Milestones in Drug Therapy
MDT

Series Editors

Erythropoietins, Erythropoietic Factors, and Erythropoiesis

Molecular, Cellular, Preclinical, and Clinical Biology

2nd Revised and Extended Edition

Edited by
Steven G. Elliott, MaryAnn Foote, and Graham Molineux

Birkhäuser
Basel · Boston · Berlin

Editors

Steven G. Elliott
Amgen Inc.
One Amgen Center Drive
Mailstop 29-1-A
Thousand Oaks, CA 91320
USA

MaryAnn Foote
MA Foote Associates
Westlake Village, CA 91362
USA

Graham Molineux
Amgen Inc.
One Amgen Center Drive
Mailstop 15-2-A
Thousand Oaks, CA 91320
USA

Library of Congress Control Number: 2008942739

Bibliographic information published by Die Deutsche Bibliothek
Die Deutsche Bibliothek lists this publication in the Deutsche Nationalbibliografie; detailed bibliographic data is available in the Internet at <http://dnb.ddb.de>.

ISBN 978-3-7643-8694-8 Birkhäuser Verlag, Basel - Boston - Berlin

Part of Springer Sciebnce+Business Media
Printed on acid-free paper produced from chlorine-free pulp. TCF ∞
Cover illustration: see p. 285, with friendly permission of Tara Arvedson.
Printed in Germany

ISBN 978-3-7643-8694-8

e-ISBN: 978-3-7643-8698-6

9 8 7 6 5 4 3 2 1

www.birkhauser.ch

Contents

Erythropoietin receptor

Clinical utility

List of contributors

Tara L. Arvedson, Amgen Inc., One Amgen Center Drive, M/S 15-2-A, Thousand Oaks, California 91320, USA; e-mail: taraa@amgen.com
Don H. Catlin, Anti-Doping Research, Inc., 3873 Grand View Blvd, Los Angeles, CA 90066, USA; e-mail: dcatlin@antidopingresearch.org
Andrew T. Chow, Department of Pharmacokinetics and Drug Metabolism, Amgen, One Amgen Center Drive, Thousand Oaks, CA 91320; e-mail: chow@amgen.com
Alice S. Chuck, Seattle Genetics, Bothell, Washington, USA; e-mail: achuck@seagen.com
Stefan N. Constantinescu, Signal Transduction Unit, Ludwig Institute for Cancer Research, Christian de Duve Institute of Cellular Pathology, Avenue Hippocrate 74, UCL 75-4, 1200 Brussels, Belgium; e-mail: stefan.constantinescu@bru.licr.org
Reinhard Depping, Institute of Physiology, University of Lübeck, Lübeck, Germany; e-mail: depping@physio.uni-luebeck.de
Rohini R. Deshpande, Amgen Inc., Thousand Oaks, California, USA; e-mail: rohinid@amgen.com
Adrian Distler, Amgen Inc., Thousand Oaks, California, USA
Sameer Doshi, Department of Pharmacokinetics and Drug Metabolism, Amgen, One Amgen Center Drive, Thousand Oaks, CA 91320; e-mail: sdoshi@amgen.com
Steve G. Elliott, Amgen Inc., One Amgen Center Drive, Mailstop 29-1-A, Thousand Oaks, CA 91320, USA; e-mail: selliott@amgen.com
MaryAnn Foote, MA Foote Associates, Westlake Village, CA 91362, USA; e-mail: fmawriter@aol.com
Jeffrey S. Friedman, Department of Molecular and Experimental Medicine, The Scripps Research Institute, La Jolla, CA, USA; e-mail: friedman@scripps.edu
Saghi Ghaffari, Department of Gene and Cell Medicine, Mount Sinai School of Medicine, New York, NY, USA; e-mail: saghi.ghaffari@mssm.edu
David A. Goodkin, Goodkin Biopharma Consulting, LLC, Bellevue, WA, USA; e-mail: davidagoodkin@comcast.net
Xylina T. Gregg, Utah Cancer Specialists, 3838 South 700 East, Ste 100, Salt Lake City, UT 84106, USA
Caroline K. Hatton, Anti-Doping Research, Inc., 3873 Grand View Blvd, Los Angeles, CA 90066, USA; e-mail: ckhatton@aol.com
Graham R. Jang, Amgen, Department of Pharmacokinetics and Drug Metabolism, One Amgen Center Drive, Thousand Oaks, CA 91320; e-mail: gjang@amgen.com

Wolfgang Jelkmann, Institute of Physiology, University of Lübeck, Lübeck, Germany; e-mail: jelkmann@physio.uni-luebeck.de

Harvey F. Lodish, Whitehead Institute for Biomedical Research, Cambridge, MA and Department of Biology, Massachusetts Institute of Technology, Cambridge, MA, USA; e-mail: lodish@wi.mit.edu

Eric Metzen, Institute of Physiology, Duisburg-Essen, Essen, Germany; e-mail: eric.metzen@uni-due.de

David R. Mole, University of Oxford, Headington Campus, Roosevelt Drive, Oxford OX3 7BN, UK; e-mail: drmole@well.ox.ac.uk

Graham Molineux, Amgen Inc., One Amgen Center Drive, Mailstop 15-2-A, Thousand Oaks, CA 91320, USA; e-mail: grahamm@amgen.com

Gregg Nyberg, Amgen Inc., Longmont, CO, USA

Timothy Osslund, Analytical and Formulation Science, Amgen Inc., One Amgen Center Drive, Thousand Oaks, CA 91230, USA; e-mail: osslund@amgen.com

Juan Jose Perez Ruixo, Amgen Inc, Department of Pharmacokinetics and Drug Metabolism, Spain; e-mail: juanjose@amgen.com

Josef T. Prchal, Professor of Medicine, Pathology and Genetics, 30 North 1900 East, Rm 5C310 SOM, University of Utah, Salt Lake City, Utah 84132-2408, USA; e-mail: josef.prchal@hsc.utah.edu

Peter J. Ratcliffe, University of Oxford, Headington Campus, Roosevelt Drive, Oxford OX3 7BN, UK; e-mail: pjr@well.ox.ac.uk

Shane A. Sander, Amgen Inc., Thousand Oaks, California, USA

Barbra J. Sasu, Amgen Inc., One Amgen Center Drive, M/S 15-2-A, Thousand Oaks, California 91320, USA; e-mail: bajohnso@amgen.com

Huub Schellekens, Departments of Pharmaceutical Sciences and Innovation Studies, Utrecht University, Utrecht, The Netherlands; e-mail: h.schellekens@uu.nl

James E. Seely, Amgen Inc., Longmont, CO, USA

Angus M. Sinclair, Amgen Inc., One Amgen Center Drive, Thousand Oaks, CA 91320, USA

Merav Socolovsky, Department of Pediatrics and Department of Cancer Biology, University of Massachusetts Medical School, Worcester, MA, USA; e-mail: merav.socolovsky@umassmed.edu

Wei Tong, Department of Pediatrics, University of Pennsylvania School of Medicine and Division of Hematology, Children's Hospital of Philadelphia, Philadelphia, PA, USA; e-mail: tongw@email.chop.edu

Bruce E. Torbett, Department of Molecular and Experimental Medicine, The Scripps Research Institute, La Jolla, CA, USA; e-mail: betorbet@scripps.edu

Jing Zhang, McArdle Laboratory for Cancer Research, University of Wisconsin, Madison, WI, USA; e-mail: zhang@oncology.wisc.edu

Preface to the first edition

Research on, and interest in, red blood cell formation spans several centuries and was thought to have peaked in the 1980s with the cloning of the erythropoietin (EPO) gene. In the years subsequent to the cloning of EPO and its expression as a recombinant protein, much was written about EPO. Although much has been learned and published, new, exciting data are becoming available on almost a daily basis. *Erythropoietins and Erythropoiesis: Molecular, Cellular, Preclinical, and Clinical Biology* compiles both pertinent historical and very recent research on this molecule and its clinical utility.

The book is divided into two sections: Background and Basic Science and Clinical Uses of Recombinant Erythropoietins. To begin, Israels and Israels describe the biology of red cells, the hierarchy of erythropoietic progenitor cells, their development to mature cells, and the effects of endogenous EPO on their development. Foote summarizes the historical interest in, and search for, an erythropoietic factor. Once EPO was identified, cloned, and expressed, the path was set for the study of other aspects of EPO biology both within erythropoiesis and other cellular systems.

The structures of recombinant human erythropoietin (rHuEPO) and its receptor (EPOR) have been studied and modeled using X-ray crystallography and other techniques, and the chapter by Osslund provides three-dimensional structural information. Activation of EPOR by EPO is essential for the survival, proliferation, and differentiation of red blood cells. EPOR is also expressed in many organs, including the brain, heart, endothelium, and ovaries, and may have physiological roles in these organs. Although studies are underway to establish the role of EPOR signaling in various organs, it is becoming increasingly apparent that red cells are not the only targets of EPO. Dame provides further data on the effects of both endogenous EPO and rHuEPO on hematologic and non-hematologic tissues. EPO has long been known to have a direct effect on the formation of red blood cells, and more recent work suggests that it may have a myriad of diverse effects that may allow the use of rHuEPO in clinical settings of neurological, cardiac, neonatal care, and as well as in other settings. Heatherington presents detailed information on the pharmacokinetics of EPO and rHuEPO in various patient populations. The pharmacokinetic properties of rHuEPO are some of the major factors that determine dosing regimens and mode of administration, and this literature review offers extensive information. Molineux reviews the basic biology of EPO and rHuEPO. The effects of treatment with rHuEPO are wide-ranging, especially in patients with degenerating kidney performance, suggesting effects beyond mere replacement of the missing endogenous EPO.

This section ends with a description by Chuck et al. of the production techniques for rHuEPO, from establishment of a cell bank to purification of the final clinical product.

Several clinical settings are discussed in detail, including treatment of the anemias of nephrology (Macdougall), oncology (Glaspy), and chronic diseases (Means) and also use in surgery (Cushner). Glaspy discusses the design of clinical trials using rHuEPO and offers insight into some published clinical data with rHuEPO and darbepoetin alfa, an erythropoiesis-stimulating protein that persists in the circulation three times longer than rHuEPO. The anemias of chronic diseases are among the most common syndromes in clinical medicine, but they often are under-recognized and under-treated. Clinical studies have demonstrated the efficacy of rHuEPO in the management of anemias of chronic diseases and have established a role for iron therapy as an adjunct to rHuEPO in this syndrome. Blood loss is inherent to the surgical setting. Using the orthopedic surgery model, Cushner focuses on the use of rHuEPO not only to decrease allogeneic transfusions but also to maximize blood parameters, such as hematocrit and hemoglobin concentration, during the peri-operative period.

Because rHuEPO is so effective in stimulating production of red blood cells, it has the potential for abuse by athletes, particularly Olympic athletes in endurance sports, such as running, cycling, and cross-country skiing. Catlin et al. describe how laboratories test for the illegal use of rHuEPO and how athletes are monitored and charged.

In 2001 and 2002, numerous cases of pure red cell aplasia, a rare but potentially life-threatening condition, were suddenly reported to health authorities worldwide. These cases of pure red cell aplasia were first noted in patients with kidney disease who were receiving rHuEPO. Mayeux and Casadevall describe this adverse event and how they test patients for the presence of anti-EPO antibodies, and they provide a possible explanation for the occurrence of pure red cell aplasia.

Finally, Elliott discusses current approaches in construction of EPO analogs to stabilize or increase activity, chemical modification of rHuEPO, gene delivery, and development of slow-release formulations. One successful strategy discussed is glycoengineering of rHuEPO, which involves construction of glycosylation analogs with increased content of sialic acid-containing carbohydrate. One such glycoengineered molecule, darbepoetin alfa, has been approved for marketing in the United States, the European Union, Australia, and Canada. Clearly, more than 20 years after the initial identification, isolation, cloning, and expression of the gene for EPO, research continues.

This book contains much information on erythropoiesis and red blood cell production, its regulation, and areas of continued or possible research, and is a resource for new and veteran researchers. Since different perspectives allow readers to arrive at their own conclusions and serve to stimulate scientific thought, we have not removed areas of controversy or overlap among chapters. We hope that this book proves useful and we invite your comments.

We have tried to acquire the necessary permissions and authorizations before publication, and great care has been taken in the preparation of the chapters. Nevertheless, errors cannot always be avoided. The editors and publishers, therefore, cannot accept responsibility for any errors or omissions that have inadvertently occurred. The views and opinions expressed in the book are those of the participating individuals and do not necessarily reflect the views of the editors, the publisher, Amgen Inc., or any other manufacturer of pharmaceutical products named herein. The package insert should be consulted before administration of any pharmaceutical product.

Graham Molineux, PhD
MaryAnn Foote, PhD
Steven G. Elliott, PhD

Thousand Oaks, California
February 2003

Preface to the second edition

Cloning of the erythropoietin (EPO) gene was accomplished in the 1980s, allowing a breakthrough in the treatment of anemia. Now more than 2 decades later, important new understandings on erythropoiesis and EPO are revealed as a consequence of continued research in this area. The first edition of this book provided a review of the field in 2003. This new edition continues the examination of this branch of learning with new information, updates, and changes.

To begin, Torbett and Friedman provide an overview of both normal and abnormal erythropoiesis as a basis for understanding various disease states and the role and action of recombinant erythropoiesis-stimulating agents (ESA). Mole and Ratcliffe discuss the production of endogenous EPO by defining *cis*-acting regulatory sequences that control *EPO* gene expression. The biology of EPO is covered in the chapter by Molineux and Sinclair. Prchal and Gregg have provided a new chapter for this edition concerning genetic abnormalities in erythropoiesis, and Constantinescu has written a new chapter on the mechanism of EPO receptor activation. Another new chapter was provided by Doshi, Perez-Ruixo, Jang, and Chow that contains updated information on the pharmacokinetics and clearance of ESA.

In the clinical section of the book, the use and abuse of ESA is described. Catlin and Hatton provide an update on the abuse of ESA by athletes. Goodkin has written a new chapter on the use of ESA in the setting of renal disease. Arvedson and Sasu have contributed a new chapter that discusses the roles iron and regulation of iron metabolism in both normal erythropoiesis and disease states. Another new chapter is by Jelkmann, Depping, and Metzen on the non-hematopoietic effects of ESA.

As with the first edition, we have allowed different perspectives and some overlap of material to allow readers to arrive at their own conclusions and to stimulate scientific thought and discussion. The views and opinions expressed in the book are those of the participating individuals and do not necessarily reflect the views of the editors, the publisher, Amgen Inc., or any other manufacturer of pharmaceutical products named herein. We have tried to acquire the necessary permissions and authorizations before publication, and care has been exercised in the preparation of the chapters. Nevertheless, errors cannot always be avoided. The editors and publishers, therefore, cannot accept responsibility for any errors or omissions that have inadvertently occurred.

The book is a review of ESA which are widely used in the treatment of anemia. The views and information contained herein are not meant to be a guide in the use of these agents. Instead, the package insert should be consulted before administration of ESA or any other pharmaceutical products.

Steven G. Elliott, PhD
MaryAnn Foote, PhD
Graham Molineux, PhD

Thousand Oaks, California
November 2008

Abbreviations

ADM	adrenalmedulin
AGM	aorta-gonad-mesonephros
AIDS	acquired immunodeficiency syndrome
Akt	protein kinase B
AMPK	AMP kinase
ANP	atrial naturetic peptide
ARNT	aryl hydrocarbon nuclear receptor translocator
ASGR	asialoglycoprotein receptor
ATM	ataxia telangiectasia mutant
2,3-BPG	2, 3 biphosphoglycerate
B-CLL	B-cell chronic lymphocytic leukemia
bHLH	basic-helix-loop-helix
BFU–E	burst-forming unit–erythroid
BPA	burst-promoting activities
BTG 1	B-cell translocation gene 1
CAD	*C*-terminal activation domain
CAIX	carbonic anhydrase-9
CAS	Court of Arbitration for Sport
CEPO	carbamoylated EPO
CFU–E	colony-forming unit–erythroid
CFU–GEMM	colony-forming unit–granulocyte, erythroid, macrophage, megakaryocte
CHMP	Committee for Medicinal Products for Human Use
CHO	Chinese hamster ovary
CHr	mean corpuscular hemoglobin
CKD	chronic kidney disease
CP	Chuvash polycythemia
CR1	complement receptor 1
CSF	cerebrospinal fluid
DFO	desferrioximine
DHFR	dihydrofolate reductase
DOPPS	Dialysis Outcomes and Practice Pattern Study
DMT1	divalent metal transporter 1
DcytB	duodenal cytochrome b
eEPO	endogenous EPO
EMEA	European Medicines Evaluation Agency
EMP	EPO mimetic peptide
eNOS	endothelial nitric-oxide synthase

EPAR	European Public Assessment Report
EPO	erythropoietin
EPOR	erythropoietin receptor
ERK	extracellular signal-regulated kinase
EryP-CFC	primitive erythroid colony-forming cell
ESA	erythropoiesis-stimulating agent
ET	essential thrombocythemia
EU	European Union
Fab	fragment of antibody
FGF	fibroblast growth factor
FIH-1	factor-inhibiting HIF-1
FIS	International Ski Federation
FOB	follow-on biologics
FRET	fluorescence resonance energy transfer
Gas6 protein	growth arrest-specific 6 protein
GC-MS	gas chromatography-mass spectrometry
G-CSF	granulocyte colony-stimulating factor
GFR	glomerular filtration rate
GLUT	glucose transporter
GMP	Good Manufacturing Processes
HCP	hematopoietic cell phosphatase
HCP-1	heme carrier protein-1
HIF	hypoxia inducible factor
HNF	hepatic nuclear factor
HO-1	hemoxygenase-1
HRE	hypoxia-responsive element
HUVEC	human umbilical vein endothelial cell
IEF	isoelectric focusing
IFN	interferon
IL	interleukin
INN	International Nonproperity Name
IOC	International Olympic Committee
iPAS	inhibitory PAS
IRS	insulin receptor substrate
IU	international unit
JAK2	Janus-type tyrosine kinase
JAK-STAT	Janus kinase/signal transducers and activators of transcription
JH	JAK homology
kDa	kilodalton
KDOQI	Kidney Disease Outcomes Quality Initiative
LC-MS-MS	liquid chromatography coupled to tandem mass spectrometry
LDHA	lactate dehydrogenase A
LPS	lipopolysacharride
MALDI	matrix-assisted laser desorption/ionisation
MAPK	Ras/mitogen-activated kinase

MCH	mean corpuscular hemoglobin
MCVr	reticulocyte mean corpuscular volume
MCV	mean corpuscular volume
MDS	myelodysplastic syndromes
MEC	minimum effective concentration
miRNA	microRNA
mTORC	mammalian target of rapamycin complex
NADO	National Anti-Doping Organization
NDRG	N-myc downstream regulated
NHCT	Normal Hematocrit Cardiac Trial
NMDA	N-methyl-D-aspartate
NMR	nuclear magnetic resonance
NO	nitrous oxide
ODDD	oxygen-dependent degradation domains
PAI	plasminogen activator inhibitor
PDGF	platelet-derived growth factor
PDK	pyruvate dehydrogenase kinase
PFCP	primary familial and congenital polycythemia
PH	pleckstrin homology
PHD1, 2, or 3	prolyl hydroxylase domain proteins 1, 2, or 3
PI3K	phosphoinositide-3 kinase
PIAS	protein inhibitor of activated stat
PKC	protein kinase C
PRCA	pure red cell aplasia
PTEC	proximal tubular epithelial cells
PV	polycythemia vera
pVHL	von Hippel-Lindau protein
rHuEPO	recombinant human erythropoietin
rHuGH	recombinant human growth hormone
RIA	radioimmune precipitation assay
ROS	reactive oxygen species
RXR	retinoic acid receptor
SAPK	c-Jun *N*-terminal kinase
SCF	stem cell factor
SDS-PAGE	sodium dodecyl sulfate-polyacrilamide gel electrophoresis
SH2	SRC homology 2
SHIP	phosphatidylinositol 3,4,5 triphosphate 5-phosphatase
siRNA	small interfering RNA
SNP	single nucleotide polymorphisms
SOCS	suppressors of cytokine signaling
STAT	signal transducer and activator of transcription
TAD	transactivating domains
sTfR	soluble transferring receptor
TGF	transforming growth factor
TIMP	tissue inhibitors of matrix metalloproteinase

TMDD	target-mediated drug disposition
TNF	tumor necrosis factor
TOF	time-of-flight
TRAIL	tumor necrosis factor-related apoptosis-inducing ligand
TRALI	transfusion-related acute lung injury
TRIM	transfusion-related immune modulation
Tsat	transferrin saturation
UCI	International Cycling Union
USADA	United States Anti-Doping Agency
VEGF	vascular epithelial growth factor
VO_2 peak	peak oxygen

Background

Erythropoietins, Erythropoietic Factors and Erythropoiesis
Edited by S.G. Elliott, M.A. Foote, and G. Molineux

Erythropoiesis: an overview

Bruce E. Torbett and Jeffrey S. Friedman

Department of Molecular and Experimental Medicine, The Scripps Research Institute, La Jolla, CA, USA

Normal erythropoiesis

Red blood cell production in vertebrates is dynamic and tightly regulated with losses due to senescence or bleeding in normal individuals being balanced by generation of new red cells. The main function of the red cells is to transport oxygen from the lungs to tissues in all parts of the body and to transport carbon dioxide back for exchange. Healthy adult humans maintain a total of $2–3 \times 10^{13}$ erythrocytes at any given time, with men having 5–6 million and woman having 4–5 million erythrocytes/microliter blood [1]. The normal life span of an erythrocyte is approximately 120 days in the blood requiring a daily replacement of about 1% of the circulating erythrocytes [1]. To put the numbers of red cells needed daily into perspective, in an adult human maintenance of steady-state red cell numbers requires the generation of 2×10^6 erythrocytes every second. Moreover, basal oxygen consumption is 4 mL/kg/min and body stores are approximately five times greater, emphasizing the need to maintain a stable red cell mass while having the capacity to increase red cells during times of tissue hypoxia [1, 2]. Changes in the production of erythrocytes from a steady-level will occur in response to blood loss, increased oxygen demand, altered erythrocyte lifespan, or diseases.

Erythropoiesis is a regulated process whereby hematopoietic progenitor cells give rise to committed erythroid progenitors, proliferate, and differentiate to mature red cells to maintain steady state red cell mass (Fig. 1). Erythropoiesis occurs in three defined stages: generation of erythroid committed blast cells from multipotent hematopoietic progenitors; division and differentiation of these morphologically identifiable erythroid progenitor cells; and terminal cellular morphologic changes, including enucleation, to produce reticulocytes and ultimately mature red cells [3, 4]. In adult humans, erythroblasts are produced from committed progenitors arising from hematopoietic stem cells present in the bone marrow. As the progeny of a stem cell progress through development, there is a loss of multipotency and increasing lineage restriction [3–5]. The various cellular stages in erythrocyte development are identified by their ability to form colonies in semisolid medium supplemented with specific cytokines and by cell surface markers (Fig. 1). The response to

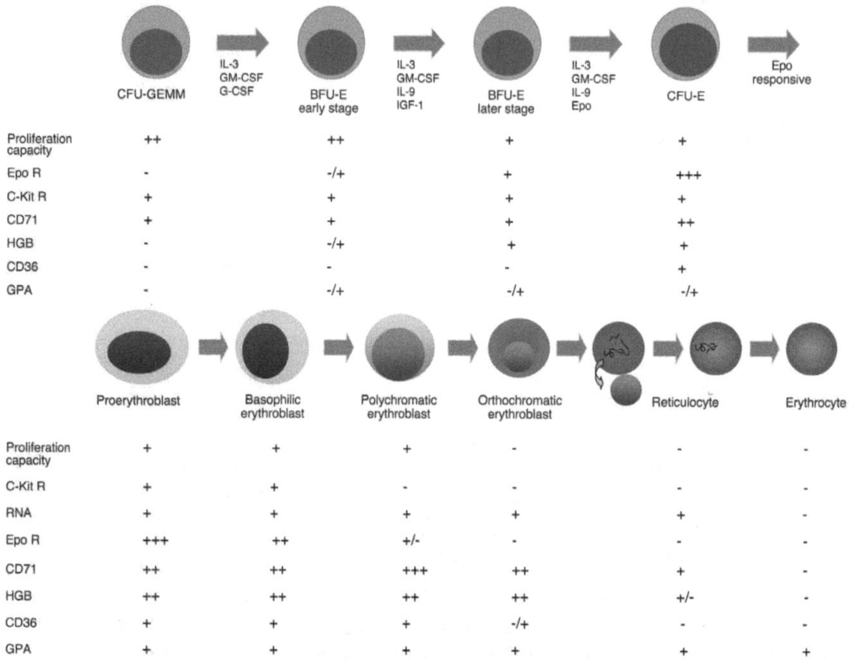

Figure 1. Lineage specific stages of erythropoiesis. Stages of erythrocyte differentiation are shown with defining cellular characteristics based on relative proliferation, cytokine responsiveness, cell surface markers, and/or presence of RNA. CFU-GEMM = colony-forming unit–granulocyte, erythroid, macrophage, megakaryocte; BFU–E = burst-forming unit–erythroid; EPOR = erythropoietin receptor; CFU–E = colony-forming unit–erythroid; C-Kit R = stem cell factor receptor, CD = CD117; CD71 = transferrin receptor; HGB = hemogloblin synthesis; CD36 = gpIIIb, thrombospondin receptor; GPA = glycophorin A present on human cells only.

selected cytokine presumably correlates to the normal environmental conditions controlling growth and differentiation of selected lineages: at early stages of development erythroid progenitors respond to multilineage cytokines; as differentiation progresses, cytokine responses become more limited and specific to the erythroid progenitors [4, 6, 7].

Hematopoietic stem cells give rise to a common myeloid progenitor that in turn, gives rise to bipotential progenitors restricted to either the granulocyte/macrophage or the erythroid/megakaryocte pathways [8–11]. A comparable stage of development defined *in vitro* using cytokine containing semisolid medium is the colony-forming unit – granulocyte, erythroid, macrophage, megakaryocte (CFU-GEMM) precursor that gives rise to bipotential progenitors restricted to either granulocyte/macrophage or erythroid/megakaryocte pathways [8–11]. Progenitors committed to the erythroid/megakaryocte pathway express erythropoietin receptor (EPOR) and are responsive to erythropoietin (EPO), whereas cells committed to the myeloid pathway do not express EPOR. The most immature erythroid-restricted progenitor is the burst-forming unit–erythroid (BFU–E) which gives rise to colony-forming unit–erythroid (CFU–E) [12, 13]. Early BFU–E are blast-like cells, highly proliferative, and give rise to clustered burst colonies of up to 20,000 cells (Fig. 1) [12, 13]. BFU–E express low amounts of EPOR but daughter cells derived from them express high amounts of EPOR, become EPO responsive, transferrin receptor positive, and begin to express hemoglobin [1, 12]. CFU–E are highly responsive to EPO, generate smaller colonies, and express many of the gene products required for definitive erythroid development (Fig. 1). EPO has been found to stimulate division and prevent apoptosis of late-stage erythroid cells generated from CFU–E [6, 7, 12]. Moreover, in tissue culture systems at this stage of development cells begin hemoglobin synthesis and acquire cytoskeletal proteins and cellular adhesion molecules, steps that help define them as nucleated erythroblasts [4, 6, 7].

The next steps of erythroid differentiation involve morphologically identifiable nucleated precursors that progress from the proerythroblast to basophilic, polychromatophilic, and orthochromatic cell forms leading to the reticulocyte (Fig. 1) [1, 13]. During these distinct stages of erythroid differentiation, four major cellular processes ensue: the accumulation of hemoglobin that participates in driving the basophilic to acidophilic cytoplasmic changes seen during maturation, limited erythroblast expansion, a continued decrease in cell size, and nuclear condensation and finally enucleation [1, 14]. Erythroblasts mature in a microenvironmental niche in bone marrow termed 'erythroblast islands' where they are closely associated with macrophages that serve as stromal or nurse cells [15–17].

In the final stages of erythroid development, reticulocytes exit the marrow and enter the circulation where they mature into erythrocytes. This stage of development is characterized by disassembly of ribosomes, golgi bodies, and other cellular machinery, removal of organelles, enucleation, changes in the cytoskeleton leading to the classic biconcave discoid shape, and then release into the circulation [1]. The last steps of differentiation again take place in the erythroblast islands, where macrophages participate in terminal maturation of erythrocytes [15–17]. The mature erythrocyte persists in the circulation for up to 120 days until 'senescent', when changes in surface antigen expression or physical characteristics of the cell trigger removal from the blood by macrophages of the reticuloendothelial system [18].

During mammalian development from embryo to adult, the body compartment responsible for hematopoiesis changes, thus providing precise regulation in specialized niches depending on developmental age [3, 4, 19–22]. In the mouse, primitive red cells arise in the yolk sac blood islands very soon after gastrulation [23, 24]. Erythroblasts originate from unique, developmental stage-specific progenitors, termed primitive erythroid colony-forming cells (EryP-CFC), that are distinguishable by their growth, maturation properties, and globin gene expression patterns [25, 26]. This cell population is short lived during development and provides red cells until hematopoiesis is established in the fetal liver by hematopoietic stem cells that give rise to definitive erythropoiesis

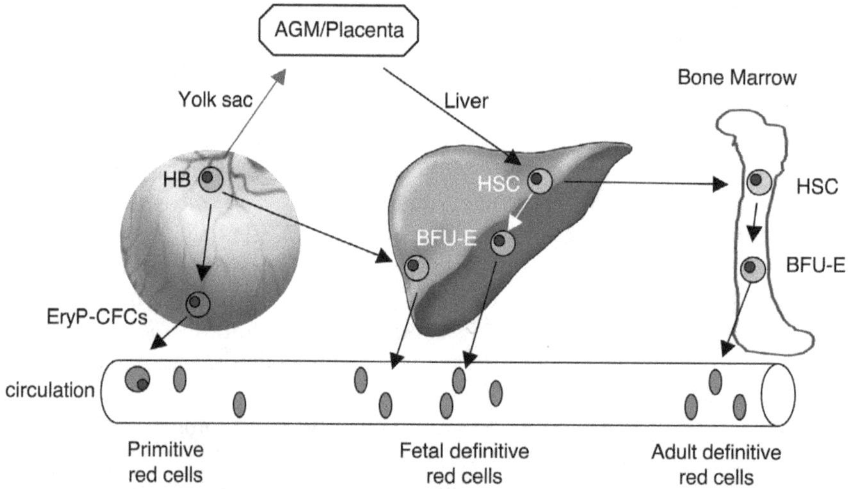

Figure 2. Model of primitive and definitive erythroid development in the mammalian embryo. The current findings in the field support a model whereby three distinct phases of erythroid progenitors emerge in the mammalian embryo. The first phase consists of primitive, nucleated erythrocytes, derived from erythrocyte precursor colony-forming cells (EryP-CFC), generated from yolk sac hemangioblasts (HB). These cells mature in the circulation to become enucleated red blood cells. The second phase is the seeding of the fetal liver by yolk sac-derived, definitive BFU–E that gives rise to red cells through the developmental pathway presented in Figure 1. In the last phase aorta-gonad-mesonephros (AGM) and placental derived HSC seed the fetal liver, which in turn, give rise to the HSC-erythroid developmental program as described in Figure 1. It should be noted that HSC that seed the fetal liver may originate in the yolk sac, migrate to the AGM where they mature into HSC, then seed the fetal liver. Determining the origin of HSC that give rise to the definitive hematopoietic system is an area of active investigation. At the end of gestation, long-term repopulating HSC from the fetal liver seed the bone marrow that then becomes the site of definitive erythropoiesis for fetal and adult red cell production.

(Fig. 2) [27–29]. Since the first primitive erythroid cells arise with endothelial cells during primitive erythropoiesis in the embryonic yolk sac, a common progenitor, the hemangioblast, has long been proposed to be the precursor. Only recently has this concept been verified experimentally in the mouse [30].

Adult hematopoiesis is characterized by the continuous generation of mature blood cells from the long-term repopulating hematopoietic stem cells. In the mouse, the developmental origin of long-term repopulating hematopoietic stem cells is associated with the appearance of clusters of cells arising from the dorsal aorta in the aorta-gonad-mesonephros (AGM) region [31, 32]. It appears that the placenta also functions as a site of hematopoietic stem cell expansion and perhaps origin [21, 33]. Within a short time after appearance of hematopoietic stem cells in the AGM/placenta, they are found in the fetal liver [32, 34, 35]. Two reports using mice bring into question the origin of the stem cells that seed the fetal liver establishing definitive hematopoiesis, and ultimately erythropoiesis [36, 37]. Both studies reply on different gene disrupted mice that severely limit dissemination of hematopoietic cells from the yolk sac to AGM/placenta and other hematopoietic sites, thus providing a developmental window to observe tissue seeding of hematopoietic stem cells before the embryo dies [36, 37]. Given the caveats of the experimental design as to whether hematopoietic cells can develop in or seed a dying embryo, the studies might suggest that the stem cells originate in the yolk sac and then must migrate to the AGM to undergo maturation before seeding the fetal liver and other tissue spaces. Lastly, these studies do not formally exclude the development of hematopoietic stem cells in the AGM and placenta.

Interestingly, definitive erythroid potential has been reported to be present in the embryo before long-term engrafting stem cells are present in the fetal liver [25, 26]. It has been reported that in mice, BFU–E arise before circulation, and after circulation formation are found in increasing numbers in the liver [25, 26]. These findings might suggest that the yolk sac produces an additional wave of immature erythroid progenitors that enter the circulation and take up residence in fetal liver [25, 26]. A similar developmental pattern has been described for the developing human embryo [20]. The findings in the mouse and human embryo would support the premise that yolk sac-derived BFU–E migrate to the fetal liver where they complete development to produce the first definitive erythroid cells in the embryo (Fig. 2) [25, 26]. During the end of gestation, hematopoiesis and definitive erythropoiesis begin to take up residence in the bone marrow [4, 5]. The current findings in the field support the model of erythroid ontogeny as depicted in Figure 2. These exciting findings underscore that there is still much to understand about hematopoietic development, which remains an area of active investigation. The reader is referred to a number of excellent classical and current reviews of the field to provide more information on the ontogeny of hematopoiesis and erythropoiesis than can be provided herein [4, 5, 21, 22, 38].

Erythropoietin

EPO, an approximately 30–34 kDa glycoprotein hormone, is the primary hematopoietic growth factor governing the rate of red cell production. The action of EPO depends upon the presence of an adequate number of responsive progenitor cells, availability of sufficient iron and cofactors (folate, B12) to support heme and globin synthesis, and an appropriate microenvironment to support erythroid development. The rate of red cell production is dynamically responsive to demand in order to replace cells lost through bleeding, destruction, and senescence. EPO concentrations are inversely related to hemoglobin concentration, ranging from approximately 10 mU/mL in nonanemic conditions up to 10,000 mU/mL in severe anemia [39, 40]. In response to EPO stimulation after blood loss, marrow production of red cells can increase 3- to 5-fold in the presence of an adequate iron supply [41]; in the case of a more severe hemolytic insult, production may increase even further (7- to 8-fold) due to more efficient recycling of iron from lysed red cells [42].

Interestingly, while EPO values are closely tied to the rate of red cell production, studies in knockout mice lacking EPO show that it is not required for development of BFU–E, the earliest committed red cell progenitors (that express only a small number of EPOR on their cell surface) [43] (Fig. 1). As such, EPO does not recruit cells to the erythroid lineage nor does it specify cell fate. Rather, EPO exerts its dynamic control over red cell formation by supporting the growth, survival, and differentiation of the progeny of erythroid-committed progenitor cells [6, 44], with the highest amounts of EPOR being found on late progenitor cells such as CFU–E [45].

Production

In adult mammals, the major physiologic site of EPO production is the kidney, although *EPO* gene transcription can be detected in liver [46], and at lower amounts in many other tissues, including brain. Within the kidney, *in situ* hybridization for EPO mRNA localizes production to peritubular interstitial fibroblasts [47, 48]. While expression in the adult liver echoes the role of this organ as the major source of EPO during embryogenesis, expression in the brain and other tissues suggest nonhematopoietic functions for this hormone/growth factor, including possible autocrine or paracrine effects in cells and tissues that express both EPO and EPOR.

The mechanism of regulation of EPO expression at the molecular level involves an ingenious coupling between oxygen sensors (prolyl hydroxylase domain proteins 1, 2, or 3 [PHD1, 2, or 3] [49] and/or the asparagine hydroxylase, factor-inhibiting HIF [FIH-1] [50]) and the transcription factor primarily responsible for *EPO* gene activity, hypoxia inducible factor [HIF]. HIF are heterodimeric transcription factors composed of a constitutively expressed β subunit and a labile α subunit [51, 52]. The O_2-dependent enzymatic activity of PDH2 is directed toward proline residues 402 and 564 of HIF-1α. Under normoxic conditions, PDH2 constitutively hydroxylates HIF-1α promoting the binding of the von Hippel-Lindau protein (pVHL), leading to the ubiquitination and degradation of HIF-1α [53] (Fig. 3). In a parallel regulatory pathway, the O_2-dependent enzymatic activity of FIH is directed toward asparagine residue 803 of HIF-1α, blocking interaction with the transcriptional coactivators p300 and CBP [54]. Thus, under normoxic conditions there are two tonic inhibitory influences on HIF-1 activity – one affecting the transcriptional activity of HIF-1α by FIH mediated hydroxylation of asparagine 803, and one affecting protein stability of HIF-1α through PDH2-mediated hydroxylation of prolines 402 and/or 564. A specific mutation in pVHL (R200W) is causal for familial erythrocytosis type 2 (also known as Chuvash polycythemia) [55, 56], and a number of additional mutations in *VHL* resulting in erythrocytosis have been described [57].

When tissue oxygenation decreases, these tonic inhibitory influences are relieved – with stabilization of transcription competent HIF-1α protein, dimer-

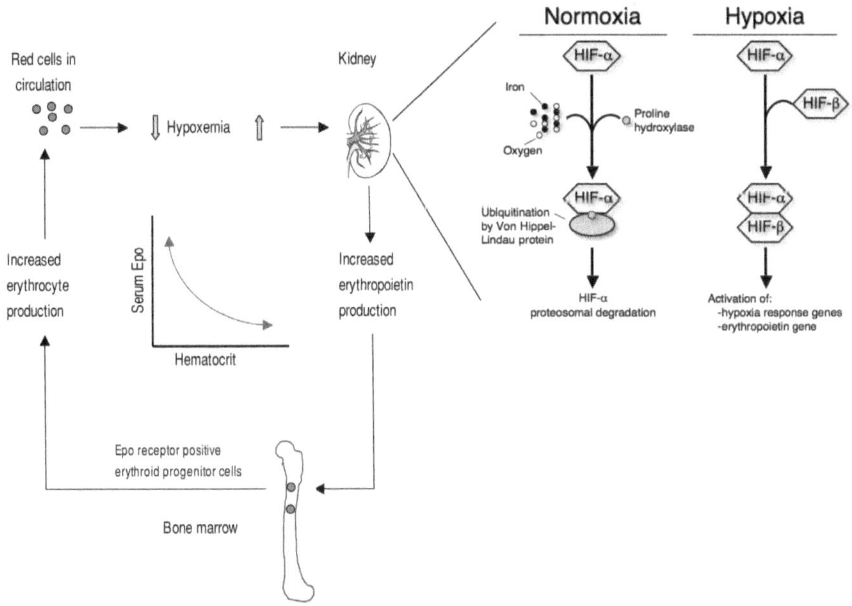

Figure 3. Regulation of erythropoietin production during normoxia and hypoxia. In adults, hypoxia resulting from decreased oxygen transport by red cells is primarily sensed in the kidneys, and to a low level in the liver, leading to an increase in EPO production. Renal produced EPO stimulates proliferation of erythroid committed precursor cells in the bone marrow resulting in the production of increased numbers of mature red cells capable of carrying oxygen and correcting tissue hypoxia. The inset graph depicts the relationship between serum EPO concentrations and the hematocrit as an indicator of hypoxia. The inset diagram to the right of the kidney depicts the roles of HIF in regulating activation of EPO and hypoxia response genes. In hypoxic situations the α-subunit is stabilized, and forms a transcriptionally competent heterodimer with the constitutively expressed HIF-1α subunit. HIF then interacts with additional nuclear factors to activate expression of erythropoietin and hypoxia response genes. The α-subunit is rate limiting in the transcription complex, and its destruction by proteosomal degradation is controlled by the amount of cellular oxygen. During normoxia, hydroxylation of two specific proline residues in HIF-1α occurs by three cellular prolyl hydoxylases in an iron, oxygen, and 2-oxglutarate-dependent fashion. This posttranslational modification, in combination with acetylation of a specific lysine residue, permits binding of the pVHL to HIF-1α promoting ubiquitination and degradation. In addition, oxygen-dependent hydroxylation of an arginine in HIF-1α blocks binding of transcriptional co-factors required for gene transcription. While both HIF-1α and HIF-2α can regulate erythropoietin gene expression, recent evidence suggests that HIF-2α is the physiologic regulator in adult mammals.

ization with HIF-1β, and transcriptional activation of responsive promoters. Two hypotheses have been put forward to explain the loss of PDH2 enzymatic activity with hypoxia: one suggests substrate limitation (O_2 and/or α-ketoglutarate) [49], the other suggests oxidation of Fe (II) in the catalytic center of PDH2 by mitochondrial-derived reactive oxygen species [58–60].

HIF affects transcription of hundreds of genes, coordinating adaptive tissue responses to hypoxia, directing cellular metabolic shifts to favor glycolysis, stimulating growth of new vessels (through induction of growth factors such as vascular endothelial growth factor [VEGF]), and regulating mitochondrial autophagy [61] in addition to increasing oxygen-carrying capacity through induction of erythropoietin production in the kidney. Two additional HIF-1α homologues exist in mammals – HIF-2α and HIF-3α, as well as two additional prolyl-hydroxylase domain proteins, PDH1 and PDH3 [49, 62]. While there is evidence of overlapping function and crosstalk between these proteins, PDH2 is the primary regulator of HIF-1α stability [63]. The importance of PDH2 as a regulator of HIF and subsequent erythropoiesis in a clinical setting has been demonstrated by the discovery of mutations (Pro317Arg and Arg371His) in PDH2 in families with hereditary erythrocytosis (now designated familial erythrocytosis type 3) [64, 65]. The functional activity and regulation of HIF-2α and HIF-3α are being investigated. Two studies demonstrate that HIF-2α is an important regulator of EPO production by the liver [66], and that postnatal ablation of HIF-2α results in anemia, suggesting that HIF-2α is the physiologically important regulator of EPO production in adult mammals [67].

Receptor

Erythroid-lineage precursor cells express EPOR as a homodimeric transmembrane protein with a ligand-binding extracellular domain, and a cytoplasmic domain constitutively associated with the Janus-type tyrosine kinase JAK2 [68]. Upon ligand binding, EPOR undergoes a conformational change, triggering the activation and transphosphorylation of JAK2, which then phosphorylates multiple tyrosine residues in the EPOR cytoplasmic tail [69]. These p-Y residues serve as docking sites for assembly, activation, and propagation of mitogenic and prosurvival signals mediated by multiple effector pathways including STAT5, PI3-kinase/AKT, and Ras/MAPK [70]. Knockout of JAK2 in mice produces a phenotype similar to that observed for knockout of EPO [71] or EPOR [43], namely embryonic lethality at approximately E12.5 with BFU–E and some CFU–E present, but unable to undergo maturation [72, 73]. Thus, EPO, EPOR, and JAK2 form a linear pathway essential for erythroid expansion and maturation, but appear to have no role in lineage specification.

EPOR signaling is negatively regulated through interaction with the SH2-domain containing protein tyrosine phosphatases, primarily SHP1, which binds to p-Y429 of the Epo R, and can downregulate the activity of JAK2 [74]. The significance of this interaction is suggested both by cell culture experiments showing EPO hyper-responsiveness when Y429 is mutated, and by clinical studies of patients with familial erythrocytosis type 1, in which truncation mutations (lacking Y429) of the c-terminal portion of the receptor have been described [75–77]. Acquired mutations in the JAK2 kinase have been discovered to play a causal role in myeloproliferative disorders, most prominently in the chronic red cell disorder polycythemia vera [78–81]. The most common mutation, JAK2 Val617Phe, occurs in the autoinhibitory domain of JAK2 and results in a constitutively active kinase that confers cytokine hypersensitivity and cytokine-independent growth to hematopoietic cells [82]. Additional negative regulatory circuits exist, most notably interaction between EPOR and

CIS1, a member of the SOCS family (suppressors of cytokine signaling). Other SOCS family members (SOCS-1, SOCS-3) may inhibit EPO stimulation through direct interactions with the JAK-2 kinase [70]. Novel strategies involving analysis of transcriptional changes in cells with normal or mutated EPOR exposed to cytokine stimulation have identified additional proteins/genes involved in mediating and regulating EPO responses [83].

Abnormal erythropoiesis

While the spectrum of abnormalities of EPO are legion, Figure 4 presents a brief sampling of disorders with increased (polycythemic) and decreased (anemic) red cell production, and describes the relative endogenous EPO concentrations observed. Polycythemia (also known as erythrocytosis) can be divided into those that are primary or secondary. In primary erythrocytosis, progenitor cells harbor genetic defects conferring either hypersensitivity to EPO or cell autonomous growth, e.g., truncation of EPOR or JAK2 mutations. Secondary erythrocytosis results largely from a normal physiologic response to tissue hypoxemia, and is seen in conditions where oxygen delivery to tissue is impaired (e.g., high altitude, cyanotic heart disease, and pulmonary disease). However, two types of familial erythrocytosis (mutations affecting pVHL or PDH2) have been described in which HIF degradation is reduced, resulting in abnormal oxygen-dependent regulation of EPO production. Finally, secondary erythrocytosis can occur as a result of 'recreational use' of EPO (usually in the context of a particular yearly cycling event in France) or as a result of blood doping, i.e., the collection and subsequent transfusion of one's own red cells (*Ibid*, cycling event in France). In such cases, endogenous EPO concentrations will be appropriately low.

Anemias can be divided into those with reduced red cell production (proliferative defects), those with red cell maturation defects (ineffective erythropoiesis), and those with accelerated loss or destruction of red cells. EPO deficiency is the cause of anemia in the context of significant renal disease, i.e., due to loss or dysfunction of cells responsible for the bulk of EPO synthesis. However, in other hypoproliferative anemias, such as anemia of chronic disease [84], anemia associated with malignancy [2], and anemia of aging, EPO values may be within or above the normal range yet fail to adequately stimulate red cell production. One of the limitations on EPO bioactivity in the setting of chronic disease is the availability of iron, which is regulated by the 25-amino acid peptide hepcidin, which in turn controls degradation of the cellular iron export channel protein, ferroportin [85, 86]. Hepcidin production by the liver is responsive to proinflammatory cytokines, particularly interleukin (IL)-6, and increased hepcidin concentrations in the context of infection or inflammation restrict iron availability, suppressing erythropoiesis [87]. The role played by hepcidin in the regulation of iron availability in normal and disordered erythropoiesis is an area of active investigation.

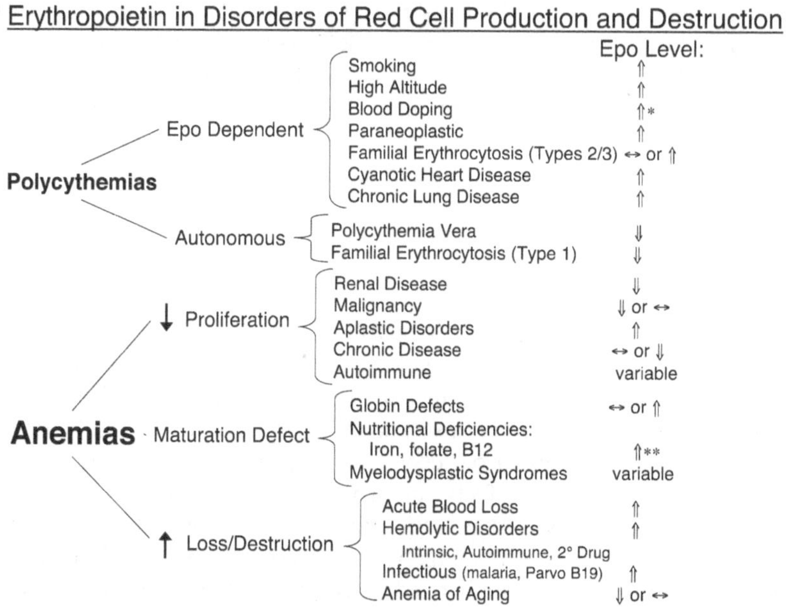

Figure 4. Erythropoietin in selected polycythemias and anemias. The increased red cell mass seen in polycythemia can be EPO dependent or cell autonomous at the level of erythroid progenitors. In many instances, increased EPO production is a normal response to tissue hypoxemia, such as that seen in individuals living at high altitude or those with cyanosis due to cardiac or pulmonary disease. Increased concentrations of EPO can occur in cancer (paraneoplastic) when produced by tumor cells, most notably in renal cell carcinoma. Mutations in pVHL or PDH2 (see text) affect regulation of EPO production, responsible for two types of familial erythrocytosis. Mutations affecting EPO signal transduction (in EPOR or the JAK2 kinase) result in erythrocytosis in the absence of increased amounts of EPO. EPO concentrations in anemic conditions can be quite variable. Robust increases in EPO concentrations are expected in normal individuals after blood cell loss or destruction. In the context of chronic disease, malignancy, and autoimmune disorders, EPO may be inappropriately low for the degree of anemia observed. In addition, the response of developing erythroid cells to EPO in these conditions may be blunted. Reduced EPO production may result from the effect of inflammatory cytokines on transcription of the *EPO* gene [106], while reduced EPO responsiveness may be a consequence of cytotoxic effects of the same cytokines on progenitor cells, as well as functional iron deficiency secondary to increased amounts of hepcidin. Increased EPO concentrations in hypoproliferative disorders may indicate marrow failure. * Blood doping, here defined as any method to boost the O_2-carrying capacity of blood. If done by transfusion, EPO concentration will be decreased; if stimulated by EPO, rHuEPO will be increased, with endogenous EPO suppressed. ** The relationship between iron regulation and regulation of EPO production is complex, with iron chelation activating HIF [107], possibly through effects on iron-dependent activity of PDH2. HIF in turn regulates hepcidin and ferroportin production, in addition to EPO [108].

A bright horizon or dark clouds?

In many nonclinical models, EPO has been shown to have a protective effect against ischemic or cytotoxic insults. Although the mechanism(s) of this effect remains unknown, it has been suggested that signaling through EPOR blocks apoptotic cell death [88, 89] in a fashion similar to that observed in erythroid progenitor cells [90, 91]. EPO signaling resulting in tissue protection appears to differ from the canonical model presented above, as the functional receptor in (many) nonerythroid cells is a heterodimer of EPOR and the common beta cytokine receptor subunit [92]. That the observed cytoprotective effects of EPO administration can be distinguished from effects on red cell production, has been shown through use of EPO derivatives (such as carbamylated or desialylated EPO) that remain cytoprotective, but do not stimulate red cell production [93–96]. Additional evidence for nonerythroid roles for EPO/EPOR comes from the study of knockout mice, demonstrating a role for EPO/EPOR signaling in brain [97] and cardiac development [98]. Studies with conditional knockout of EPOR or with 'erythroid rescued' EPOR-knockout animals have demonstrated that an intact EPO/EPOR pathway mediates neuroprotection and cardioprotection in response to a variety of insults [99]. These nonclinical studies represent the bright horizon – and provide the basis for a series of ongoing clinical studies exploring the potential of EPO and derivatives as tissue protective agents in a variety of neurologic and cardiovascular disorders, and in a critical care setting.

Despite the bright outlook, there are clouds on the horizon in the form of recent clinical studies that highlight known or potential side effects of EPO therapy, namely an increased risk of thrombosis in many patient groups, and a question of tumor promotion when recombinant human EPO (rHuEPO) is given to patients with certain types of cancer. Two published meta-analyses, one involving patients with kidney disease [100] and the other patients with cancer [101] highlight the risk of increased thrombosis in patients treated with rHuEPO, and emphasize that risk increases with increasing target hematocrit, concomitant therapy with other known thrombogenic agents, or the presence of an underlying prothrombotic condition. Perhaps even more concerning are reports from placebo-controlled studies in patients with head-and-neck cancer [102], breast cancer [103], lymphoid malignancies, and nonsmall cell lung cancer [104] linking rHuEPO therapy with decreased survival, disease progression, or both. These published results, as well as reports of safety concerns from additional studies, have led to updated recommendations on the clinical use of rHuEPO [105]. Equally important, these findings point to the need for a more thorough investigation of the relationship between the cytoprotective effects of EPO and the possibility of enhanced tumor progression.

During the past 5 years, developmental biologists have begun unraveling the intricacies of erythropoiesis, as evidenced by the many reports on ontogeny ranging from defining the roles of the hemangioblast in early development to specific requirements for yolk sac and AGM in primitive *versus* definitive erythropoiesis. These ongoing studies are providing new insights into the precise regulation of stage-specific erythroid differentiation, and may reveal additional pathways for manipulation of red cell production without the side effects of thrombosis or tumor promotion. Insights into the cellular and molecular consequences of EPO signaling in nonhematopoietic cells should provide information as to the extent and risk of EPO protective effects for selected neurologic and cardiovascular disorders.

Acknowledgements
This is TSRI manuscript number 19494. The authors are supported by the National Institutes of Health Grants R01 HL091219 (B.E.T.) and R21 DK075763 (J.S.F.), the Stein Endowment Fund (both), and the Joyce Klein Stock Gift (B.E.T.)

We dedicate this chapter to Dr. Ernest Beutler, an outstanding hematologist/scientist and mentor. While Dr. Beutler had broad scientific interests, he was particularly fond of red cells and was a pioneer in elucidating the biochemical and genetic basis for a number of red cell metabolic disorders. Through his enthusiasm and insights, we have also come to appreciate the elegance, simplicity and medical import of the erythropoietic system.

References

1. Lichtman M, Beutler E, Kipps T et al. Williams hematology: 2007;McGraw-Hill.
2. Spivak JL. The anaemia of cancer: death by a thousand cuts. *Nat Rev Cancer* 2005;5:543–555.
3. Baron MH, Fraser ST. The specification of early hematopoiesis in the mammal. *Curr Opin Hematol* 2005;12:217–221.
4. McGrath K, Palis J. Ontogeny of erythropoiesis in the mammalian embryo. *Curr Top Dev Biol* 2008;82:1–22.
5. Dzierzak E, Speck NA. Of lineage and legacy: the development of mammalian hematopoietic stem cells. *Nat Immunol* 2008;9:129–136.
6. Koury MJ, Bondurant MC. Erythropoietin retards DNA breakdown and prevents programmed death in erythroid progenitor cells. *Science* 1990;248:378–381.
7. Migliaccio AR, Migliaccio G. Human embryonic hemopoiesis: control mechanisms underlying progenitor differentiation *in vitro*. *Dev Biol* 1988;125:127–134.
8. Akashi K, Traver D, Miyamoto T et al. A clonogenic common myeloid progenitor that gives rise to all myeloid lineages. *Nature* 2000;404:193–197.
9. Debili N, Coulombel L, Croisille L et al. Characterization of a bipotent erythro-megakaryocytic progenitor in human bone marrow. *Blood* 1996;88:1284–1296.
10. McLeod DL, Shreve MM, Axelrad AA. Induction of megakaryocyte colonies with platelet formation *in vitro*. *Nature* 1996;88:492–494.
11. Suda T, Suda J, Ogawa M. Single-cell origin of mouse hemopoietic colonies expressing multiple lineages in variable combinations. *Proc Natl Acad Sci USA* 1983;80:6689–6693.
12. Heath DS, Axelrad AA, McLeod DL et al. Separation of the erythropoietin-responsive progenitors BFU–E and CFU–E in mouse bone marrow by unit gravity sedimentation. *Blood* 1976;47:777–792.
13. Stephenson JR, Axelrad AA, McLeod DL et al. Induction of colonies of hemoglobin-synthesizing cells by erythropoietin *in vitro*. *Proc Natl Acad Sci USA* 1971;68:1542–1546.
14. Granick S, Levere RD. Heme synthesis in erythroid cells. Prog Hematol 194;27:1–47.
15. Bessis M, Mize C, Prenant M. Erythropoiesis: comparison of *in vivo* and *in vitro* amplification. *Blood Cells* 1978;4:155–174.
16. Manwani D, Bieker JJ. The erythroblastic island. *Curr Top. Dev Biol* 2008;82:23–53.
17. Rhodes MM, Kopsombut P, Bondurant MC et al. Adherence to macrophages in erythroblastic islands enhances erythroblast proliferation and increases erythrocyte production by a different mechanism than erythropoietin. *Blood* 2008;111:1700–1708.
18. Gifford SC, Derganc J, Shevkoplyas SS et al. A detailed study of time-dependent changes in human red blood cells: from reticulocyte maturation to erythrocyte senescence. *Br J Haematol* 2006;135:395–404.
19. McGrath KE, Palis J. Hematopoiesis in the yolk sac: more than meets the eye. *Exp Hematol* 2005;33:1021–1028.
20. Migliaccio G, Migliaccio AR, Petti S et al. Human embryonic hemopoiesis. Kinetics of progenitors and precursors underlying the yolk sac – liver transition. *J Clin Invest* 1986;78:51–60.
21. Mikkola HK, Gekas C, Orkin SH et al. Placenta as a site for hematopoietic stem cell development. *Exp Hematol* 2005;33:1048–1054.
22. Palis J, Yoder MC. Yolk-sac hematopoiesis: the first blood cells of mouse and man. *Exp Hematol* 2001;29:927–936.
23. Haar JL, Ackerman GA. A phase and electron microscopic study of vasculogenesis and erythropoiesis in the yolk sac of the mouse. *Anat Rec* 1971;170:199–223.

24. Silver L, Palis J. Initiation of murine embryonic erythropoiesis: a spatial analysis. *Blood* 197;89:1154–1164.
25. Palis J, Robertson S, Kennedy M et al. Development of erythroid and myeloid progenitors in the yolk sac and embryo proper of the mouse. *Development* 1999;126:5073–5084.
26. Wong PM, Chung SW, Chui DH et al. Properties of the earliest clonogenic hemopoietic precursors to appear in the developing murine yolk sac. *Proc Natl Acad Sci USA* 1986;83:3851–3854.
27. Brotherton TW, Chui DH, Gauldie J et al. Hemoglobin ontogeny during normal mouse fetal development. *Proc Natl Acad Sci USA* 1979;76:2853–2857.
28. Kingsley PD, Malik J, Fantauzzo KA et al. Yolk sac-derived primitive erythroblasts enucleate during mammalian embryogenesis. *Blood* 2004;104:19–25.
29. Steiner R, Vogel H. On the kinetics of erythroid cell differentiation in fetal mice. I. Microspectrophotometric determination of the hemoglobin content in erythroid cells during gestation. *J Cell Physiol* 1973;81:323–338.
30. Huber TL, Kouskoff V, Fehling HJ et al. Haemangioblast commitment is initiated in the primitive streak of the mouse embryo. *Nature* 2004;432:625–630.
31. de Bruijn MF, Speck NA, Peeters MC et al. Definitive hematopoietic stem cells first develop within the major arterial regions of the mouse embryo. *EMBO J* 2000;19:2465–2474.
32. Muller AM, Medvinsky A, Strouboulis J et al. Development of hematopoietic stem cell activity in the mouse embryo. *Immunity* 1994;1:291–301.
33. Zeigler BM, Sugiyama D, Chen M et al. The allantois and chorion, when isolated before circulation or chorio-allantoic fusion, have hematopoietic potential. *Development* 2006;133:4183–4192.
34. Ema H, Nakauchi H. Expansion of hematopoietic stem cells in the developing liver of a mouse embryo. *Blood* 2000;95:2284–2288.
35. Kumaravelu P, Hook L, Morrison AM et al. Quantitative developmental anatomy of definitive haematopoietic stem cells/long-term repopulating units (HSC/RUs): role of the aorta-gonad-mesonephros (AGM) region and the yolk sac in colonisation of the mouse embryonic liver. *Development* 2002;129:4891–4899.
36. Ghiaur G, Ferkowicz MJ, Milsom MD et al. Rac1 is essential for intraembryonic hematopoiesis and for the initial seeding of fetal liver with definitive hematopoietic progenitor cells. *Blood* 2008;111:3313–3321.
37. Lux CT, Yoshimoto M, McGrath K et al. All primitive and definitive hematopoietic progenitor cells emerging before E10 in the mouse embryo are products of the yolk sac. *Blood* 2008;111:3435–3438.
38. Palis J. Developmental biology: no red cell is an island. *Nature* 2004;432:964–965.
39. Jelkmann W. Erythropoietin: structure, control of production, and function. *Physiol Rev* 1992;72:449–489.
40. Adamson JW. The erythropoietin-hematocrit relationship in normal and polycythemic man: implications of marrow regulation. *Blood* 1968;32:597–609.
41. Hillman RS, Henderson PA. Control of marrow production by the level of iron supply. *J Clin Invest* 1969;48:454–460
42. Crosby WH, Akeroyd JH. The limit of hemoglobin synthesis in hereditary hemolytic anemia; its relation to the excretion of bile pigment. *Am J Med* 1952;13:273–283.
43. Lin CS, Lim SK, D'Agati V et al. Differential effects of an erythropoietin receptor gene disruption on primitive and definitive erythropoiesis. *Genes Dev* 1996;10:154–164.
44. Spivak JL, Pham T, Isaacs M et al. Erythropoietin is both a mitogen and a survival factor. *Blood* 1991;77:1228–1233.
45. Sawyer ST, Penta K. Erythropoietin cell biology. *Hematol Oncol Clin North Am* 1994;8:895–911.
46. Koury ST, Bondurant MC, Koury MJ et al. Localization of cells producing erythropoietin in murine liver by *in situ* hybridization. *Blood* 1991;77:2497–2503.
47. Lacombe C, DaSilva JL, Bruneval P et al. Peritubular cells are the site of erythropoietin synthesis in the murine hypoxic kidney. *J Clin Invest* 1998;81:620–623.
48. Bachmann S, Le Hir M, Eckardt KU. Co-localization of erythropoietin mRNA and ecto-5'-nucleotidase immunoreactivity in peritubular cells of rat renal cortex indicates that fibroblasts produce erythropoietin. *J Histochem Cytochem* 1993;41:335–341.
49. Epstein AC, Gleadle JM, McNeill LA et al. C. elegans EGL-9 and mammalian homologs define a family of dioxygenases that regulate HIF by prolyl hydroxylation. *Cell* 2001;107:43–54.
50. Mahon PC, Hirota K, Semenza GL. FIH-1: a novel protein that interacts with HIF-1alpha and VHL to mediate repression of HIF-1 transcriptional activity. *Genes Dev* 2001;15:2675–2686.

51. Wang GL, Semenza GL. Desferrioxamine induces erythropoietin gene expression and hypoxia-inducible factor 1 DNA-binding activity: implications for models of hypoxia signal transduction. *Blood* 1993;82:3610–3615.
52. Wang GL, Semenza GL. Characterization of hypoxia-inducible factor 1 and regulation of DNA binding activity by hypoxia. *J Biol Chem* 1993;268:21513–21518.
53. Semenza GL. HIF-1, O(2), and the 3 PHDs: how animal cells signal hypoxia to the nucleus. *Cell* 2001;107:1–3.
54. Peet D, Linke S. Regulation of HIF: asparaginyl hydroxylation. *Novartis Found Symp* 2006;272:37–49; discussion 49–53, 131–140.
55. Ang SO, Chen H, Gordeuk VR et al. Endemic polycythemia in Russia: mutation in the VHL gene. *Blood Cells Mol Dis* 2002;28:57–62.
56. Ang SO, Chen H, Hirota K et al. Disruption of oxygen homeostasis underlies congenital Chuvash polycythemia. *Nat Genet* 2002;32:614–621.
57. Bento MC, Chang KT, Guan Y et al. Congenital polycythemia with homozygous and heterozygous mutations of von Hippel-Lindau gene: five new Caucasian patients. *Haematologica* 2005;90:128–129.
58. Brunelle JK, Bell EL, Quesada NM et al. Oxygen sensing requires mitochondrial ROS but not oxidative phosphorylation. *Cell Metab* 2005;1:409–414.
59. Guzy RD, Hoyos B, Robin E et al. Mitochondrial complex III is required for hypoxia-induced ROS production and cellular oxygen sensing. *Cell Metab* 2005;1:401–408.
60. Mansfield KD, Guzy RD, Pan Y et al. Mitochondrial dysfunction resulting from loss of cytochrome c impairs cellular oxygen sensing and hypoxic HIF-alpha activation. *Cell Metab* 2005;1:393–399.
61. Zhang H, Bosch-Marce M, Shimoda LA et al. Mitochondrial autophagy is a HIF-1-dependent adaptive metabolic response to hypoxia. *J Biol Chem* 2008;283:10892–10903.
62. Semenza GL. Hypoxia-inducible factor 1 (HIF-1) pathway. *Sci STKE* 2007;2007:cm8.
63. Berra E, Benizri E, Ginouves A et al. HIF prolyl-hydroxylase 2 is the key oxygen sensor setting low steady-state levels of HIF-1alpha in normoxia. *EMBO J* 2003;22:4082–4090.
64. Percy MJ, Zhao Q, Flores A et al. From the Cover: A family with erythrocytosis establishes a role for prolyl hydroxylase domain protein 2 in oxygen homeostasis. *Proc Natl Acad Sci USA* 2006;103:654–659.
65. Percy MJ, Furlow PW, Beer PA et al. A novel erythrocytosis-associated PHD2 mutation suggests the location of a HIF binding groove. *Blood* 2007;110:2193–2196.
66. Rankin EB, Biju MP, Liu Q et al. Hypoxia-inducible factor-2 (HIF-2) regulates hepatic erythropoietin *in vivo*. *J Clin Invest* 2007;117:1068–1077.
67. Gruber M, Hu CJ, Johnson RS et al. Acute postnatal ablation of Hif-2alpha results in anemia. *Proc Natl Acad Sci USA* 2007;104:2301–2306.
68. Witthuhn BA, Quelle FW, Silvennoinen O et al. JAK2 associates with the erythropoietin receptor and is tyrosine phosphorylated and activated following stimulation with erythropoietin. *Cell* 1993;74:227–236.
69. Wilson IA, Jolliffe LK. The structure, organization, activation and plasticity of the erythropoietin receptor. *Curr Opin Struct Biol* 1999;9:696–704.
70. Richmond TD, Chohan M, Barber DL. Turning cells red: signal transduction mediated by erythropoietin. *Trends Cell Biol* 2005;15:146–155.
71. Wu H, Liu X, Jaenisch R et al. Generation of committed erythroid BFU–E and CFU–E progenitors does not require erythropoietin or the erythropoietin receptor. *Cell* 1995;83:59–67.
72. Neubauer H, Cumano A, Muller M et al. Jak2 deficiency defines an essential developmental checkpoint in definitive hematopoiesis. *Cell* 1998;93:397–409.
73. Parganas E, Wang D, Stravopodis D et al. Jak2 is essential for signaling through a variety of cytokine receptors. *Cell* 1998;93:385–395.
74. Klingmuller U, Lorenz U, Cantley LC et al. Specific recruitment of SH-PTP1 to the erythropoietin receptor causes inactivation of JAK2 and termination of proliferative signals. *Cell* 1995;80:729–738.
75. Kralovics R, Indrak K, Stopka T et al. Two new EPO receptor mutations: truncated EPO receptors are most frequently associated with primary familial and congenital polycythemias. *Blood* 1997;90:2057–2061.
76. Sokol L, Prchal JF, D'Andrea A et al. Mutation in the negative regulatory element of the erythropoietin receptor gene in a case of sporadic primary polycythemia. *Exp Hematol* 1994;22:447–453.

77. de la Chapelle A, Traskelin AL, Juvonen E. Truncated erythropoietin receptor causes dominantly inherited benign human erythrocytosis. *Proc Natl Acad Sci USA* 1993;90:4495–4499.
78. Levine RL, Wadleigh M, Cools J et al. Activating mutation in the tyrosine kinase JAK2 in polycythemia vera, essential thrombocythemia, and myeloid metaplasia with myelofibrosis. *Cancer Cell* 2005;7:387–397.
79. James C, Ugo V, LeCouedic JP et al. A unique clonal JAK2 mutation leading to constitutive signalling causes polycythaemia vera. *Nature* 2005;434:1144–1148.
80. Baxter EJ, Scott LM, Campbell PJ et al. Acquired mutation of the tyrosine kinase JAK2 in human myeloproliferative disorders. *Lancet* 2005;365:1054–1061.
81. Kralovics R, Passamonti F, Buser AS et al. A gain-of-function mutation of JAK2 in myeloproliferative disorders. *N Engl J Med* 2005;352:1779–1790.
82. Lu X, Levine R, Tong W et al. Expression of a homodimeric type I cytokine receptor is required for JAK2V617F-mediated transformation. *Proc Natl Acad Sci USA* 2005;102:18962–18967.
83. Sathyanarayana P, Dev A, Fang J et al. EPO receptor circuits for primary erythroblast survival. *Blood* 2008;blood-2007–2010–119743.
84. Spivak JL. The blood in systemic disorders. *Lancet* 2000;355:1707–1712.
85. Nemeth E, Tuttle MS, Powelson J et al. Hepcidin regulates cellular iron efflux by binding to ferroportin and inducing its internalization. *Science* 2004;306:2090–2093.
86. Ganz T. Molecular control of iron transport. *J Am Soc Nephrol* 2007;18:394–400.
87. Nemeth E, Rivera S, Gabayan V et al. IL-6 mediates hypoferremia of inflammation by inducing the synthesis of the iron regulatory hormone hepcidin. *J Clin Invest* 2004;113:1271–1276.
88. Maiese K, Li F, Chong ZZ. New avenues of exploration for erythropoietin. *JAMA* 2005;293:90–95.
89. Ghezzi P, Brines M. Erythropoietin as an antiapoptotic, tissue-protective cytokine. *Cell Death Differ* 2004;11 Suppl 1:S37–44.
90. Gregory T, Yu C, Ma A et al. GATA-1 and erythropoietin cooperate to promote erythroid cell survival by regulating bcl-xL expression. *Blood* 1999;94:87–96.
91. Orkin SH, Weiss MJ. Apoptosis. Cutting red-cell production. *Nature* 1999;401:433, 435–436.
92. Brines M, Grasso G, Fiordaliso F et al. Erythropoietin mediates tissue protection through an erythropoietin and common beta-subunit heteroreceptor. *Proc Natl Acad Sci USA* 2004;101:14907–14912.
93. Erbayraktar S, Grasso G, Sfacteria A et al. Asialoerythropoietin is a nonerythropoietic cytokine with broad neuroprotective activity *in vivo*. *Proc Natl Acad Sci USA* 2003;100:6741–6746.
94. Schmidt RE, Green KG, Feng D et al. Erythropoietin and its carbamylated derivative prevent the development of experimental diabetic autonomic neuropathy in STZ-induced diabetic NOD-SCID mice. *Exp Neurol* 2008;209:161–170.
95. Fiordaliso F, Chimenti S, Staszewsky L et al. A nonerythropoietic derivative of erythropoietin protects the myocardium from ischemia-reperfusion injury. *Proc Natl Acad Sci USA* 2005;102:2046–2051.
96. Leist M, Ghezzi P, Grasso G et al. Derivatives of erythropoietin that are tissue protective but not erythropoietic. *Science* 2004;305:239–242.
97. Yu X, Shacka JJ, Eells JB et al. Erythropoietin receptor signalling is required for normal brain development. *Development* 2002;129:505–516.
98. Wu H, Lee SH, Gao J, Liu X, Iruela-Arispe ML. Inactivation of erythropoietin leads to defects in cardiac morphogenesis. *Development* 1999;126:3597–3605.
99. Arcasoy MO. The non-haematopoietic biological effects of erythropoietin. *Br J Haematol* 2008;141:14–31.
100. Phrommintikul A, Haas SJ, Elsik M et al. Mortality and target haemoglobin concentrations in anaemic patients with chronic kidney disease treated with erythropoietin: a meta-analysis. *Lancet* 2007;369:381–388.
101. Bohlius J, Wilson J, Seidenfeld J et al. Recombinant human erythropoietins and cancer patients: updated meta-analysis of 57 studies including 9353 patients. *J Natl Cancer Inst* 2006;98:708–714.
102. Henke M, Laszig R, Rube C et al. Erythropoietin to treat head and neck cancer patients with anaemia undergoing radiotherapy: randomised, double-blind, placebo-controlled trial. *Lancet* 2003;362:1255–1260.
103. Leyland-Jones B, Semiglazov V, Pawlicki M et al. Maintaining normal hemoglobin levels with epoetin alfa in mainly nonanemic patients with metastatic breast cancer receiving first-line

chemotherapy: a survival study. *J Clin Oncol* 2005;23:5960–5972.
104. Wright JR, Ung YC, Julian JA et al. Randomized, double-blind, placebo-controlled trial of erythropoietin in non-small-cell lung cancer with disease-related anemia. *J Clin Oncol* 2007;25:1027–1032.
105. Rizzo JD, Somerfield MR, Hagerty KL et al. Use of epoetin and darbepoetin in patients with. *Cancer* 2007 American Society of Hematology/American Society of Clinical Oncology clinical practice guideline update. *Blood* 2008;111:25–41.
106. Jelkmann W, Pagel H, Wolff M et al. Monokines inhibiting erythropoietin production in human hepatoma cultures and in isolated perfused rat kidneys. *Life Sci* 1992;50:301–308.
107. Bianchi L, Tacchini L, Cairo G. HIF-1-mediated activation of transferrin receptor gene transcription by iron chelation. *Nucleic Acids Res* 1999;27:4223–4227.
108. Peyssonnaux C, Zinkernagel AS, Schuepbach RA et al. Regulation of iron homeostasis by the hypoxia-inducible transcription factors (HIFs). *J Clin Invest* 2007;117:1926–1932.

Erythropoietins, Erythropoietic Factors and Erythropoiesis
Edited by S.G. Elliott, M.A. Foote, and G. Molineux

Regulation of endogenous erythropoietin production

David R. Mole and Peter J. Ratcliffe

University of Oxford, Headington Campus, Roosevelt Drive, Oxford OX3 7BN, United Kingdom

Hypoxic regulation of erythropoietin

The principal function of the red cell is to convey oxygen from the lungs to the tissues through the oxygen transport molecule, haemoglobin, a concept finally established in the 19th Century after major advances in biology and chemistry over the preceding two centuries. The description of the blood circulation by William Harvey (1578–1657) in *De Motu Cordis et Sanguinis in Animalibus* in 1628 framed the question as to the purpose of moving such large volumes of fluid around the body. Richard Lower (1631–1691) working in Oxford with Robert Hooke (1635–1702) noted that whereas the blood leaving the heart for the lungs was blue, that returning from the lungs to the heart was red, concluding that 'Nitrous spirit of the air, vital to life is mixed with the blood during transit through the lungs'. After the first consistent measurements of oxygen in blood, by Gustav Magnus, the role of the blood circulation in delivering oxygen to the tissues was confirmed by showing that there was more oxygen in arterial than venous blood [1]. Finally, a specific role for erythrocytes in oxygen transport was established by the demonstration of reversible binding of oxygen to the pigmented haemoglobin that accounted for the colour change [2, 3].

The idea of hormonal regulation of erythropoiesis, the production of new erythrocytes, was first formulated by Carnot and Deflandre in 1906 [4] only a few years after the concept of hormones was first suggested by Henri Brown-Sequard in 1889. However, Allan Erslev (1919–2003) is generally credited with providing definitive proof of the existence of erythropoietin in 1953, by transfusing large quantities of plasma from anaemic rabbits into normal animals, which responded by generating a consistent reticulocytosis, followed by a major increase in hematocrit [5]. With remarkable foresight, he also postulated "Conceivably isolation and purification of this factor would provide an agent useful in the treatment of conditions associated with erythropoietic depression, such as chronic infection and chronic renal disease." Much of the early work on the effects of reduced oxygen levels (hypoxia) on erythropoiesis was undertaken at altitude, with the earliest description generally attributed to the Jesuit priest, Father Joseph De Acosta. Both Journadet and Bert, considered the pioneer of altitude medicine, described an association between altitude and polycythaemia. Induction of erythropoiesis was noted to be both rapid and sensitive to small changes in altitude [6]. Subsequently, after the recognition of erythropoietin (EPO) as the regulatory hormone, measurement of EPO concentrations revealed an extremely rapid response to ascent, with endogenous EPO peaking at 1–3 days after arrival at a new altitude and decreasing close to baseline by day 10 [7].

Renal oxygenation and erythropoietin

The kidneys are the main source of circulating EPO in adults, as initially suggested by the lack of an EPO response to haemorrhage in nephrectomised rabbits and rats [8], and subsequently confirmed through the demonstration of EPO activity in isolated serum-free perfused kidneys from hypoxic rabbits [9]. Further confirmation was by the isolation of EPO mRNA from the kidneys of hypoxic rodents [10]. More sensitive analyses of mRNA demonstrated that the liver, spleen, lung, bone marrow, and brain also express lesser amounts of EPO mRNA [11], although the contribution to EPO production and the biologic role of gene expression at these sites is unclear. Within the kidneys, the EPO-producing cells reside within the interstitium of the renal cortex, and co-localise with fibroblast markers, defining the interstitial fibroblast as the source of renal EPO [12, 13].

Given that the teleological purpose of EPO is to maintain the blood haemoglobin concentration in the normal range, and that the main function of haemoglobin is to carry and deliver oxygen from the lungs to the respiring tissues, it is fitting that the sensed parameter in the feedback loop is tissue oxygenation. Given that the kidney, however, has no direct role in erythropoiesis or oxygen transport, it is not immediately apparent why the kidney is so well suited to sensing oxygen.

A potential answer to this paradox lies in the unusual anatomy of the renal circulation, in which the filtering glomeruli and reabsorbing tubules are supplied in series by the same network of vessels. Given that most of the glomerular filtrate is reabsorbed, the renal oxygen demand for tubular reabsorption will vary in proportion with the glomerular filtration rate. Thus, reductions in renal perfusion are predicted to result in a parallel reduction in oxygen supply and oxygen demand (proportional to filtration/reabsorption) so that renal oxygen balance is little affected by the rate of blood flow and more accurately reflects changes in oxygen content of the blood (as determined by haemoglobin concentration). A second interesting feature of the renal vascular anatomy is the tendency for arteries and veins to run countercurrent. This set up allows direct 'shunt' diffusion of oxygen from artery to vein so that the tips of the countercurrent loops are particularly hypoxic, which is most marked in the renal medulla where the innermost regions are profoundly hypoxic even under basal physiologic conditions. Similar effects, however, are also observed in the renal cortex, where EPO is produced, with cortical oxygen tension decreasing with increasing cortical depth. Interestingly, under basal conditions, EPO production is confined to a few cells near the corticomedullary junction, but as the haemoglobin concentration decreases in progressive anaemia, so the hypoxic regions of the kidney propagate outwards, and increasing numbers of peritubular fibroblasts are recruited to produce EPO in a 'march' from deep to superficial cortex (Fig. 1) [14]. The induction of EPO in this manner shows a strikingly high gain response, with even modest changes in haemoglobin leading to large changes in EPO over several orders of magnitude.

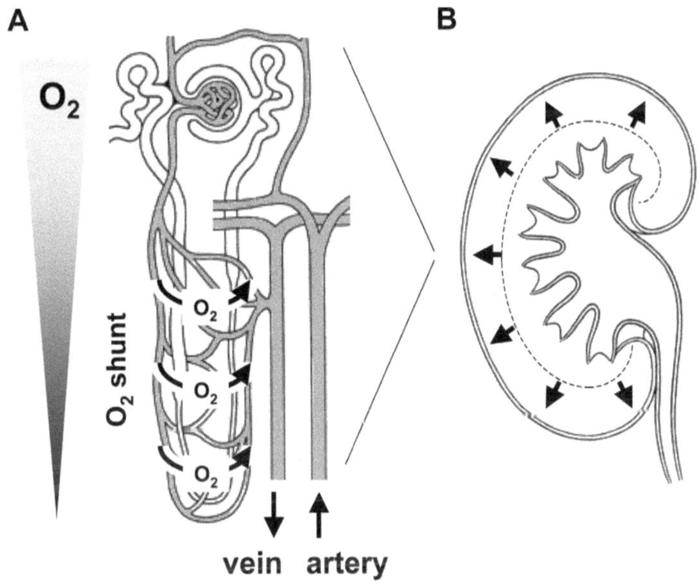

Figure 1. (A) The highly specialized microvascular anatomy of the kidney, illustrating the same network of vessels, in series, supplying both the filtering glomerulus and the reabsorbing tubule, of each nephron. Tubular reabsorption varies in proportion with glomerular filtration such that tubular oxygen requirements parallel blood flow, rendering renal oxygenation insensitive to changes in blood flow, while remaining sensitive to changes in oxygen content (and hence haemoglobin concentration) of the blood. Countercurrent tubular flow is paralleled by countercurrent blood flow, with consequent shunting of oxygen, generating an oxygen gradient within the kidney. (B) As the haemoglobin concentration decreases in progressive anaemia, so the hypoxic regions of the kidney propagate outwards, and increasing numbers of peritubular fibroblasts are recruited to produce EPO in a 'march' from deep to superficial cortex.

cis-acting regulatory elements of the erythropoietin gene

Identification of the *EPO* gene provided the opportunity to study the pathways controlling its expression by defining its *cis*-acting regulatory sequences (non-coding sequences at a gene locus that direct expression by interacting with transcription factors) (Fig. 2).

EPO contains five exons spanning a region of DNA of approximately 2.9 kb [15, 16]. Comparison of primate and mouse sequences showed strong homol-

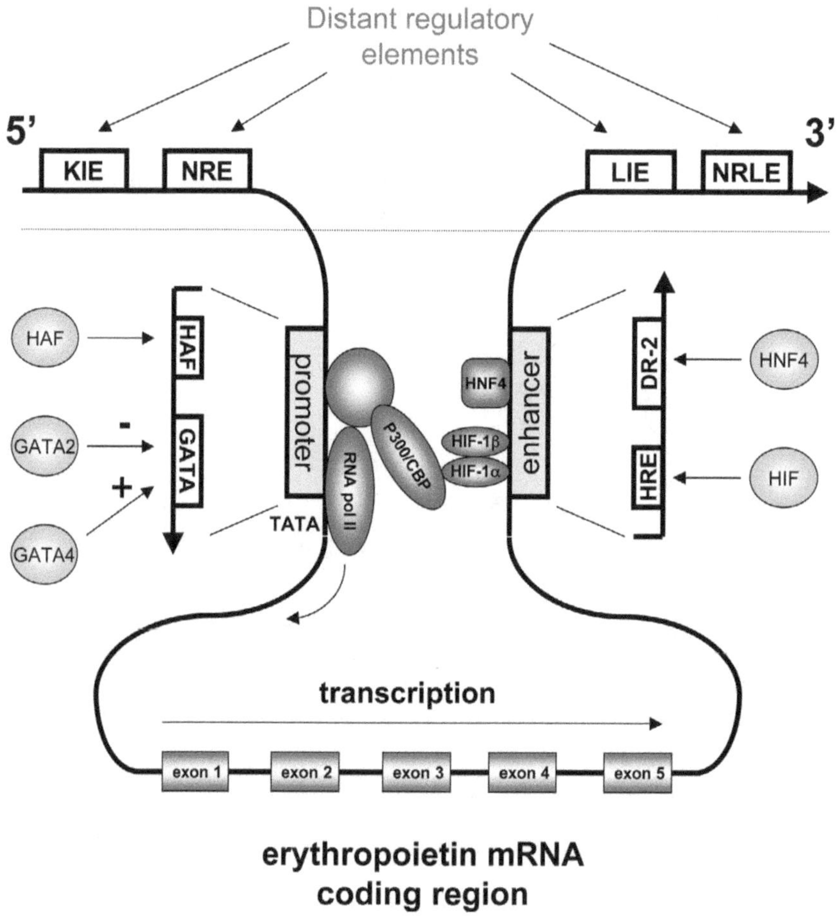

Figure 2. Regulatory elements of the *EPO* gene. This schematic representation of *EPO* shows the promoter region, the five exons and the 3' EPO enhancer. Additional *cis*-acting regulatory are required for tissue specific and developmental regulation. The kidney inducible element (KIE) confers expression in interstitial peritubular cells. The negative regulatory element (NRE) suppresses transcription. Two *cis*-acting regulatory elements 3' to the RPO-coding sequence, the liver-inducible element (LIE), and the negative regulatory liver element (NRLE) promote and enhance gene expression in the liver. At the promoter, 5' sequences contain a "GATA" motif that regulates opposing effects of diverse GATA factors. While the EPO promoter is a weak promoter, it does contribute to hypoxic regulation through a hypoxia-associated factor (HAF) binding site. However, the 3' enhancer containing a hypoxia-responsive element (HRE) that binds hypoxia-inducible factor (HIF) is responsible for most of the fold-induction of the *EPO* gene in response to hypoxia. In addition, the direct repeat of two steroid hormone receptor half-sites (DR2) confers regulation through binding of hepatic nuclear factor 4 (HNF-4).

ogy, not only in the coding sequence of *EPO*, but also within the 5' flanking region, the first intron, and the 3' flanking region, suggesting that one or more sequences within these regions might function in regulation. Both studies in transgenic mice and transfection studies in cultured cells have been used to define important *cis*-acting sequences within *EPO* that lead to tissue-specific, developmental, and hypoxia-inducible control of *EPO* expression.

The first studies of EPO transgenic mice, containing a 4 kb DNA fragment that included 0.4 kb of the 5'-flanking sequence and 0.7 kb of the 3'-flanking sequence revealed anaemia inducible expression of the transgene in the liver, indicating the existence of at least one sequence that could interact with the hypoxia-signalling apparatus within these sequences. The widespread unregulated expression in other tissues that do not normally express EPO, however, indicated that additional, more distant, elements are required for correct tissue-specific and developmental regulation. The widespread anomalous expression of EPO was reduced in transgenes containing a further 6 kb of the 5'-flanking sequence, suggesting that this region contains one or more negative regulatory elements. In these mice, however, inducible EPO expression occurred in the liver, rather than kidney, in contrast to normal adult mice. Extension of the 5'-flanking sequence still further, identified a kidney inducible element that mediated inducible expression of the EPO transgene in interstitial fibroblasts of the kidney. These mice still manifest polycythaemia as a result of excessive hepatic EPO expression. Regulation of the transgene by physiologic and tissue specific signals only mirrored expression of the endogenous gene when the 3'-flanking sequence was extended to 7 kb defining a further negative regulatory element restricting hepatic expression. Thus, these studies have demonstrated the existence of distinct sequences that promote or restrict EPO expression in different tissues. Of particular interest, however, was evidence for local sequences that could mediate responses to anaemia in the liver. Mapping of a DNAse hypersensitivity site within this region suggested the possibility of a regulatory site lying 3' to the gene. Further analysis by transfection of EPO-producing hepatoma cell lines (HepG2 and Hep3b), identified a hypoxia-inducible control sequence termed the EPO 3' enhancer. In both the mouse and human *EPO*, this enhancer lies in the highly conserved region 120 bp 3' to the polyadenylation site. Analysis of the EPO 3' enhancer in hepatoma cells defined three functional sites. Critical for the regulation by hypoxia is a DNA motif, termed the hypoxia-responsive element (HRE) that was found to bind an inducible protein complex termed hypoxia-inducible factor (HIF), the transcription factor that is responsible for hypoxic regulation of EPO expression [17].

Two other sites were shown to be necessary for proper function of the enhancer, a second motif (CACA), 7 bp 3' to the HRE, and a third motif consisting of a direct repeat of two steroid hormone receptor half-sites separated by 2 bp (DR-2). In the setting of the EPO enhancer, this motif binds hepatic nuclear factor (HNF)-4. These multiple sites are typical of transcriptional enhancers and are presumed to enable cooperative assembly of transcriptional complexes. How these proteins cooperate with HIF bound to the core HRE is not fully understood, but they may modulate the response to hypoxia and contribute to the tissue specificity of EPO expression. For example, HNF-4, which is expressed in the liver and renal cortex, may contribute to developmental and tissue specific regulation of endogenous EPO expression. In this respect, it is also interesting that binding of retinoic acid receptor (RXRα) to the DR-2 site during early embryonic development is superseded by HNF-4. This switch corresponds temporally with a change in the characteristics of EPO expression from responsiveness to retinoic acid to responsiveness to hypoxia [18].

DNA methylation may also have a role in regulation of endogenous *EPO* expression. The consensus HIF-binding site (G/ACGTG) contains a CpG dinucleotide known to be a potential target of cytosine methylation. Methylation at this site abolishes binding of HIF to the HRE, recruits inhibitory methyl-CpG-binding proteins, and abrogates hypoxic regulation of reporter gene expression [19, 20]. Overall, these studies define a specific point of interaction between the oxygen-sensitive signalling system and the *EPO* locus. Unexpectedly, however, transfection studies of the 3' EPO enhancer also demonstrated oxygen-regulated activity after introduction into a wide variety of cells, regardless of whether the cells produced EPO, or were derived from an EPO-producing organ. Thus, it became clear that the highly specific and sensitive response to hypoxia that was manifest in the EPO-producing tissues was, in fact, a general property of mammalian cells [21]. The biologic importance of this oxygen-sensing mechanism is further emphasised by the discovery that the HIF system is conserved across the entire animal kingdom, even in primitive species without specialised oxygen delivery systems, such as blood vessels.

EPO – a paradigm for widespread hypoxic gene regulation

The discovery that the oxygen-sensing mechanism operated in all cells, irrespective of their ability to express EPO, led to the realisation that the range of genes regulated by oxygen and HIF extended far beyond EPO, and was involved in the regulation of many biologic processes that both facilitate oxygen delivery and reduce oxygen demand (Fig. 3) [22]. Effective erythropoiesis requires not only stimulation of the bone marrow by EPO, but also coordinated provision of iron, supplied through its transporter, transferrin, and its cellular uptake mechanism, the transferrin receptor, and a range of other mechanisms, many of which are now recognised to be regulated by HIF.

Increased oxygen delivery is mediated not only through erythropoiesis, but also through changes in the vasculature. During localised hypoxia, vasomotor tone is controlled through HIF-mediated transcriptional regulation of genes encoding vasoactive factors such as endothelin 1, endothelial nitric-oxide synthase (eNOS), hemoxygenase-1 (HO-1), and atrial naturetic peptide (ANP), and also a range of ion channels that may operate to alter vascular tone and responsiveness.

Another important vascular response to hypoxia is the regulation of angiogenesis through angiogenic growth factors. Both vascular epithelial growth factor (VEGF) and its receptor Flt-1 are transcriptionally activated by HIF, and

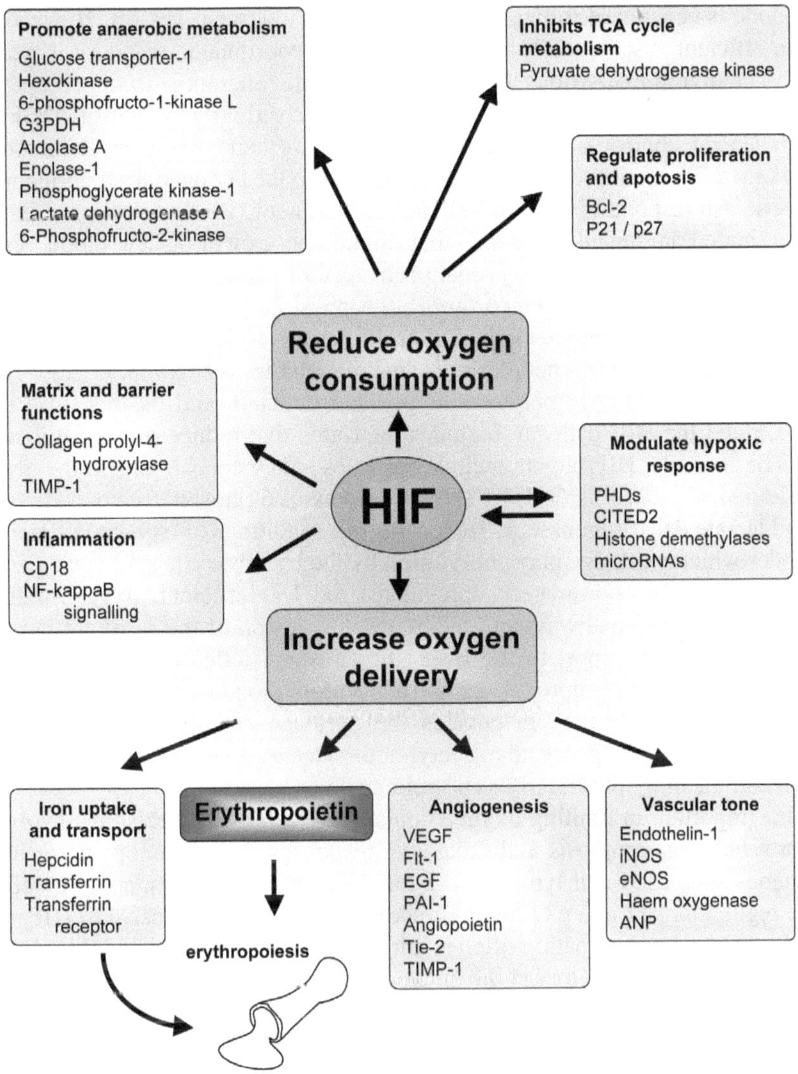

Figure 3. Transcriptional targets of HIF. A representation of the increasing number of genes regulated by the HIF oxygen-sensing transcriptional pathway is illustrated. Broadly these act to increase oxygen delivery systemically by promoting erythropoiesis and iron delivery, and at the tissue level by promoting angiogenesis and controlling vascular tone, in addition to reducing oxygen consumption by inhibiting TCA (tricarboxylic acid) cycle metabolism, promoting anaerobic glycolysis, and modulating cell proliferation and apoptosis. In addition it is now recognised that several components of the HIF regulatory pathway itself are also transcriptionally regulated by HIF, leading to autoregulatory feedback loops. Glyceraldehyde-3-phosphate dehydrogenase (G3PHD), B-cell CLL/lymphoma 2 (Bcl-2). vascular endothelial growth factor (VEGF), fms-related tyrosine kinase 1/VEGF receptor (Flt-1), epidermal growth factor (EGF), plasminogen activator inhibitor-1 (PAI-1), endothelium-specific receptor tyrosine kinase 2 (Tie-2), tissue inhibitor of matrix metalloproteinase-1 (TIMP-1), inducible nitric oxide synthase (iNOS), endothelial nitric oxide synthase (eNOS), atrial natriuretic peptide (ANP), prolyl hydroxylase domain containing (PHD), CBP/p300 interacting transactivator with ED-rich tail (CITED2).

this alone is capable of initiating angiogenesis in quiescent vessels. However, for an efficient vasculature to be formed a more coordinated response is necessary, involving many other factors. For example, angiopoietin, its receptor Tie-2, fibroblast growth factor (FGF), platelet-derived growth factor (PDGF), and balanced control of matrix metalloproteinases and tissue inhibitors of matrix metalloproteinases (TIMP) all contribute to the endogenous angiogenic response. All respond to the HIF system. Activation of HIF therefore promotes a coordinated angiogenic response that directs the growth of new blood vessels into areas of impaired oxygenation to restore adequate oxygen delivery. This activation is of importance during the physiologic processes of growth and development, wound healing, and proliferation, as well as in pathological responses, resulting from neoplasia, ischaemia, and inflammation. Local oxygen tension is a balance between supply and demand, and the response to hypoxia and the HIF pathway includes measures that reduce oxygen utilisation. For example, HIF targets include enzymes at key checkpoints in the glycolytic pathway such as GLUT1 (glucose uptake), 6-phosphofructo-1-kinase L, and lactate dehydrogenase A. Hence, during conditions of oxygen deficiency under which oxidative phosphorylation by the tricarboxylic acid cycle cannot proceed, HIF coordinately upregulates the less efficient, oxygen-independent, glycolytic pathway, and facilitates conversion of the resultant pyruvate to lactate, for export to the liver (the "Pasteur" effect). In addition, in hypoxia HIF directly upregulates pyruvate dehydrogenase kinase (PDK), which phosphorylates and inactivates the pyruvate dehydrogenase enzyme complex that converts pyruvate to acetyl-coenzyme A, thereby inhibiting pyruvate metabolism by the tricarboxylic acid cycle.

Also important in limiting oxygen demand are cell-based decisions involving proliferation, apoptosis and protective responses such as autophagy, with key genes such as B-cell lymphoma-2 family members (Bcl-2), and the cell-cycle regulators p21 and p27 being directly or indirectly responsive to HIF.

Infection and inflammation often result in lowered oxygen tension. Myeloid cells, and other key effectors of the innate immune response, have evolved several HIF-dependent responses such as enhanced glycolysis, inhibition of NF-κB dependent apoptosis, and enhanced CD18 dependent diapedesis of neutrophils into hypoxic regions.

Hypoxia inducible-factor (HIF)

Hypoxia-inducible factor was first isolated by DNA affinity chromatography using the EPO 3' enhancer sequences and shown to be a heterodimeric transcription factor composed of α and β subunits [23]. The HIFβ subunit also functions as the aryl hydrocarbon nuclear receptor translocator (ARNT) in other transcription pathways, while HIFα subunits are specific to the hypoxic response and are highly regulated by the availability of oxygen [23–25]. Overall, three HIFα subunits have been identified, though HIF-1α and HIF-2α are the best studied to date. Both HIF-1α and HIF-2α have a similar domain structure and contain, in their *N*-terminal regions, both a basic-helix-loop-helix (bHLH) motif, common to many transcription factors, and a PAS (Per AHR/ARNT Sim) domain, defined by its presence in the *Drosophila* Per and Sim proteins and in the mammalian ARNT and AHR proteins (Fig. 4) [23]. While this *N*-terminal portion of the HIF subunits is necessary for dimerisation and DNA binding, it is the *C*-terminal region of the HIFα subunits that is responsible for responsiveness to oxygen. In addition to oxygen-dependent degradation domains (ODDD) that regulate protein stability, *C*-terminal sequences of HIF-1α and HIF-2α also contain transactivating domains (TAD) that control the hypoxically inducible transactivating ability of the protein [26], in part through binding to p300/CBP [27]. Thus, the control of HIF by oxygen involves the regulation of two independent properties of HIFα subunits, stability and activity.

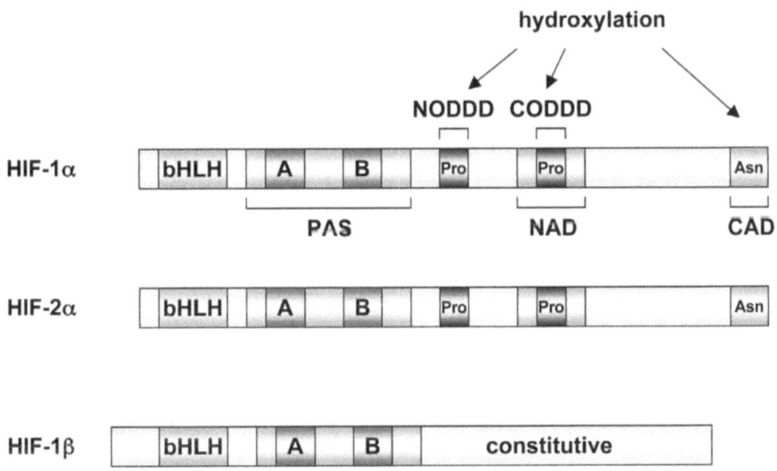

Figure 4. The domain structure of HIF-1α, HIF-2α, and HIF-1α/ARNT subunits. The figure shows the basic helix-loop-helix (bHLH) domains and the PAS domains responsible for dimerisation and DNA binding. The *C*-terminal portion of HIF-1α and HIF-2α contains the regulatory domains: the amino-terminal oxygen-dependent degradation domain (NODDD) and the carboxy- terminal oxygen-dependent degradation domain (CODDD), responsible for regulating HIFα stability and the amino-terminal activation domain (NAD) and the carboxy-terminal activation domain (CAD) involved in regulating transactivating ability. These regulatory properties involve enzymatic hydroxylation of specific residues in these regions. PAS domain = PER, AHR, ARNT, SIM domain; PER = periodic circadian protein, AHR = aryl-hydrocarbon receptor, ARNT = aryl-hydrocarbon receptor nuclear translocator, SIM = single-minded protein.

Differential functions of the HIF isoforms

The identification of multiple HIF isoforms raises questions about the existence and extent of distinct, *versus* overlapping, roles. At present the precise function of HIF-3α remains unknown, although a splice variant, inhibitory

PAS (iPAS) can antagonise responses to the other two HIFα isoforms [28]. To date, study has concentrated mostly on HIF-1α and HIF-2α. These isoforms have largely concordant, although not identical, mechanisms of regulation. However, differing patterns of organ expression and cellular distribution indicate differing roles for HIF-1α and HIF-2α. Originally, HIF-2α was thought to be largely restricted to the mouse endothelium, while HIF-1α had an almost ubiquitous distribution. Better-characterised patterns of expression have been reported by immunohistochemistry [29]. For example, in the hypoxic kidney HIF-2α is seen in the EPO-producing interstitial fibroblasts as well as capillary endothelial cells, whereas HIF-1α is the dominant form expressed in tubular epithelial cells [30].

Further evidence of distinct roles for HIF-1α and HIF-2α comes from the contrasting phenotypes seen in mouse knockout models. HIF-1α knockout mice have early embryonic lethality with enhanced mesenchymal cell death and defects in angiogenesis that involve both the yolk sac and embryonic tissues [31, 32]. By contrast, the phenotype of HIF-2α knockout mice is more variable, depending on the genetic background, with some homozygotes surviving into the perinatal period, and manifesting a range of abnormalities. Thus, defects in vascular remodelling, catecholamine production, fatty acid oxidation, lung maturation, and surfactant production, and an anaemia that can be rescued by exogenous EPO have variously been described in different reports of HIF-2α knockout mice [33–37].

HIF-1α and HIF-2α exhibit similar functional domain structures, containing DNA binding, dimerisation, and regulatory domains with a high degree of homology. Furthermore, each dimerises with HIF-1β and recognises identical DNA sequences. It was therefore initially suggested that HIF-1α and HIF-2α had unique roles in hypoxic responses as a function of their distinct expression patterns. However, with the development of microarray technology and the ability to inactivate each isoform in specific cells using small interfering RNA (siRNA), it has become apparent that HIF-1α and HIF-2α also differ in the range of target genes that each is able to activate [38, 39]. For example, lactate dehydrogenase A (LDHA), carbonic anhydrase-9 (CAIX), and many glycolytic genes, are exclusively regulated by HIF-1α in response to hypoxia [39–44]. Other HIF regulated targets, such as the Pou transcription factor Oct-4, cyclin D1, TWIST1, transforming growth factor-β (TGF-β) and EPO are dominantly or exclusively controlled by HIF-2α [43, 45–49]. Furthermore, both HIF-1α and HIF-2α are capable of regulating genes such as vascular endothelial growth factor (VEGF), facilitated glucose transporter-1 (GLUT-1), adipose differentiation-related protein, adrenomedulin (ADM) and N-myc downstream regulated 1 (NDRG-1).

Chromatin immunoprecipitation experiments have shown that both HIFα subunits can bind to the endogenous HRE of hypoxia-responsive genes irrespective of whether the gene is functionally regulated by HIF-1α, HIF-2α, or both factors, indicating that selective HIF target gene activation is not based on the unique DNA-binding abilities of these factors [50, 51]. In contrast, domain-exchange experiments have demonstrated *C*-terminal regions of each isoform that are required to confer target gene specificity.

Several other transcription factors are known to physically interact with HIF-2α, but not HIF-1α and have been proposed to cooperate with HIF-2α in activating a number of target genes [52]. For example, HIF-2α interacts, through its *C*-terminus, with Ets-1 and synergistically activates expression of VEGF receptor-2 (VEGFR-2 or Flk-1) [52]. Elk-1, another Ets family member, cooperates with HIF-2α to activate a number of HIF-2α-specific targets genes, including *EPO* [50, 53]. Furthermore, several other HIF-2α-specific targets contain Ets-binding sites adjacent to HREs, suggesting that this may be a common mechanism for regulating target gene selectivity by HIF-2α [53].

Regulation of HIF by oxygen

Despite evidence for disparate roles of HIF-1α and HIF-2α, the mechanisms regulating the abundance and activity of the two isoforms in response to hypoxia are largely concordant, and this is reflected in similar domain structures [54] (Fig. 4). The amounts of HIF-1α and HIF-2α are regulated by proteolytic degradation, dependent upon distinct *N*-terminal and *C*-terminal oxygen-dependent degradation domains (NODDD and CODDD), located in the central region of the molecule [24–26, 55, 56]. In addition, both isoforms also possess two transactivation domains, required for recruiting cofactors that mediate transcription (such as the coactivator p300 [27]) – an *N*-terminal activation domain (NAD) that overlaps with the CODDD, and a *C*-terminal activation domain (CAD).

Both the proteolytic destruction and the transactivation domains are under oxygen-dependent regulation and in each case oxygen-dependence is conferred by enzymatic hydroxylation of specific amino-acid residues (Fig. 5). In the case of the NODDD and the CODDD, hydroxylation of PRO residues permits specific binding to an E3 ubiquitin ligase, which then targets the molecule for rapid degradation by the ubiquitin-proteasome pathway [57–59]. The apparent K_mO_2 (concentration of oxygen that supports a half maximal initial rate of catalysis) is reported to be 230–250 μM, well above the oxygen concentration in tissues (in the range 10–30 μM) [60]. Consequently, oxygen is rate-limiting for HIFα hydroxylation, such that the reaction rate varies with oxygen concentration over the physiologic range of tissue oxygen tensions. Because this step is rate limiting for the whole degradation pathway, this allows HIFα levels to reflect oxygen concentration in a graded manner, essential for the system to act as an oxygen sensor.

In a similar process, under oxygenated conditions, hydroxylation of an asparaginyl residue within the CAD blocks its interaction with the CH1 domain of p300, impeding its ability to initiate transcription [61]. Additionally, evidence has been reported for other levels of oxygen-dependent control. For example, oxygen-dependent regulation of HIF-1α nuclear/cytoplasmic locali-

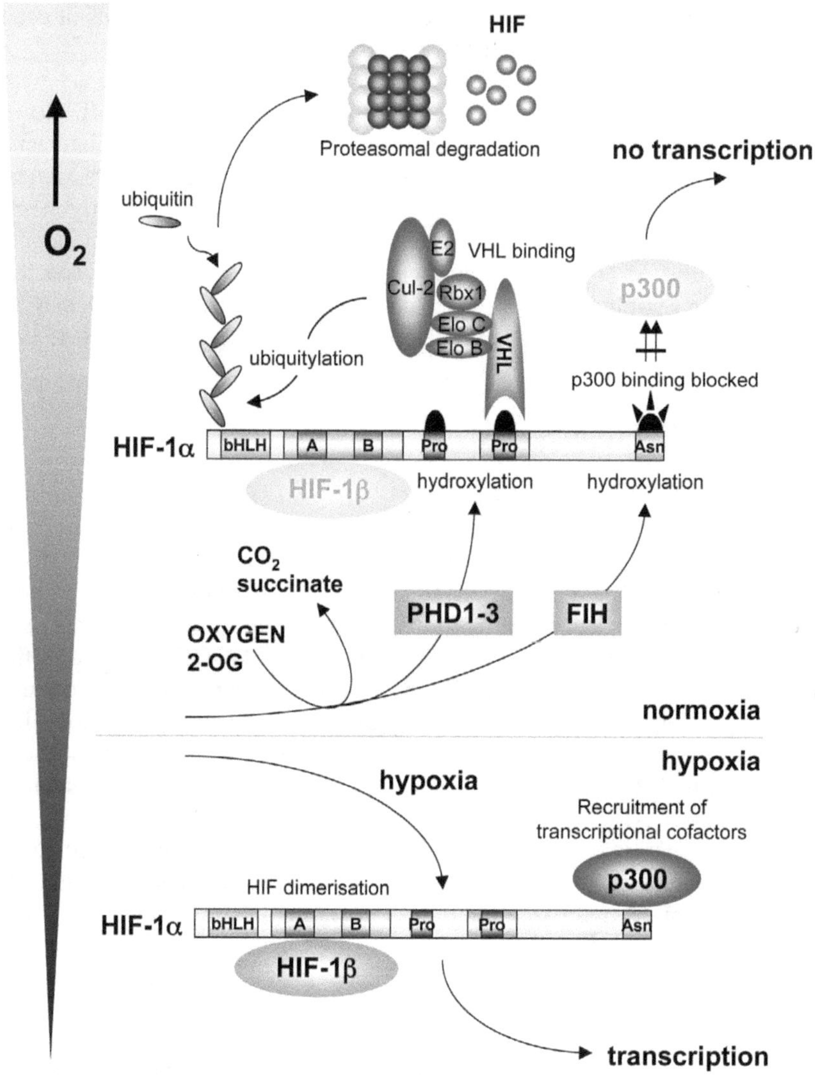

Figure 5. Oxygen-dependent regulation of HIF signalling. Under hypoxic conditions HIFα subunits dimerise with HIF-1α, bind DNA sequences containing hypoxia response elements (HREs) and recruit transcriptional cofactors such as p300 to effect transcription of target genes. When oxygen is available, enzymatic hydroxylation of residues within the HIFα subunit is effected by the PHD and FIH enzymes. Hydroxylation of a specific ASP residue within the CAD blocks recruitment of the cofactor p300, while hydroxylation of specific PRO residues in the NODDD and CODDD allows binding of the VHL E3 ubiquitin ligase complex. This tags the HIFα subunit with a multiubiquitin chain that targets it for proteasomal destruction.

sation has been observed in a number of systems [62, 63]. For HIF-3α prolyl hydroxylase-dependent proteolytic regulation that involves a region homologous to the CODDD is conserved [64]. Interestingly, the alternative splicing process that generates the inhibitory protein IPAS is also reported to be regulated by hypoxia, though the mechanism is currently unknown [28, 65].

In addition, a number of non-oxygen regulated controls of HIF stability and activity have been reported though it is unclear what effect they have on the physiologic regulation of EPO. In addition to ubiquitylation, HIFα subunits can be sumoylated, though it unclear whether this promotes stability or instability. Mice that are deficient in Sumo-specific protease-1 after targeted inactivation develop anaemia due to failure to stabilise HIFα. Enzymatic hydroxylation of residues within HIFα subunit proved to be an unprecedented mechanism of signal transduction, although such modifications are well established, and have structural significance in the post-translational processing of collagen molecules [66]. To date, all prolyl and asparaginyl hydroxylases, including the HIF-modifying enzymes, have been found to be members of the super-family of 2-oxoglutarate dependent dioxygenases [67]. In addition to oxygen, these enzymes also require 2-oxoglutarate, as a cosubstrate, and nonheme iron (Fe^{2+}), as a cofactor. Furthermore, ascorbate (vitamin C) is required for full enzymatic activity, possibly to maintain the catalytic iron centre in the reduced state, or to facilitate the availability of Fe^{2+} in a more general way. The ease with which iron in these enzymes can be removed by metal chelators, or substituted by other transition metals, explains the classical observations that metal ions, such as Co^{2+}, and iron chelators can induce EPO. It also raises important questions as to what extent HIF-hydroxylase activity is affected by physiologic or pathologic variation in iron or ascorbate availability. Experiments in tissue culture suggest that such effects could be important, particularly in cancer [68, 69].

In mammalian cells, three closely related hydroxylases termed prolyl hydroxylase domain PHD1, 2, and 3 catalyse hydroxylation of the prolyl residues within the NODDD and CODDD [70, 71], and one asparaginyl hydroxylase, termed factor inhibiting HIF (FIH), is responsible for modifying the CAD [72, 73]. As with HIFα isoforms, the presence of multiple HIF hydroxylases raises questions about their possible redundancy and diverse roles. Although all three PHDs contribute to the regulation of HIF, inactivation of each PHD individually using siRNA, has shown that loss of PHD2 alone is sufficient to increase HIF-1α levels in oxygenated cells, leading to the proposal that PHD2 is the most important isoform in oxygen sensing [74]. This likely reflects the relative high abundance of PHD2, which is the most abundant isoform in most cell types. In support of the importance of PHD2 is the embryonic lethality seen in PHD2 knockout mice, while those lacking PHD1 and PHD3 survive relatively normally [75]. However, these enzymes show differential patterns of organ expression, intracellular localisation, inducibility by exogenous stimuli, and substrate selectivity for HIF-1α *versus* HIF-2α, suggesting that each has distinct, if overlapping, functions in modulating responses to hypoxia [59, 76–81].

While the activity of the HIF hydroxylases depends directly on the ambient oxygen concentration, the levels of various hydroxylase enzymes themselves are also under the control of oxygen, adding another layer of control in the response to hypoxia. Hypoxia leads to increased transcription of PHD2 and PHD3 mRNA, in a HIF-dependent manner that acts, by a negative feedback loop, to limit the increase in HIF observed during prolonged hypoxia [70, 74]. In addition, the stability of PHD1 and PHD3 decreases in hypoxic cells, as a result of increased expression of another E3 ubiquitin ligase, Siah2, that targets these isoforms for proteasomal degradation and acts to enhance HIFα stability [79]. Indeed Siah2$^{-/-}$ mice exhibit an impaired hypoxic EPO response. Furthermore, an additional negative-feedback mechanism acts to autoregulate transcriptional activity of the HIF DNA-binding complex. CITED2 (CBP/p300 interacting transactivator with ED-rich tail), itself a HIF regulated gene, binds p300/CBP, in hypoxia, inhibiting its recruitment to the HIF transcriptional complex, and limiting the ability of HIF to activate gene transcription [82].

Von Hippel-Lindau disease

While hydroxylation of prolyl residues within the HIFα isoforms marks them for destruction, it is the von Hippel-Lindau (VHL) tumour suppressor protein that provides the specific recognition component that links the hydroxylated proteins to the constitutive ubiquitination machinery, and accounts for the dramatic upregulation of the HIF system after VHL inactivation in certain types of cancer.

VHL disease is a tumour syndrome, inherited as an autosomal dominant trait, affecting 1 in 36,000 of the population, and associated with (often multiple) clear cell renal cell carcinoma; hemangioblastomas (especially of the cerebellum, spinal cord, and retina); pheochromocytoma; endolymphatic-sac tumours of the inner ear; and cysts of the kidney, pancreas, and epididymis. Tumours are associated with somatic loss or inactivation of the remaining wild-type allele in individuals heterozygote for germline mutation in the VHL tumour-suppressor gene, in accordance with the Knudson ‘two-hit’ hypothesis. Sporadic tumours are associated with somatic loss or inactivation of both alleles within the same cell. Tumours are characteristically highly vascular and overexpress both HIFα and the target gene *VEGF*. The *VHL* gene, located on chromosome 3p25–26, was first identified in 1993 [83], and shown to comprise three exons coding for a protein of 213 amino acids with a second isoform of 150 residues resulting from internal translational initiation at an in frame ATG at codon 54. pVHL, the product of the *VHL* gene, forms a larger multiprotein complex, including elongins B and C, CUL2, and Rbx1, and forms the specific recognition component of an E3 ubiquitin-ligase that covalently tags HIFα subunits with a polyubiquitin chain. The labelled protein is taken up by the 26S proteasome, where it is rapidly destroyed [84–88]. X-ray crystallography has demonstrated that the formation of two hydrogen bonds between the alcohol of the hydroxylated prolines on HIFα and a binding pocket within pVHL stabilises the interaction by at least three orders of magnitude, permitting exquisite discrimination between the hydroxylated and nonhydroxylated forms of HIFα [89, 90]. Loss of functional pVHL in tumour cells blocks oxygen-dependent degradation of the HIFα subunits. Substantial evidence supports a causal role for the consequent upregulation of HIF and hypoxia pathways in the promotion of VHL-associated renal cell carcinoma.

Integration of hypoxic, metabolic, and redox signals

The involvement of two sub-families of 2-oxoglutarate dependent dioxygenases at discrete points in the oxygen-sensing pathway raises questions as to whether such enzymes are especially suited to physiologic oxygen-sensing, and whether their unusually extensive cosubstrate and cofactor requirements play a role in a shaping responses to hypoxia. Indeed the cosubstrate 2-oxoglutarate is an intermediate metabolite of the TCA cycle. Furthermore, both succinate (a product of the hydroxylase reaction) and fumarate are also products of TCA cycle enzymes, and can both compete with 2-oxoglutarate to inhibit PHD enzyme activity (Fig. 6) [91, 92]. It is possible, therefore, that in addition to sensing oxygen, these enzymes are also in some way acting as metabolic sensors, able to detect stresses on oxygen-requiring TCA cycle metabolism. Clinically, normoxic upregulation of HIF is seen in tumours lacking succinate dehydrogenase or fumarate hydratase activity.

Other factors that influence HIF-activation independently of oxygen concentrations include many growth factors and can be broadly grouped, according to their ability to induce common cellular signalling pathways, in particular those involving kinases, reactive oxygen species (ROS) or both. Such stimuli may pre-prime the cell to anticipate an increase in oxygen requirement consequent on the increased cell growth. More specifically, kinases such as p38 mitogen-activated protein kinase (p38), p42/p44 extracellular signal-regulated kinse 1 and 2, and phosphatidyliositol 3-kinase (PI3K) have all been reported to enhance normoxic HIF levels. In addition, phosphorylation of the tuberose sclerosis gene, *TSC2*, by AMP kinase (AMPK), consequent upon energy starvation, leads to inhibition of mammalian target of rapamycin (mTOR). In turn, mTOR affects HIFα levels independent of oxygen, and may provide another means of coupling nutrient availability to oxygen sensing [93].

The effect of hypoxia on levels of ROS is complex and somewhat controversial. However, ROS do have the potential to interfere with HIF hydroxylation by altering the redox state of the active iron-binding enzymatic site, leading to HIFα accumulation, as observed in JunD$^{-/-}$ cells [94]. A ROS of particular interest is nitric oxide (NO), which has been reported to variably stabilise or destabilise HIFα under normoxic conditions. These discrepancies may result from interactions of NO with mitochondrial respiration and consequence effects on intracellular oxygen concentrations, from temporal observations of a transient effect and/or the ability of NO to affect other transcription factors, such as GATA-2 that are involved in the regulation of HIF target genes such as *EPO*.

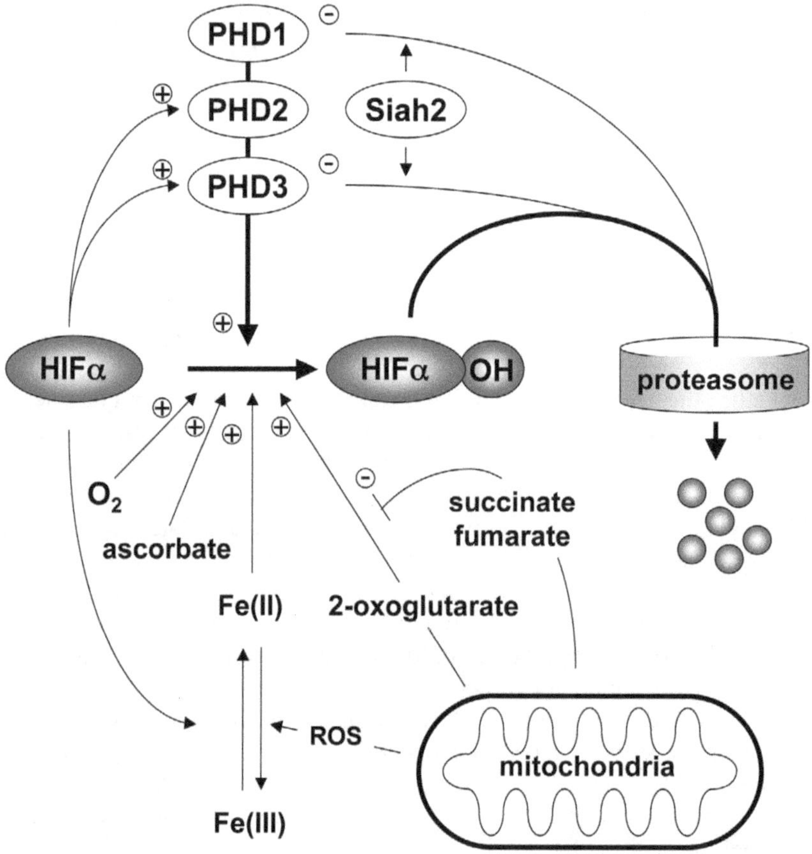

Figure 6. Integration of hypoxic, metabolic, and redox signals. In addition to oxygen, hydroxylation of HIFα subunits requires a number of cosubstrates, availability of which can modulate hydroxylase activity. While intracellular free-iron is tightly regulated, through HIF-dependent and HIF-independent mechanisms, reactive oxygen species (ROS) have the potential to interfere with the redox state of the active iron-binding enzymatic site, leading to HIFα accumulation by interfering with HIF hydroxylation. 2-Oxoglutarate, a tricarboxylic acid cycle (TCA) intermediate, is required for hydroxylation, and several other intermediates can antagonise this action, such that HIF also has the potential to respond to metabolic changes in the cell. Different isoforms of the HIF prolyl hydroxylases (PHDs) are variably transcriptionally activated by HIF, or hypoxically downregulated by Siah2 mediated proteasomal degradation, adding another layer of complexity.

Renal anaemia and erythropoietin regulation

Many factors likely contribute to the anaemia of chronic renal failure, including hematinic deficiency, lack of RPO, and accumulation of circulating inhibitors of erythropoiesis. It has been postulated that increased levels of uremic toxins may antagonise the bone marrow response to EPO. Indeed, it is known that dialysis alone improves anaemia in patients with renal failure.

It is clear, however, that in chronic renal impairment, the EPO response to a lowered hematocrit is dramatically blunted and that inappropriately low EPO concentrations are the main drive for anaemia. Given the very substantial capacity of the normal kidney to upregulate EPO production it seems unlikely that the failure of the kidney to generate an adequate EPO response to anaemia can be attributed to loss of renal mass alone. Several studies have shown that the failure to produce EPO in renal disease is relative rather than absolute. When subject to more severe hypoxic stimulation, individuals with renal disease can often generate an EPO response that would be sufficient to correct the anaemia if it were sustained. After renal injury, the number of fibroblast-like interstitial cells is increased in the kidney. The cells assume a myofibroblastoid phenotype and their potential for EPO expression is reduced [95]. The molecular mechanisms underlying this change in phenotype are unknown and despite much progress in understanding the molecular regulation of EPO production, the reason(s) why generation of EPO is inappropriately low in diseased kidneys remain unclear. The decrease in glomerular filtration rate (GFR) associated with chronic renal failure might be predicted to reduce the need for sodium reabsorption in the tubule, and hence oxygen consumption, leading to downregulation of the hypoxic drive for erythropoietin production. Owing to the portal nature of the renal circulation, however, glomerular sclerosis also reduces blood flow to the tubule, and intrarenal oxygen tensions in chronic renal failure are, in fact, lowered. Inappropriately low EPO concentrations have also been described in anaemias associated with chronic inflammatory conditions, such as acquired immunodeficiency syndrome (AIDS), and rheumatoid arthritis, as well as in cancer. Using animal and cell culture models, proinflammatory cytokines such as tissue necrosis factor-α (TNF-α) and interleukin-1 (IL-1) have been shown to suppress hypoxia induced EPO production, although the unexpected concomitant increase in HIF-1α activity indicates that the effects must be mediated by other transcription factors, possibly by GATA-2 and NF-κB, acting on the EPO promoter [96, 97].

Pharmacologic manipulation of the HIF response

The central role of HIF in a wide range of common pathologies, including cancer and ischaemia as well as renal anaemia has made it an attractive target for pharmacologic manipulation. The enzymatic basis of its regulation together with the absolute requirement for a number of co-factors and cosubstrates has identified the PHDs and FIH as key objectives for such a strategy. The experimental approaches of iron chelation or substitution are likely to prove too toxic for clinical use. Indeed in renal anaemia iron deficiency can often confound the anaemia. Approaches based on the use of 2-oxoglutarate analogues as competitive inhibitors and related compounds that are able to bind to the hydroxylases and hold greater promise. Existing knowledge of HIF pathways and of this class of enzyme suggests that obtaining sufficient specificity for clinical use in anaemia therapy will be challenging [98]. Bioinformatic predictions suggest that there may be up to 60 2-oxoglutarate-dependent oxygenases encoded by the human genome, some of which are as yet uncharacterised and thus have the potential to generate unexpected off-target effects [99]. The pleiotropic actions of HIF also have the potential to generate unwanted effects, even with inhibitors that are entirely specific for the HIF hydroxylases. For example, HIF activation has the potential to increase both erythropoiesis and angiogenesis. Promotion of erythropoiesis, leading to polycythaemia, may provide an unwanted side-effect of treatments aimed at supporting angiogenesis and *vice versa*. However, dissociating these actions may simply be a matter of delivering the drug to the required organ. For example, systemic hypoxia at altitude predominantly affects erythropoiesis, with little in the way of systemic angiogenesis, whereas localised hypoxia, in healing or neoplastic in tissues incapable of expressing EPO, produces localised angiogenesis. In addition, drugs that preferentially target different HIF hydroxylases or different HIFα isoforms could further aid specificity.

Summary

Analysis of pathways regulating *EPO* expression, in accordance with oxygen availability, have revealed unprecedented signal systems involving the post-translational hydroxylation of specific prolyl and asparaginyl residues in HIF, the master regulator of transcriptional responses to hypoxia. These pathways are conserved throughout the animal kingdom and mediate diverse cellular and systemic responses to hypoxic challenge. Therapeutic manipulation has the potential to augment the natural cellular response to hypoxia in a coordinated way, and may be of benefit in ischaemic/hypoxic diseases including anaemia.

References

1. Magnus G. Ueber die im Blute enthaltenen Gase, Sauerstoff, Stickstoff und Kohlensaeure. *Ann Physik* 1837;12:583–606.
2. Hoppe F. Ueber das Verhalten des Blutfarbstoffes im Spectrum des Sonnenlichtes. *Virchows Arch* 1862;23:446–449.
3. Stokes GG. On the reduction and oxidation of the colouring matter of blood. *Proc Roy Soc London* 1864;13:355–364.
4. Carnot P, Deflandre C. Sur l'activité hémopoiétique du sérum au cours de la régénération du sang. *CR Acad Sci Paris* 1906;143:384–386.
5. Erslev A. Humoral regulation of red cell production. *Blood* 1953;8:349–357.
6. FitzGerald MP. The changes in the breathing and the blood at various high altitudes. *Phil Trans R Soc Lond B*, Containing Papers of a Biological Character. 1913;203:351–371.
7. Abbrecht PH, Littell JK. Plasma erythropoietin in men and mice during acclimatization to different altitudes. *J Appl Physiol* 1972;32:54–58.
8. Jacobson LO, Goldwasser E, Fried W et al. Role of the kidney in erythropoiesis. *Nature* 1957;179:633–634.
9. Erslev AJ. *In vitro* production of erythropoietin by kidneys perfused with a serum-free solution. *Blood* 1974;44:77–85.

10. Beru N, McDonald J, Lacombe C et al. Expression of the erythropoietin gene. *Mol Cell Biol* 1986;6:2571–2575.
11. Fandrey J, Bunn HF. *In vivo* and *in vitro* regulation of erythropoietin mRNA: measurement by competitive polymerase chain reaction. *Blood* 1993;81:617–623.
12. Bachmann S, Le Hir M, Eckardt KU. Co-localization of erythropoietin mRNA and ecto-5'-nucleotidase immunoreactivity in peritubular cells of rat renal cortex indicates that fibroblasts produce erythropoietin. *J Histochem Cytochem* 1993;41:335–341.
13. Maxwell PH, Osmond MK, Pugh CW et al. Identification of the renal erythropoietin producing cells using transgenic mice. *Kidney Int* 1993;44:1149–1162.
14. Koury ST, Koury MJ, Bondurant MC et al. Quantitation of erythropoietin-producing cells in kidneys of mice by *in situ* hybridization: correlation with hematocrit, renal erythropoietin mRNA, and serum erythropoietin concentration. *Blood* 1989;74:645–651.
15. Jacobs K, Shoemaker C, Rudersdorf R et al. Isolation and characterization of genomic and cDNA clones of human erythropoietin. *Nature* 1985;313:806–810.
16. Lin FK, Suggs S, Lin CH et al. Cloning and expression of the human erythropoietin gene. *Proc Natl Acad Sci USA* 1985;82:7580–7584.
17. Semenza GL, Wang GL. A nuclear factor induced by hypoxia via *de novo* protein synthesis binds to the human erythropoietin gene enhancer at a site required for transcriptional activation. *Mol Cell Biol* 1992;12:5447–5454.
18. Makita T, Hernandez-Hoyos G, Chen TH et al. A developmental transition in definitive erythropoiesis: erythropoietin expression is sequentially regulated by retinoic acid receptors and HNF4. *Genes Dev* 2001;15:889–901.
19. Wenger RH, Kvietikova I, Rolfs A et al. Oxygen-regulated erythropoietin gene expression is dependent on a CpG methylation-free hypoxia-inducible factor-1 DNA-binding site. *Eur J Biochem* 1998;253:771–777.
20. Yin H, Blanchard KL. DNA methylation represses the expression of the human erythropoietin gene by two different mechanisms. *Blood* 2000;95:111–119.
21. Maxwell PH, Pugh CW, Ratcliffe PJ. Inducible operation of the erythropoietin 3' enhancer in multiple cell lines: evidence for a widespread oxygen-sensing mechanism. *Proc Natl Acad Sci USA* 1993;90:2423–2427.
22. Wenger RH, Stiehl DP, Camenisch G. Integration of oxygen signaling at the consensus HRE. *Sci STKE* 2005;2005:re12.
23. Wang GL, Jiang BH, Rue EA et al. Hypoxia-inducible factor 1 is a basic-helix-loop-helix-PAS heterodimer regulated by cellular O_2 tension. *Proc Natl Acad Sci USA* 1995;92:5510–5514.
24. Salceda S, Caro J. Hypoxia-inducible factor 1alpha (HIF-1alpha) protein is rapidly degraded by the ubiquitin-proteasome system under normoxic conditions. Its stabilization by hypoxia depends on redox-induced changes. *J Biol Chem* 1997;272:22642–22647.
25. Wiesener MS, Turley H, Allen WE et al. Induction of endothelial PAS domain protein-1 by hypoxia: characterization and comparison with hypoxia-inducible factor-1alpha. *Blood* 1998;92:2260–2268.
26. Pugh CW, O'Rourke JF, Nagao M et al. Activation of hypoxia-inducible factor-1; definition of regulatory domains within the alpha subunit. *J Biol Chem* 1997;272:11205–11214.
27. Arany Z, Huang LE, Eckner R et al. An essential role for p300/CBP in the cellular response to hypoxia. *Proc Natl Acad Sci USA* 1996;93:12969–12973.
28. Makino Y, Kanopka A, Wilson WJ et al. Inhibitory PAS domain protein (IPAS) is a hypoxia-inducible splicing variant of the hypoxia-inducible factor-3alpha locus. *J Biol Chem* 2002;277:32405–32408.
29. Wiesener MS, Jurgensen JS, Rosenberger C et al. Widespread hypoxia-inducible expression of HIF-2alpha in distinct cell populations of different organs. *FASEB J* 2003;17:271–273.
30. Rosenberger C, Mandriota S, Jurgensen JS et al. Expression of hypoxia-inducible factor-1alpha and -2alpha in hypoxic and ischemic rat kidneys. *J Am Soc Nephrol* 2002;13:1721–1732.
31. Iyer NV, Kotch LE, Agani F et al. Cellular and developmental control of O_2 homeostasis by hypoxia-inducible factor 1 alpha. *Genes Dev* 1998;12:149–162.
32. Carmeliet P, Dor Y, Herbert JM et al. Role of HIF-1alpha in hypoxia-mediated apoptosis, cell proliferation and tumour angiogenesis. *Nature* 1998;394:485–490.
33. Scortegagna M, Ding K, Oktay Y et al. Multiple organ pathology, metabolic abnormalities and impaired homeostasis of reactive oxygen species in Epas1–/– mice. *Nat Genet* 2003;35:331–340.
34. Tian H, Hammer RE, Matsumoto AM et al. The hypoxia-responsive transcription factor EPAS1 is

essential for catecholamine homeostasis and protection against heart failure during embryonic development. *Genes Dev* 1998;12:3320–3324.
35. Peng J, Zhang L, Drysdale L et al. The transcription factor EPAS-1/hypoxia-inducible factor 2alpha plays an important role in vascular remodeling. *Proc Natl Acad Sci USA* 2000;97:8386–8391.
36. Compernolle V, Brusselmans K, Acker T et al. Loss of HIF-2alpha and inhibition of VEGF impair fetal lung maturation, whereas treatment with VEGF prevents fatal respiratory distress in premature mice. *Nat Med* 2002;8:702–710.
37. Gruber M, Hu CJ, Johnson RS et al. Acute postnatal ablation of Hif-2alpha results in anemia. *Proc Natl Acad Sci USA* 2007;104:2301–2306.
38. Elvidge GP, Glenny L, Appelhoff RJ et al. Concordant regulation of gene expression by hypoxia and 2-oxoglutarate-dependent dioxygenase inhibition: the role of HIF-1alpha, HIF-2alpha, and other pathways. *J Biol Chem* 2006;281:15215–15226.
39. Wang V, Davis DA, Haque M et al. Differential gene up-regulation by hypoxia-inducible factor-1alpha and hypoxia-inducible factor-2alpha in HEK293T cells. *Cancer Res* 2005;65:3299–3306.
40. Sowter HM, Raval RR, Moore JW et al. Predominant role of hypoxia-inducible transcription factor (Hif)-1alpha *versus* Hif-2alpha in regulation of the transcriptional response to hypoxia. *Cancer Res* 2003;63:6130–6134.
41. Greijer AE, van der Groep P, Kemming D et al. Up-regulation of gene expression by hypoxia is mediated predominantly by hypoxia-inducible factor 1 (HIF-1). *J Pathol* 2005;206:291–304.
42. Hu CJ, Wang LY, Chodosh LA et al. Differential roles of hypoxia-inducible factor 1alpha (HIF-1alpha) and HIF-2alpha in hypoxic gene regulation. *Mol Cell Biol* 2003;23:9361–9374.
43. Raval RR, Lau KW, Tran MG et al. Contrasting Properties of Hypoxia-Inducible Factor 1 (HIF-1) and HIF-2 in von Hippel-Lindau-Associated Renal Cell Carcinoma. *Mol Cell Biol* 2005;25:5675–5686.
44. Grabmaier K, MC AdW, Verhaegh GW et al. Strict regulation of CAIX(G250/MN) by HIF-1alpha in clear cell renal cell carcinoma. *Oncogene* 2004;23:5624–5631.
45. Warnecke C, Zaborowska Z, Kurreck J et al. Differentiating the functional role of hypoxia-inducible factor (HIF)-1alpha and HIF-2alpha (EPAS-1) by the use of RNA interference: erythropoietin is a HIF-2alpha target gene in Hep3B and Kelly cells. *FASEB J* 2004;18:1462–1464.
46. Baba M, Hirai S, Yamada-Okabe H et al. Loss of von Hippel-Lindau protein causes cell density dependent deregulation of CyclinD1 expression through hypoxia-inducible factor. *Oncogene* 2003;22:2728–2738.
47. Gunaratnam L, Morley M, Franovic A et al. Hypoxia inducible factor activates the transforming growth factor-alpha/epidermal growth factor receptor growth stimulatory pathway in VHL(–/–) renal cell carcinoma cells. *J Biol Chem* 2003;278:44966–44974.
48. Covello KL, Kehler J, Yu H et al. HIF-2alpha regulates Oct-4: effects of hypoxia on stem cell function, embryonic development, and tumor growth. *Genes Dev* 2006;20:557–570.
49. Gort EH, van Haaften G, Verlaan I et al. The TWIST1 oncogene is a direct target of hypoxia-inducible factor-2alpha. *Oncogene* 2008;27:1501–1510.
50. Hu CJ, Sataur A, Wang L et al. The N-terminal transactivation domain confers target gene specificity of hypoxia-inducible factors HIF-1alpha and HIF-2alpha. *Mol Biol Cell* 2007;18:4528–4542.
51. Lau KW, Tian YM, Raval RR et al. Target gene selectivity of hypoxia-inducible factor-alpha in renal cancer cells is conveyed by post-DNA-binding mechanisms. *Br J Cancer* 2007;96:1284–1292.
52. Elvert G, Kappel A, Heidenreich R et al. Cooperative interaction of hypoxia-inducible factor-2alpha (HIF-2alpha) and Ets-1 in the transcriptional activation of vascular endothelial growth factor receptor-2 (Flk-1). *J Biol Chem* 2003;278:7520–7530.
53. Aprelikova O, Wood M, Tackett S et al. Role of ETS transcription factors in the hypoxia-inducible factor-2 target gene selection. *Cancer Res* 2006;66:5641–5647.
54. Schofield CJ, Ratcliffe PJ. Oxygen sensing by HIF hydroxylases. *Nat Rev Mol Cell Biol* 2004;5:343–354.
55. O'Rourke JF, Tian YM, Ratcliffe PJ et al. Oxygen-regulated and transactivating domains in endothelial PAS protein 1: comparison with hypoxia-inducible factor-1alpha. *J Biol Chem* 1999;274:2060–2071.
56. Huang LE, Gu J, Schau M et al. Regulation of hypoxia-inducible factor 1alpha is mediated by an O_2-dependent degradation domain via the ubiquitin-proteasome pathway. *Proc Natl Acad Sci USA*

1998;95:7987–7992.
57. Jaakkola P, Mole DR, Tian YM et al. Targeting of HIF-alpha to the von Hippel-Lindau ubiquitylation complex by O_2-regulated prolyl hydroxylation. *Science* 2001;292:468–472.
58. Ivan M, Kondo K, Yang H et al. HIFalpha targeted for VHL-mediated destruction by proline hydroxylation: implications for O_2 sensing. *Science* 2001;292:464–468.
59. Masson N, Willam C, Maxwell PH et al. Independent function of two destruction domains in hypoxia-inducible factor-alpha chains activated by prolyl hydroxylation. *EMBO J* 2001;20:5197–5206.
60. Hirsila M, Koivunen P, Gunzler V et al. Characterization of the human prolyl 4-hydroxylases that modify the hypoxia-inducible factor HIF. *J Biol Chem* 2003;278:30772–30780.
61. Lando D, Peet DJ, Whelan DA et al. Asparagine hydroxylation of the HIF transactivation domain a hypoxic switch. *Science* 2002;295:858–861.
62. Kallio PJ, Okamoto K, O'Brien S et al. Signal transduction in hypoxic cells: inducible nuclear translocation and recruitment of the CBP/p300 coactivator by the hypoxia-inducible factor-1alpha. *EMBO J* 1998;17:6573–6586.
63. Groulx I, Lee S. Oxygen-dependent ubiquitination and degradation of hypoxia-inducible factor requires nuclear-cytoplasmic trafficking of the von Hippel-Lindau tumor suppressor protein. *Mol Cell Biol* 2002;22:5319–5336.
64. Maynard MA, Qi H, Chung J et al. Multiple splice variants of the human HIF-3 alpha locus are targets of the von Hippel-Lindau E3 ubiquitin ligase complex. *J Biol Chem* 2003;278:11032–11040.
65. Hara S, Hamada J, Kobayashi C et al. Expression and characterization of hypoxia-inducible factor (HIF)-3alpha in human kidney: suppression of HIF-mediated gene expression by HIF-3alpha. *Biochem Biophys Res Commun* 2001;287:808–813.
66. Myllyharju J. Prolyl 4-hydroxylases, the key enzymes of collagen biosynthesis. *Matrix Biol* 2003;22:15–24.
67. Schofield CJ, Zhang Z. Structural and mechanistic studies on 2-oxoglutarate-dependent oxygenases and related enzymes. *Curr Opin Struct Biol* 1999;9:722–731.
68. Knowles HJ, Mole DR, Ratcliffe PJ et al. Normoxic stabilization of hypoxia-inducible factor-1alpha by modulation of the labile iron pool in differentiating U937 macrophages: effect of natural resistance-associated macrophage protein 1. *Cancer Res* 2006;66:2600–2607.
69. Knowles HJ, Raval RR, Harris AL et al. Effect of ascorbate on the activity of hypoxia-inducible factor in cancer cells. *Cancer Res* 2003;63:1764–1768.
70. Epstein AC, Gleadle JM, McNeill LA et al. C. elegans EGL-9 and mammalian homologs define a family of dioxygenases that regulate HIF by prolyl hydroxylation. *Cell* 2001;107:43–54.
71. Bruick RK, McKnight SL. A conserved family of prolyl-4-hydroxylases that modify HIF. *Science* 2001;294:1337–1340.
72. Hewitson KS, McNeill LA, Riordan MV et al. Hypoxia-inducible factor (HIF) asparagine hydroxylase is identical to factor inhibiting HIF (FIH) and is related to the cupin structural family. *J Biol Chem* 2002;277:26351–26355.
73. Lando D, Peet DJ, Gorman JJ et al. FIH-1 is an asparaginyl hydroxylase enzyme that regulates the transcriptional activity of hypoxia-inducible factor. *Genes Dev* 2002;16:1466–1471.
74. Berra E, Benizri E, Ginouves A et al. HIF prolyl-hydroxylase 2 is the key oxygen sensor setting low steady-state levels of HIF-1{alpha} in normoxia. *EMBO J* 2003;22:4082–4090.
75. Takeda K, Ho VC, Takeda H et al. Placental but not heart defects are associated with elevated hypoxia-inducible factor alpha levels in mice lacking prolyl hydroxylase domain protein 2. *Mol Cell Biol* 2006;26:8336–8346.
76. Metzen E, Berchner-Pfannschmidt U, Stengel P et al. Intracellular localisation of human HIF-1alpha hydroxylases: implications for oxygen sensing. *J Cell Sci* 2003;116:1319–1326.
77. Appelhoff RJ, Tian Y-M, Raval RR et al. Differential function of the prolyl hydroxylases, PHD1, 2 and 3 in the regulation of Hypoxia inducible factor (HIF). *J Biol Chem* 2004;279:38458–38465.
78. Cioffi CL, Qin Liu X, Kosinski PA et al. Differential regulation of HIF-1alpha prolyl-4-hydroxylase genes by hypoxia in human cardiovascular cells. *Biochem Biophys Res Commun* 2003;303:947–953.
79. Nakayama K, Frew IJ, Hagensen M et al. Siah2 regulates stability of prolyl-hydroxylases, controls HIF1alpha abundance, and modulates physiological responses to hypoxia. *Cell* 2004;117:941–952.
80. Hopfer U, Hopfer H, Jablonski K et al. The novel WD-repeat protein Morg1 acts as a molecular

scaffold for hypoxia-inducible factor prolyl hydroxylase 3 (PHD3). *J Biol Chem* 2006;281:8645–8655.
81. Lieb ME, Menzies K, Moschella MC et al. Mammalian EGLN genes have distinct patterns of mRNA expression and regulation. *Biochem Cell Biol* 2002;80:421–426.
82. Bhattacharya S, Michels CL, Leung MK et al. Functional role of p35srj, a novel p300/CBP binding protein, during transactivation by HIF-1. *Genes Dev* 1999;13:64–75.
83. Latif F, Tory K, Gnarra J et al. Identification of the von Hippel-Lindau disease tumor suppressor gene. *Science* 1993;260:1317–1320.
84. Maxwell PH, Wiesener MS, Chang GW et al. The tumour suppressor protein VHL targets hypoxia-inducible factors for oxygen-dependent proteolysis. *Nature* 1999;399:271–275.
85. Cockman ME, Masson N, Mole DR et al. Hypoxia inducible factor-alpha binding and ubiquitylation by the von Hippel-Lindau tumor suppressor protein. *J Biol Chem* 2000;275:25733–25741.
86. Kamura T, Sato S, Iwai K et al. Activation of HIF1alpha ubiquitination by a reconstituted von Hippel-Lindau (VHL) tumor suppressor complex. *Proc Natl Acad Sci USA* 2000;97:10430–10435.
87. Ohh M, Park CW, Ivan M et al. Ubiquitination of hypoxia-inducible factor requires direct binding to the beta-domain of the von Hippel-Lindau protein. *Nat Cell Biol* 2000;2:423–427.
88. Tanimoto K, Makino Y, Pereira T et al. Mechanism of regulation of the hypoxia-inducible factor-1 alpha by the von Hippel-Lindau tumor suppressor protein. *EMBO J* 2000;19:4298–4309.
89. Hon WC, Wilson MI, Harlos K et al. Structural basis for the recognition of hydroxyproline in HIF-1 alpha by pVHL. *Nature* 2002;417:975–978.
90. Min JH, Yang H, Ivan M et al. Structure of an HIF-1alpha -pVHL complex: hydroxyproline recognition in signaling. *Science* 2002;296:1886–1889.
91. Selak MA, Armour SM, MacKenzie ED et al. Succinate links TCA cycle dysfunction to oncogenesis by inhibiting HIF-alpha prolyl hydroxylase. *Cancer Cell* 2005;7:77–85.
92. Isaacs JS, Jung YJ, Mole DR et al. HIF overexpression correlates with biallelic loss of fumarate hydratase in renal cancer: Novel role of fumarate in regulation of HIF stability. *Cancer Cell* 2005;8:143–153.
93. Brugarolas J, Kaelin WG Jr, Dysregulation of HIF and VEGF is a unifying feature of the familial hamartoma syndromes. *Cancer Cell* 2004;6:7–10.
94. Gerald D, Berra E, Frapart YM et al. JunD reduces tumor angiogenesis by protecting cells from oxidative stress. *Cell* 2004;118:781–794.
95. Maxwell PH, Ferguson DJ, Nicholls LG et al. The interstitial response to renal injury: fibroblast-like cells show phenotypic changes and have reduced potential for erythropoietin gene expression. *Kidney Int* 1997;52:715–724.
96. Faquin WC, Schneider TJ, Goldberg MA. Effect of inflammatory cytokines on hypoxia-induced erythropoietin production. *Blood* 1992;79:1987–1994.
97. Jelkmann W, Pagel H, Wolff M et al. Monokines inhibiting erythropoietin production in human hepatoma cultures and in isolated perfused rat kidneys. *Life Sci* 1992;50:301–308.
98. Hewitson KS, Schofield CJ. The HIF pathway as a therapeutic target. *Drug Discov Today* 2004;9:704–711.
99. Elkins JM, Hewitson KS, McNeill LA et al. Structure of factor-inhibiting hypoxia-inducible factor (HIF) reveals mechanism of oxidative modification of HIF-1 alpha. *J Biol Chem* 2003;278:1802–1806.

Erythropoietins, Erythropoietic Factors and Erythropoiesis
Edited by S.G. Elliott, M.A. Foote, and G. Molineux

Biology of erythropoietin

Graham Molineux and Angus M. Sinclair

Amgen Inc., One Amgen Center Drive, Thousand Oaks, CA 91320, USA

Introduction

Recombinant erythropoietin is a 165-amino acid glycoprotein hormone with a molecular weight of 30.4 kDa that contains 39% carbohydrate [1, 2] and has the same biologic function as the endogenous product of a human gene located on chromosome 7q22 [3, 4]. Both the endogenous material (eEPO), and recombinant human glycoprotein (rHuEPO) bind a single receptor (EPOR) [5–7]. Although data from one group challenge this concept by suggesting signaling through a heterodimeric receptor including the beta-common cytokine receptor [8], these data have yet to be confirmed by other groups. For some years it had been widely accepted that EPOR was expressed broadly within the body in both normal and diseased tissues (e.g., [9]), but as discussed elsewhere in this book, methodological flaws in almost all of the published data draw this conclusion into doubt. To date, it is possible only to conclude with certainty that EPOR is expressed at effective levels in erythroid tissues in the bone marrow and spleen, so our discussion will focus only on the function of EPOR in that context.

Manipulated overexpression or deletion of the *EPO* gene in 'transgenic' or 'knockout' mice provided insight into the physiologic role of EPO. EPO transgenic mice have high hematocrits (80%), increased blood volume, enlarged hearts, reduced exercise capacity, and premature death [10]. It is likely that the latter changes are compensatory in nature and not the result of a direct action of EPO in cardiovascular tissues. Both EPO and EPOR knockout mice lack embryonic erythropoiesis, have heart problems, and die at approximately gestational day 13.5 [11–13]. The close similarity between the EPO and EPOR knockout phenotypes is good evidence that only a single ligand exists for EPOR and only a single receptor exists for EPO. EPOR knockout mice can be rescued by expressing human EPOR within the hematopoietic system, thus suggesting EPOR is not required for normal non-hematopoietic tissue development and confirming the role of EPO and EPOR in erythropoiesis [14]. However, some brain-specific EPOR knockout studies have suggested EPOR may play a role in neural progenitor proliferation, survival and migration [15, 16], although the functional significance of these findings are unclear given the rescued mice appear normal.

Control of EPO production

The control of EPO production *in vivo* through the HIF-PH pathway has provided additional pharmacologic targets for novel therapeutic options. This finding assumes the plieotropic nature of HIF-1α stabilization can be harnessed; a significant proportion of the genome is controlled by this axis, but early signs are encouraging [17].

Pharmacologic stabilization of HIF complexes has been shown to increase the production of endogenous eEPO [17]. How this will differ from the pharmacologic administration of rHuEPO is not fully understood. For example, though eEPO has an identical amino acid sequence to rHuEPO [2] eEPO collected from urine contains more sulphate and is more negatively charged [18, 19]. How serum eEPO activated through the HIF-PH pathway corresponds to rHuEPO or urinary eEPO is unknown. Therefore, the role of HIF stabilizers in anemia treatment will require substantial investigation not only because the EPO which is liberated may have different clinical effects than rHuEpo (which we know a great deal about), but also because a large number of proteins potentially regulated by HIF could influence the pharmacologic effects of the eEPO.

A number of properties of the EPO molecule are known to affect its *in vivo* potency. However, two general attributes explain the performance of all erythropoiesis-stimulating agents (ESA) described to date: interactions with the receptor, and clearance from the body.

The first of these, receptor interaction, can be summarized by the affinity of the molecule for EPOR. The receptor probably exists on the surface of responsive cells as a preformed homodimer [20]. Two regions of EPO have been shown to interact with the receptor, one with high affinity, the other relatively low [21]. Direct modifications in these regions of EPO will obviously affect the affinity of the ligand for the receptor but destruction of either one will effectively neutralize activity [21]. The interaction between EPO and EPOR is highly dynamic. Receptor is cycled to and from the cell surface relatively quickly [22]. The physicochemical properties (e.g., charge, size, hydrophobicity) of the ESA as well as the concentration outside the cell will determine the probability of the ligand being bound to the receptor (Kd) at equilibrium although equilibrium is a consequence of both the rate of association (K_{on}) and the rate of dissociation (K_{off}) [23]. To date, all studied EPO analogs display only quantitative variations in these constants; none has been shown to present a qualitatively new type of interaction with the receptor. Should an EPO analog be bound to the receptor when the complex is internalized then its fate may be determined by a number of constants that contribute to the probability of the EPO analog being recycled or not [23]. EPOR activators described to date appear to be full agonists and their ability to transduce a signal through the receptor is dominated by affinity. An interesting class of agonists is represented by antibodies selected for their ability to activate EPOR [24]. These 'agonistic antibodies' show a unique bell-shaped dose response curve. It has been suggested that at lower concentrations, these agents work as conventional agonists by inducing the appropriate activation of EPOR dimers. However, because each arm of the antibody F_{ab} region can itself bind a receptor, at high molar ratios of antibody to receptor, two copies of the antibody can potentially bind to the EPOR dimer while neither of them binds to both receptors to transduce a signal. The agents would, therefore, become self-antagonistic at higher concentrations. This curious observation has yet to be exploited to develop a therapeutic but a number of other modifications have been introduced to create analogs and derivatives of EPO, including hyperglycosylation (to different degrees), PEGylation, glycopegylation, Fc-conjugation, and albumin-conjugation. These changes tend to reduce the affinity of the ligand for the receptor, probably by simple stearic hindrance. These modifications in general slow the clearance of EPO from the body.

The clearance of EPO from plasma is not well understood. Various mechanisms and clearance organs have been suggested, but no explanation to date is fully satisfactory. What is known is that the modifications cited have extended the *in vivo* residence time of EPO analogs and that in general has led to the creation of more 'potent' molecules, i.e., more hemoglobin response per molar equivalent administered, despite the parallel loss of affinity for the receptor. The empirical observation made from these efforts is that persistence *in vivo* is a far more significant determinant of *in vivo* activity than is affinity for the receptor; though obviously a molecule that does not bind to the receptor will not activate it. This relatively simple observation needs to be borne in mind in any assessment of ESA 'potency' – data collected *in vitro* might be swayed by the dominant effect of receptor affinity in that setting, whereas data collected *in vivo* tend to be weighted in favor of persistence in the body.

Erythropoiesis

The major function of red blood cells (erythrocytes) is to transport oxygen to the tissues of the body. The erythrocyte component represents about 40% to 45% of the blood by volume. A normal adult may have 5 L of blood and an erythrocyte count of $5 \times 10^6/\mu L$. The red-cell lifespan in such a normal adult is approximately 100–120 days. To maintain a stable number of erythrocytes, replacing those lost by normal aging alone, requires replacement of 1% of the erythrocytes per day, i.e., production of 2.5×10^{11} cells per day or more than 10^{10} cells per hour. This prodigious rate of production is maintained in most circumstances for the lifetime of the individual. Thus, erythropoiesis is a highly dynamic process. EPO has a pivotal role in this process.

As discussed elsewhere in this book, the cellular target of EPO in erythropoiesis was traditionally referred to as the erythropoietin responsive cell (ERC). This population is now known to represent progenitor cells in late blast-forming unit – erythroid (BFU–E) stage through the colony-forming unit – erythroid (CFU–E) stage and the morphologically recognizable erythroblasts. EPOR is, however, absent on reticulocytes and mature erythrocytes [25].

EPOR+ progenitor cells have somewhat different cytokine dependencies in culture, with CFU–E requiring the addition of only EPO to grow and BFU–E requiring EPO and more primitive-acting cytokines previously known collectively as burst-promoting activities (BPA). The BPA are now molecularly identified as cytokines such as stem cell factor (SCF), interleukin-3 (IL-3), granulocyte-macrophage colony-stimulating factor (GM-CSF), and IL-6 among others, or occasionally mixtures of all or some of these.

Once EPO has bound and activated EPOR, a cascade of events is set in motion including activation of the dimerized receptor [26], and signal transduction through JAK2, Stat5, MAP kinase protein kinase, PI3 kinase (reviewed in [27–29] and protein kinase C [30, 31]. The actions of EPO include promotion of the survival of sensitive progenitors through prevention of apoptotic processes [32, 33], stimulation of proliferation [34, 35], and differentiation into large numbers of hemoglobinized cells. To maintain homeostasis and supply the necessary number of erythrocytes, the degree of proliferation required is quite remarkable, and occurs in parallel with the acquisition of features of the specialized functions of erythrocytes – accumulation of hemoglobin and disposal of the cell nucleus.

As discussed, EPO does not act alone to stimulate erythropoiesis in the bone marrow. A host of other agents have been shown to affect EPO-driven erythropoiesis. The hierarchy of cell populations leading to erythrocyte production includes successively more specialized cell types. Among the primitive-acting factors, the product of the mouse Steel locus, which is known as SCF and is the functional ligand for the *c-kit* cell-surface receptor, is one of the better understood factors. In mice anemic because of defects in either SCF or *c-kit*, administration of exogenous EPO still increases hemoglobin concentration [36, 37], although to a modest degree. This result would suggest that the few CFU–E that survive functional SCF deficiency *in vivo* can still respond normally to exogenous EPO, though their numbers are reduced by limitations imposed on the cellular pathway at more primitive levels. The interaction of SCF and EPO, which is of obvious significance in these earlier populations, has been elegantly examined at the molecular level and shown to result from a subtle interplay of survival, proliferation, and differentiation signals [38].

These effects at the cellular level lead consistently to increases in hemoglobin concentration in response to increases in circulating EPO, whether exogenous rHuEPO or eEPO produced in response to hypoxia (or a pharmacologic mimic of hypoxia). The relationship between serum concentrations of EPO and response is relatively straightforward (Fig. 1). A significant temporal mismatch exists between the residence time of rHuEPO ($t_{1/2}$ = approximately 8 h), the lifespan of reticulocytes (approximately 5 days), and the lifespan of erythrocytes (approximately 120 days). A more thorough discussion of the pharmacokinetics of EPO is to be found in other chapters in this book, but in summary, an increased EPO serum concentration of relatively short duration (measured in hours or days) can produce a very large number of reticulocytes (see calculations earlier). The flux through the reticulocyte compartment is huge under

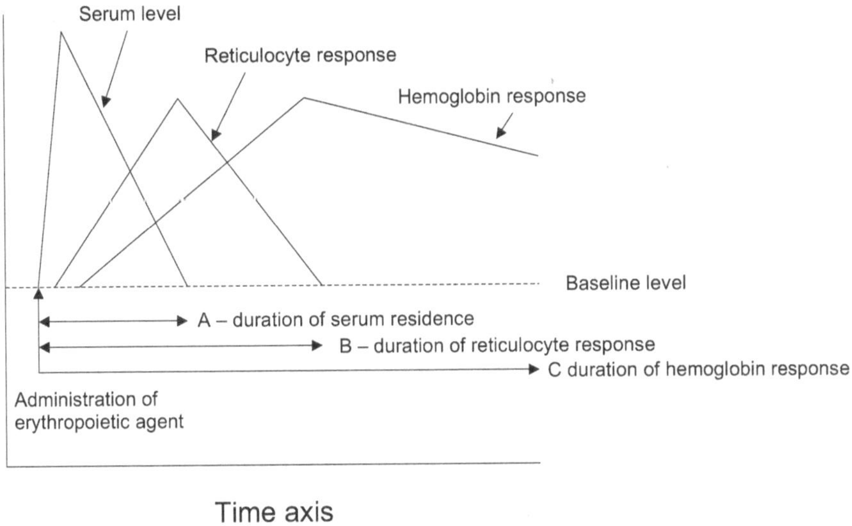

Figure 1. The relationship between residence time of EPO, or any other erythropoietic agent, reticulocyte response and hemoglobin response. The disproportionate effect on hemoglobin from increased amounts of an erythropoietic agent. The duration A is controlled largely by three factors: the dose, the route of injection that controls the gradient of the upward slope, and the rate of clearance that controls the gradient of the downward slope. The half-life of rHuEPO is approximately 8 h. The duration B of the reticulocyte response is determined by the duration of A – the time when serum concentrations of the erythropoietic agent are above baseline. Once serum levels of the erythropoietic agent return to baseline, reticulocyte numbers will begin to decline at a rate proportional to their half-life in the circulation. Reticulocyte lifespan is of the order of 5 days. The duration C is controlled again by the duration B. During the time supranormal numbers of reticulocytes are being produced the rate of production of hemoglobin exceeds the rate of loss, so hemoglobin concentration will be increasing. When reticulocyte numbers return to baseline, the hemoglobin curve will inflect and hemoglobin will be lost at a rate proportional to the lifespan of erythrocytes. Normal erythrocyte lifespan in humans is around 120 days. Overall, an erythropoietic agent with a half-life measured in hours can influence hemoglobin concentrations on a timescale measured in months. Thus changes in the duration A, influenced by dose, route of administration and rate of clearance, will have a disproportionate effect on the duration C.

conditions of accelerated erythropoiesis and delivers a large number of erythrocytes to the circulation. The erythrocytes formed, however, live a comparatively long time (on the order of 3 months). This process, in effect, leads to a disproportion between the duration of accelerated erythropoiesis (sustained by increased amounts of EPO and measured in hours) and the longevity of the hemoglobin response (measured in months). This information is of little value beyond curiosity until new therapeutic derivatives and analogs of EPO or novel dosing paradigms are considered, where an increase in circulating half-life of an erythroid molecule will have a large effect on hemoglobin – on a scale disproportionate to the increased residence time of the EPO analog (Fig. 1).

The body uses short-lived eEPO to sustain a normal blood hemoglobin concentration between 14 and 15 g/dL (in men) by maintaining an eEPO concentration of approximately 20 mU/mL in plasma. Pharmacologic dosing with an

ESA results in concentrations many fold higher than this (e.g., EPO C_{max} is approximately 1,000 mU/mL after a subcutaneously administered dose of 450 U/kg [39] and remains >20 mU/mL for approximately 2 days in humans and perhaps 4 days in rats). Ignoring for a moment disease states where response may be impaired, administration of any of these agents yields serum concentrations many fold higher than the baseline level and effectively maximizes erythroid response while [ESA] > MEC (minimum effective concentration).The essential concept that emerges from animal models of ESA dosing and its relationship to hemoglobin response is the concept of percentage cycle time coverage (%CT, cycle time being the interval between doses). On any dosing schedule, ESA plasma concentrations above the MEC will provide a stimulatory signal to erythropoiesis (accelerated erythropoiesis is 'on'). When plasma concentrations decrease below MEC, hemoglobin is in free fall (erythropoiesis is 'off', the rate of hemoglobin decline is controlled mainly by the lifespan of red blood cells). The proportion of the cycle time (%CT) when erythropoiesis in 'on' *versus* 'off' can then be related to hemoglobin response. The higher the percentage of a cycle that erythropoiesis is 'on' the greater will be the accumulation of hemoglobin over that period. Should the %CT approach 100, then obviously a maximal rate of rise in hemoglobin will be noted, until that is, something else constrains response (iron availability, blood viscosity, etc.). A lower %CT will yield a lower average rate of hemoglobin increase. At some point, any dose and schedule of ESA administration will result in a more or less stable hemoglobin concentration when loss from the compartment that occurs throughout the entire cycle is balanced by input during the period where [ESA] > MEC and red blood cells are being added to the circulating pool. Hemoglobin concentration is relatively well buffered from changes related to ESA dosing, unlike reticulocyte measurements that can wax and wane even on an ESA dose schedule that can be shown to maintain stable hemoglobin concentrations; high shortly after dosing (where [ESA] > MEC and zero later in the cycle (where [ESA] < MEC) (Fig. 2).

Consider the example of weekly dosing, an interval of 7 days between consecutive ESA administrations. In rodents, data obtained with a number of ESAs suggest approximately 40%CT is required, i.e., approximately 2.8 days of [ESA] > MEC is required to sustain a 1 g/dL increase in hemoglobin concentration. With knowledge of the clearance profile of any ESA, it is relatively straightforward to calculate the dose to give 2.8 days coverage and thus the desired dose to sustain hemoglobin concentration on a weekly schedule. The 40%CT can also be used to calculate the dose to sustain an interval of 3 days ([ESA] > MEC for 28.8 h), 2 weeks ([ESA] > MEC for 5.6 days), or any other interval.

Though this model is generally applicable, it is also oversimplified. MEC will be different for each different ESA – reflecting, among other things, affinity for the receptor. MEC also changes in disease settings where response may be constrained by multiple factors, it changes from one species to another, it may be a simple single concentration level, or may be a range through which

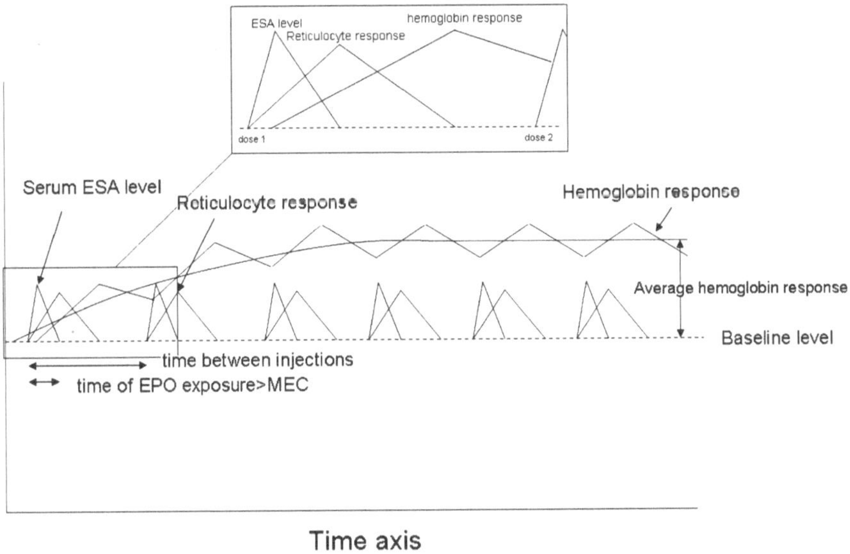

Figure 2. The relationship between exposure to ESA and stable hemoglobin response. As indicated in Figure 1, the duration of ESA exposure can be a small fraction of the time over which a hemoglobin effect can be detected. Further exploring this relationship reveals that to sustain a stable hemoglobin concentration by repeated ESA administration, the major determinant of the hemoglobin concentration attained is the fraction of the interval between administrations where ESA serum levels are above a threshold value. Irrespective of the clearance, the intrinsic potency of the ESA or the interval between administrations it is possible to calculate the percentage of the cycle time (%CT) – the fraction of the interval between administrations – that drives efficacy to a set level.

submaximal response may be obtained, in addition %CT will vary with red blood cell lifespan that is, in turn, dependent on species, disease, and numerous other variables. In general, this concept is broadly applicable to a number of ESAs and forms the basis of how ESA pharmacodynamics can be explained.

One interesting aspect of the %CT model is that essentially any ESA can sustain almost any dosing interval. A short-lived ESA, such as eEPO or rHuEPO, will require doses that increase logarithmically as dosing interval extends beyond their practically useful range. Longer acting analogs, such as darbepoetin alfa and PEGylated epoetin beta, may not suffer this logarithmic dose increase quite so soon, but in turn the maximum dose interval they can sustain will be constrained by the acceptability of their short-term effects.

Overall, much of the machinery involved in translating an increase of either endogenous or recombinant EPO into an increase in circulating hemoglobin concentration has been defined. At the opposite end of the response is the effect of EPO withdrawal. In addition to the predictable cessation of erythropoiesis when EPO is in short supply, the precipitous decrease in circulating EPO concentrations associated with a return from altitude, the return from space flight, and, perhaps, the end of a course of ESA treatment in the clinic, is associated with an accelerated rate of hemoglobin loss from the blood. The seminal work in this area has led to the proposal of a mechanism of neocytolysis, or the premature destruction of newly formed erythrocytes after interaction with endothelial cells or macrophages [40–42]. The precise role a mechanism such as neocytolysis may play in therapeutic intervention with rHuEPO has to be defined.

Assays

A number of assays were developed to detect erythropoietic activity in body fluids and to assess the potency of manufactured erythropoietic agents. In the early part of the 20th Century possibly the first bioassay of EPO was devised by Carnot when he transferred serum from anemic to normal rabbits and counted red cell number. Although preceded by measurements more or less proximal to erythropoiesis, the seminal work of Cotes illustrated the first viable technique for the quantitative estimate of EPO activity in body fluids [43]. This assay, which was later slightly modified to assess potency and well as quantity, exploited the normal physiologic response to polycythemia.

Rodents can be made polycythemic by either blood transfusion or they can be preconditioned at low oxygen tension. Under either condition the resulting plethora suppresses EPO production and therefore ongoing erythropoiesis. Under this condition, the animal becomes very sensitive to the administration of exogenous ESA and its effect can be readily measured by the incorporation of administered radioactive iron into the hemoglobin newly synthesized in response to ESA administration. This type of assay was the mainstay of rHuEPO potency assays for most of the 1990s, indeed it was even modified to account for longer acting analogs such as darbepoetin alfa [44]. This assay is perhaps still the gold standard assay for EPO activity for a number of reasons. Other assays such as single timepoint reticulocyte assays and *in vitro* cell proliferation assays (both cell lines and primary tissues) are also employed but a number of measurements from different systems need to be combined to approach the usefulness of the ^{59}Fe incorporation assay (Tab. 1).

The unit of EPO activity

Historically, assessing the activity of erythropoietin purified from an array of sources required comparison with a standard preparation, to assure comparability across investigators, sites, and studies. Preceding the adoption of a standard preparation one comparator was the result obtained in an iron incorporation assay with 5 μM cobalt chloride [45] – the so-called ‘cobalt unit’. The mechanism by which cobalt salts such as this can indirectly stabilize HIF-1α and upregulate *EPO* gene expression is described [46]. Other standards were a dried extract of rabbit plasma and a preparation of anemic sheep plasma (adopted as the first International Reference Standard; Standard A in 1961

Table 1. A comparison of assays of EPO performance

	Example format	Detects	Pros	Cons
In vitro assays				No PK component. Tend to emphasize differences in affinity for EPOR.
Immunoassay	EPO ELISA	EPO core peptide	Sensitive, specific, inexpensive	Does not measure activity. Variably applicable to different analogs but can only be used for ESA-like analogs Cannot distinguish eEPO from other ESAs
Affinity	BIACORE	Affinity for receptor	Sensitive, specific	Expensive Does not measure activity. For analogs, can read out as the inverse of *in vivo* activity
Cell line proliferation	UT/7 EPO	Proliferation or other measure of response (e.g., respiration, reporter genes)	Sensitive, can be selective, simple, inexpensive, acceptable biological relevance	Can read out as inverse of *in vivo* activity
Primary cell proliferation	Spleen cells BFU–E, CFU–E	Proliferation, or clonal growth	Good biological relevance. Can be performed with human tissue.	Can read out as inverse of *in vivo* activity. Clonal assays somewhat subjective. Labor intensive. Subject to inter-donor variability
In vivo assays				Animal use, labor intensive
^{59}Fe incorporation	Ex-hypoxic (polycythemic) mouse assay	ESA or non-ESA potency and short-term PK	Integrates activity over exposure time i.e., includes both potency and PK (requires adaptation for longer lived molecules. Sensitive, reproducible (low n).	Radioactive basis

(continued on next page)

Table 1. (Continued)

	Example format	Detects	Pros	Cons
In vivo assays				Animal use, labor intensive
Reticulocyte assay	Single dose ESA, single timepoint reticulocyte count	ESA or non-ESA potency and to a lesser degree, PK	High biological relevance Sensitive.	Highly variable, requires high n.
Single timepoint hemoglobin	Single dose ESA, single timepoint erythrocyte count	ESA potency and to a lesser degree, PK	High biological relevance	Highly variable, requires high n. Insensitive
Hemoglobin maintenance assay	Weekly administration of ESA, repeated hemoglobin measures over several weeks to months	ESA potency and short-term PK	Highly sensitive	Long-term assay, significant risk of immune reaction; hence highly variable, highest n

eEPO = endogenous EPO; EPOR = EPO receptor; ESA = erythropoiesis-stimulating agent; PK = pharmacokinetics

[47]). This material was assigned the activity of 10 units per ampoule, the unit being roughly equivalent to the 'cobalt unit'. Subsequent to this, the first standard adopted by the WHO in 1964 (the Second International Reference Standard) was purified from human urine and named Standard B [48] with an assigned activity of 10 units per ampoule. 'Activity' defined by this unit came from some form of iron incorporation assay. In the standard application of this assay, the radioactive readout represents an index of the accumulated hemoglobin synthesis across the time interval between administration of isotope and sampling of the blood. It therefore comprises both a pharmacokinetic and a pharmacodynamic component – integrating the persistence of the agent across the time interval (pharmacokinetic) with the stimulatory effects (pharmacodynamic), yet these components cannot be easily deconvoluted. A highly potent analog that is rapidly cleared from the body might read out the same as a modestly potent material which persisted for the duration of the assay. It became apparent that even though this type of assay was suitable for the level of heterogeneity found in different EPO preparations it was not readily applicable to unnatural or engineered forms such as darbepoetin alfa. In considering in some detail how activity of epoetin alfa and darbepoetin alfa might be compared Sasu et al. [49] concluded that the EPO 'unit' is not a legitimate measure of the activity of engineered derivatives. The usefulness of the standard preparation remains however, as more pharmaceutical preparations of erythropoietin become available under emerging regulatory guidelines for follow-on or biosimilar biologics. The International Standard has been updated with The First International Standard for Recombinant Derived Erythropoietin in 1990 and The Second International Standard for Recombinant Derived Erythropoietin in 2003.

Pharmacology – disease rationale

The administration of rHuEPO in a situation where every other constituent of the biologic response is in excess and EPO is the only limiting factor in erythropoiesis will yield the most marked response. Such a situation is extremely rare. Even in the case of renal disease where compromised endogenous EPO production is an issue, other factors such as uremia, general inflammation, or iron deficiency can limit the effectiveness of injected rHuEPO by placing other constraints on the response [50]. These constraints may be the demonstrated inhibition of erythropoiesis by tumor necrosis factor (TNF), IL-1, and interferon, the proinflammatory cytokines [51], or by limited supply of iron for hemoglobin synthesis. Recognizing that such a clear-cut situation is not likely to present itself, the effectiveness of administered rHuEPO will depend on a milieu of negative influences including those above in addition to anti-EPO antibodies and soluble EPOR. The latter two factors have direct consequences for rHuEPO therapy. Antibodies to EPO that are capable of neutralizing the action of administered rHuEPO have been noted in normal patients [52] and also in diseases such as systemic lupus erythematosus and HIV infection [53–55]. Antibodies may be raised in response to administered rHuEPO with serious consequences that may include refractoriness to both rHuEPO and eEPO and a long-term dependence on blood transfusions for survival. Soluble EPOR is a truncated form of the full-length membrane-bound version that has been suggested to act as a competitive inhibitor of the cell-borne form of the receptor and can limit the response to injected rHuEPO [56], although this has not been well established.

Interaction with iron

Iron is a vital component of hemoglobin and limitations on its supply can compromise the ability of erythropoietic cells to synthesize hemoglobin. Paradoxically, iron is also a poisonous metal for which most mammals, including humans, have highly evolved strategies to ameliorate its toxicities. Iron deficiency is commonly discussed, although the body's treatment of iron as a toxin can lead to its deposition as relatively inert forms (such as ferritin or hemosiderin) in various tissues [57]. This iron maldistribution situation, where the body load of iron may be excessive yet the presentation is one of functional iron deficiency, is typical of anemia of chronic disorders, the anemia of cancer, and perhaps other anemias with an inflammatory component. When deficient, and depending upon the reason for the deficiency, supplementation with oral or intravenous iron may be relatively straightforward. The move from oral administration of simple salts such as ferrous sulfate towards more complex intravenous forms of iron resulted from a series of clinical studies, although agreement is not complete on whether one is more beneficial than the other [58–61]. Emerging data suggest that when hepcidin, the key regulator of iron metabolism, is increased absorption of iron from the diet may be compromised. Assays for the detection of hepcidin are in their infancy and it remains to be seen whether circulating hepcidin affects the effectiveness of oral iron supplementation in some cases. EPO has not been shown to have a direct role in mobilizing iron from storage organs, but can act indirectly by stimulating hemoglobin synthesis and drawing iron from storage pools or perhaps other mechanisms [62]. Although iron may often be a limiting step in erythropoiesis, it may be in part corrected by supplementation or mobilization from existing body stores.

The use of EPO in different species

EPO from mammalian species as diverse as human, hamster, rat, mouse, cat, lion, dog, horse, sheep, dolphin, and pig has been cloned and sequenced and found to be remarkably conserved with a minimum identity of 79.8% between any two of these species [63, 64]. This finding has lead to the experimental use of rHuEPO in cross-species settings, for example in anemic FLV-infected cats [65], for idiopathic anemia in dolphins [66], in blood doping of racehorses [67], in transient brain ischemia in gerbils [68], for subarachnoid hemorrhage in rabbits [69], and to study erythrokinetics in sheep [70]. Even with the high degree of conservation across species, the administration of rHuEPO can raise antibodies that may cross-neutralize eEPO (e.g., in horses [71]). Despite the precise species match between recombinant and endogenous forms of EPO, instances of cross-neutralizing antibodies have been reported in humans receiving some forms of rHuEPO, but not with other forms bearing an identical amino acid sequence. It may be the case that although these antibodies bind to similar epitopes on the protein backbone of rHuEPO and endogenous EPO, their induction is dependent on other factors such as autoimmune disease or manufacturing or formulation methods resulting in protein aggregation or degradation.

Rationale for rHuEPO treatment in disease

Anemia of renal failure

As the kidney represents the major site of EPO production in the adult, the progressive destruction of this organ might be intuitively linked to anemia of an EPO limited nature. Indeed, since shortly after the gene was cloned in 1983, the anemia of renal failure has been successfully treated with rHuEPO [72, 73]. In fact, rHuEPO represents a routine treatment for patients receiving regular hemodialysis or peritoneal dialysis and for many patients not receiving dialysis (i.e., predialysis patients) [74, 75]. Biologically, this use represents a form of hormone supplementation therapy, although the benefits to the patient extend beyond the reduced need for transfusions and the associated risks of iron overload, incompatibility reactions, and viral infections; and include the benefits of increased exercise tolerance, cognitive and psychomotor changes, and relief from hyperdynamic cardiac states [76–78]. In total, these changes produce a significant impact on the quality of life of patients [79]. Beyond the surprisingly wide ramifications of correction of anemia in these patients, the biologic basis for rHuEPO therapy in patients with renal disease remains replacement of a relative hormone deficiency (in contrast to absolute deficiency) secondary to failure of the organ that produces the hormone. Other uses for rHuEPO, for example in anemias associated with chemotherapy treatment have led to extensive use outside the dialysis clinic.

Anemia in oncology

Anemia in patients with cancer results from a complex interaction of various factors [80] that make treatment somewhat less predictable than the hormone replacement-like use in renal failure states. Included in the list of contributing factors are hemodilution, bleeding, hypersplenism (and hemophagocytosis), hemolysis, nutritional deficiencies, marrow damage, chemotherapy and radiotherapy, and the anemia of cancer itself (included in the broader classification of anemia of chronic disease) [81–84].

Addressing any one of these factors alone represents a significant challenge, yet the perceived benefits of managing anemia in patients with cancer has led to a number of successful therapeutic strategies. Perhaps the most obvious rationale for rHuEPO use in anemic patients with cancer is in patients with documented low endogenous serum EPO concentrations. Concentrations of eEPO are inherently variable and use of the data is further confounded by a phenomenon widely documented in anemia of chronic disease where the absolute EPO concentration is within the normal range and is low only when considered relative to the degree of anemia – the so-called 'blunted' EPO production [85]. For the most part, studies of endogenous EPO concentrations in animals have not illustrated principles to be applied to clinical practice. For example, inappropriately low EPO concentrations in tumor-bearing mice did not confirm the earlier observations made in tumor-bearing rats [86, 87]. Later work, however, showed that eEPO concentrations were predictable and consistent with the degree of anemia [88]. In addition, the effects of chemotherapy on eEPO concentrations have been variously reported to increase, decrease, or remain unchanged [89–91].

Patients with cancer frequently have circulating eEPO amounts that are lower than would be expected for the degree of anemia. In early studies in patients with cancer, inappropriately low eEPO concentrations were reported, yet only in association with other problems such as infection or inflammation [92], while other studies showed normal levels [93, 94]. Blunted EPO production, however, has been seen in more recent work in patients with cancer without complications, in contrast to patients with iron-deficiency anemia [95–99]. Overall, a lower endogenous serum EPO concentration than might be expected with the degree of anemia has suggested that rHuEPO use is warranted in anemic cancer patients even before they receive chemotherapy.

Chemotherapy- or radiotherapy-induced anemia is the most frequently encountered anemia in patients with cancer [100]. Chemotherapy or radiotherapy cause systemic changes outside the bone marrow that result in anemia, e.g., hemolysis, blood loss, nutritional deficiency, damage to the erythroid populations in the bone marrow, or other conditions may reduce hemoglobin concentrations [101]. The eEPO concentration in such patients can be variable depending on a host of factors including, but not limited to, tumor type and chemotherapy agent. Most patients (more than 50%) respond to administered rHuEPO [102]. Whether this responsiveness results from the correction of low eEPO values or a heightened response from supernormal circulating EPO values (combining both endogenous and exogenous material) is unclear. The observation remains that the proportion of patients responding to rHuEPO after chemotherapy increases with increasing dose. In the case of limited marrow damage, the problem can be overcome by rHuEPO administration, although when marrow damage becomes extensive, it is unlikely that rHuEPO will be effective.

Other anemias

Although the most appealing indication for a recombinant protein therapeutic may be where amounts of the endogenous prototype are limited, the discussion earlier suggests that, in cancer at least, an eEPO-deficit condition is not the only time when using rHuEPO may be useful. In general in those cases where endogenous serum EPO concentrations are reduced, administration of rHuEPO is most effective. Thus, the rationale for rHuEPO therapy in iron-deficiency anemia due to limited dietary iron intake, for example, is weak. Anemic conditions range from an apparent EPO-deficit condition documented in neonates through premature erythrocyte destruction noted in elderly patients. Despite the different cause of these two types of anemia, rHuEPO appears to be effective in both. In contrast to these anemias are the hemoglobinopathies, where an rHuEPO-induced increase in hematocrit, without the accompanying conversion of hemoglobin type from diseased to fetal (e.g., with hydroxyurea therapy), would be of limited benefit. Anemia in the various intensive care patient populations responds, in general, quite well to rHuEPO and appears to have many features in common with anemia of chronic disease. Similarly, the anemia noted in patients with AIDS, despite the multifactorial causation, responds predictably to rHuEPO.

A variety of anemias exist, many of which will respond well to rHuEPO administration, but other cases exist where rHuEPO may not be a good therapeutic choice. In anemic conditions where eEPO concentrations are low or even normal, rHuEPO has a good chance of being effective. In other conditions where EPO concentrations are supernormal, the administration of rHuEPO can help where it might be reasoned that EPO is not the only limiting factor in the development of anemia, but that supplementing EPO can make up for the limitation elsewhere in erythropoiesis [36].

Non-anemic conditions

There exists a widespread literature on the effects of EPO in nonerythropoietic tissues. For these effects to be direct EPOR must be present in the target tissue. Recent studies illustrate very clearly that the presence of mRNA for EPOR is not sufficient to guarantee that functional receptor is expressed or functional [103–106]. Neither it turns out is antibody mediated detection of EPOR with the most commonly used reagents a reliable method to detect the receptor. However, it remains a point for discussion whether a direct action of ESAs via tissue borne EPOR is the only way in which an ESA could result in a biologically useful signal in target tissues. A shift in oxygen saturation associated with a hemoglobin response, changed ferrokinetics in response to ESA administration, and changes in blood pressure or coagulation parameters are examples of effects that might be mediated indirectly by ESAs.

Unfortunately, many of the studies purporting to show an effect of rHuEPO in tissue protection, angiogenesis, nerve regeneration, stroke, etc., have to be re-examined in the light of defective methods used to detect EPOR expression and function. If, as would appear to be the case, these tissues do not express physiologically relevant, functional amounts of EPOR, then the findings cannot be interpreted as being direct since there is no tenable mechanism of action.

References

1. Lai PH, Everett R, Wang FF et al. Structural characterization of human erythropoietin. *J Biol Chem* 1986;261:3116–3121.
2. Browne JK, Cohen AM, Egrie JC. Erythropoietin: Gene cloning protein structure and biological properties. *Cold Spring Harbor Symp Quant Biol* 1986;51:693–702.
3. Davis JM, Arakawa T, Strickland TW et al. Characterization of recombinant human erythropoietin produced in Chinese hamster ovary cells. *Biochemistry* 1987;26:2633–2638.
4. Law ML, Cai GY, Lin FK et al. Chromosomal assignment of the human erythropoietin gene and its DNA polymorphism. *Proc Natl Acad Sci USA* 1986;83:6920–6924.
5. D'Andrea AD, Lodish HF, Wong GG. Expression cloning of the murine erythropoietin receptor. *Cell* 1989;57:277–285.
6. Noguchi CT, Bae KS, Chin K et al. Cloning of the human erythropoietin receptor gene. *Blood* 1991;78:2548–2556.
7. Maouche L, Tournamille C, Hattab C et al. Cloning of the gene encoding the human erythropoietin receptor. *Blood* 1991;78:2557–2563.
8. Brines M, Grasso G, Fiordaliso F et al. Erythropoietin mediates tissue protection through an erythropoietin and common β-subunit heteroreceptor. *Proc Natl Acad Sci USA* 2004;101:14907–14912.
9. Juul SE, Yachnis AT, Christensen RD. Tissue distribution of erythropoietin and erythropoietin receptor in the developing human fetus. *Early Human Dev* 1998;52:235–249
10. Wagner KF, Katschinski DM, Hasegawa J et al. Chronic inborn erythrocytosis leads to cardiac dysfunction and premature death in mice overexpressing erythropoietin. *Blood* 2001;97:536–542.
11. Wu H, Liu X, Jaenisch R et al. Generation of committed erythroid BFU–E and CFU–E progenitors does not require erythropoietin or the erythropoietin receptor. *Cell* 1995;83:59–67.
12. Lin CS, Lim SK, D'Agati V et al. Differential effects of an erythropoietin receptor gene disruption on primitive and definitive erythropoiesis. *Genes Dev* 1996;10:154–164.
13. Wu H, Lee SH, Gao J et al. Inactivation of erythropoietin leads to detects in cardiac morphogenesis. *Development* 1999;126:3597–3605.
14. Suzuki N, Ohneda O, Takahashi S et al. Erythroid-specific expression of the erythropoietin receptor rescued its null mutant mice from lethality. *Blood* 2002;100:2279–2288.
15. Chen ZY, Asavaritikrai P, Prchal JT et al. Endogenous erythropoietin signaling is required for normal neural progenitor cell proliferation. *J Biol Chem* 2007;282:25875–25883.
16. Tsai PT, Ohab JJ, Kertesz N et al. A critical role of erythropoietin receptor in neurogenesis and post-stroke recovery. *J Neurosci* 2006;26:1269–1274.
17. Hsieh MM, Linde NS, Wynter A et al. HIF-prolyl hydroxylase inhibition results in endogenous erythropoietin induction, erythrocytosis, and modest fetal hemoglobin expression in rhesus macaques. *Blood* 2007;110:2140–2147.
18. Sasaki H, Bothner B, Dell A et al. Carbohydrate structure of erythropoietin expressed in Chinese hamster ovary cells by a human erythropoietin cDNA. *J Biol Chem* 1987;262:12059–12076.
19. Kawasaki N, Haishima Y, Ohta M et al. Structural analysis of sulfated N-linked oligosaccharides in erythropoietin. *Glycobiology* 2001;11:1043–1049.

20. Lu X, Gross AW, Lodish HF. Active conformation of the erythropoietin receptor: Random and cysteine-scanning mutagenesis of the extracellular juxtamembrane and transmembrane domains. *J Biol Chem* 2006;281:7002–7011.
21. Cheetham JC, Smith DM, Aoki KH et al. NMR structure of human erythropoietin and a comparison with its receptor bound conformation. *Nat Struct Biol* 2006;5:861–866.
22. Sawyer ST, Hankins WD. The functional form of the erythropoietin receptor is a 78-kDa protein: correlation with cell surface expression, endocytosis and phosphorylation. *Proc Nat Acad Sci USA* 1993;90:6849–6853.
23. Gross AW, Lodish HF. Cellular trafficking and degradation of erythropoietin and novel erythropoiesis stimulating protein (NESP). *J Biol Chem* 2006;281:2024–2032.
24. Liu Z, Stoll VS, DeVries PJ et al. A potent erythropoietin-mimicking human antibody interacts through a novel binding site. *Blood* 2007;110:2408–2413.
25. Broudy VC, Lin N, Brice M et al. Erythropoietin receptor characteristics on primary human erythroid cells. *Blood* 1991;77:2583–2590.
26. Constantinescu SN, Keren T, Socolovsky M et al. Ligand-independent oligomerization of cell-surface erythropoietin receptor is mediated by the transmembrane domain. *Proc Natl Acad Sci USA* 2001;98:4379–4384.
27. Frank SJ. Receptor dimerization in GH and erythropoietin action – it takes two to tango, but how? *Endocrinology* 2002;143:2–10.
28. Cheung JY, Miller BA. Molecular mechanisms of erythropoietin signaling. *Nephron* 2001;87:215–222.
29. Wojchowski DM, Gregory RC, Miller CP et al. Signal transduction in the erythropoietin receptor system. *Exp Cell Res* 1999;253:143–156.
30. Chen C, Sytkowski AJ. Erythropoietin activates two distinct signaling pathways required for the initiation and the elongation of c-myc. *J Biol Chem* 2001;276:38518–38526.
31. von Lindern M, Parren-van Amelsvoort M, van Dijk T et al. Protein kinase C alpha controls erythropoietin receptor signaling. *J Biol Chem* 2000;275:34719–34727.
32. Koury MJ, Bondurant MC. Erythropoietin retards DNA breakdown and prevents programmed death in erythroid progenitor cells. *Science* 1990;248:378–381.
33. Koury MJ, Bondurant MC. Control of red cell production: the roles of programmed cell death (apoptosis) and erythropoietin. *Transfusion* 1990;30:673–674.
34. Spivak JL, Pham T, Isaacs M et al. Erythropoietin is both a mitogen and a survival factor. *Blood* 1991;77:1228–1233.
35. Krantz SB. Erythropoietin. *Blood* 1991;77:419–434.
36. Cynshi O, Satoh K, Higuchi M et al. Effects of recombinant human erythropoietin on anaemic W/Wv and Sl/Sld mice. *Br J Haematol* 1990;75:319–324.
37. Pharr PN, Hofbauer A, Worthington RE et al. Residual erythroid progenitors in W/W mice respond to erythropoietin in the absence of steel factor signals. *Int J Hematol* 2000;72:178–185
38. Kapur R, Zhang L. A novel mechanism of cooperation between c-Kit and erythropoietin receptor. Stem cell factor induces the expression of Stat5 and erythropoietin receptor, resulting in efficient proliferation and survival by erythropoietin. *J Biol Chem* 2001;276:1099–1106.
39. Woo S, Jusko WJ. Interspecies comparisons of pharacokinetics and pharmacodynamics of recombinant human erythropoietin. *Drug Metab Dispos* 2007;35:1672–1678.
40. Rice L, Ruiz W, Driscoll T et al. Neocytolysis on descent from altitude: a newly recognized mechanism for the control of red cell mass. *Ann Intern Med* 2001;134:652–656.
41. Rice L, Alfrey CP. Modulation of red cell mass by neocytolysis in space and on Earth. *Pflügers Arch* 2000;441:R91–R94.
42. Trial J, Rice L, Alfrey CP. Erythropoietin withdrawal alters interactions between young red blood cells, splenic endothelial cells, and macrophages: an *in vitro* model of neocytolysis. *J Investig Med* 2001;49:335–345.
43. Cotes PM, Bangham DR. Bio-assay of erythropoietin in mice made polycythaemic by exposure to air at reduced pressure. *Nature* 1961;191:1065–1067.
44. Elliott S, Egrie J, Browne J et al. Control of rHuEPO biological activity: The role of carbohydrate. *Exp Hematol* 2004;32:1146–1155.
45. Goldwasser E, White WF. *Fed Proc* 1959;18:236
46. Ebert BL, Bunn HF. Regulation of the erythropoietin gene. *Blood* 1999;94:1864–1877
47. Jacobson LO, Goldwasser E, Gurney CW. Transfusion induced polycythemia as a model for studying factors influencing erythropoiesis. In: *CIBA Foundation Symposium on Haemopoiesis.*

GEW Wolstenhome & M O'Connor, Eds:1960;423. Little Brown and Co Boston, Mass.
48. Cotes PM, Bangham DR. An international reference preparation for bioassay of erythropoietin. *Bull Wld Hlth Org* 1966;35:751
49. Sasu BJ, Hartley C, Schultz H et al. Comparison of Epoetin alfa and darbepoetin alfa biological activity under different administration schedules in normal mice. *Acta Haematologica* 2005;113:163–174.
50. Macdougall IC. Role of uremic toxins in exacerbating anemia in renal failure. *Kidney Int Suppl* 2001;78:S67–S72.
51. Macdougall IC, Cooper A. The inflammatory response and epoetin sensitivity. *Nephrol Dial Transplant* 2002;17:48–52.
52. Casadevall N, Dupuy E, Molho-Sabatier P et al. Autoantibodies against erythropoietin in a patient with pure red-cell aplasia. *N Engl J Med* 1996;334:630–633.
53. Voulgarelis M, Kokori SI, Ioannidis JP et al. Anaemia in systemic lupus erythematosus: aetiological profile and the role of erythropoietin. *Ann Rheum Dis* 2000;59:217–222.
54. Tzioufas AG, Kokori SI, Petrovas CI et al. Autoantibodies to human recombinant erythropoietin in patients with systemic lupus erythematosus: correlation with anemia. *Arthritis Rheum* 1997;40:2212–2216.
55. Sipsas NV, Kokori SI, Ioannidis JP et al. Circulating autoantibodies to erythropoietin are associated with human immunodeficiency virus type 1-related anemia. *J Infect Dis* 1999;180:2044–2047.
56. Kuramochi S, Ikawa Y, Todokoro K. Characterization of murine erythropoietin receptor genes. *J Mol Biol* 1990;216:567–575.
57. Hercberg S, Preziosi P, Galan P. Iron deficiency in Europe. Public Health Nutr 2001;4:537–545.
58. Bayoumeu F, Subiran-Buisset C, Baka NE et al. Iron therapy in iron deficiency anemia in pregnancy: intravenous route *versus* oral route. *Am J Obstet Gynecol* 2002;186:518–522.
59. Stoves J, Inglis H, Newstead CG. A randomized study of oral *versus* intravenous iron supplementation in patients with progressive renal insufficiency treated with erythropoietin. *Nephrol Dial Transplant* 2001;16:967–974.
60. Bayoumeu F, Subiran-Buisset C, Baka NE et al. Iron therapy in iron deficiency anemia in pregnancy: intravenous route *versus* oral route. *Am J Obstet Gynecol* 2002;186:518–522.
61. Stoves J, Inglis H, Newstead CG. A randomized study of oral *versus* intravenous iron supplementation in patients with progressive renal insufficiency treated with erythropoietin. *Nephrol Dial Transplant* 2001;16:967–974.
62. Pak M, Lopez MA, Gabayan V et al. Suppression of hepcidin during anemia requires erythropoietic activity. *Blood* 2006;108:3730–3735.
63. Wen D, Boissel JP, Tracy TE et al. Erythropoietin structure-function relationships: high degree of sequence homology among mammals. *Blood* 1993;82:1507–1516.
64. David RB, Blom AK, Sjaastad OV et al. The porcine erythropoietin gene: cDNA sequence, genomic sequence and expression analyses in piglets. *Domest Anim Endocrinol* 2001;20:137–147.
65. Arai M, Darman J, Lewis A et al. The use of human hematopoietic growth factors (rhGM-CSF and rhEPO) as a supportive therapy for FIV-infected cats. *Vet Immunol Immunopathol* 2000;77:71–92.
66. Manire CA, Rhinehart HL. Use of human recombinant erythropoietin for the treatment of nonregenerative anemia in a rough-toothed dolphin (Steno bredanensis). *J Zoo Wildl Med* 2000;31:57–163.
67. McKeever KH. Erythropoietin: a new form of blood doping in horses. *Proceedings of the 11th International Conference of Racing Analysts and Veterinarians.* 1996;79–84. R&W Press, Newmarket, UK.
68. Catania MA, Marciano MC, Parisi A et al. Erythropoietin prevents cognition impairment induced by transient brain ischemia in gerbils. *Eur J Pharmacol* 2002;437:147–150.
69. Grasso G, Passalacqua M, Sfacteria A et al. Does administration of recombinant human erythropoietin attenuate the increase of S-100 protein observed in cerebrospinal fluid after experimental subarachnoid hemorrhage? *J Neurosurg* 2002;96:565–570.
70. Chapel SH, Veng-Pedersen P, Schmidt RL et al. A pharmacodynamic analysis of erythropoietin-stimulated reticulocyte response in phlebotomized sheep. *J Pharmacol Exp Ther* 2000;295:346–351.
71. Piercy RJ, Swardson CJ, Hinchcliff KW. Erythroid hypoplasia and anemia following administration of recombinant human erythropoietin to two horses. *J Am Vet Med Assoc* 1998;212:244–247.

72. Winearls CG, Oliver DO, Pippard MJ et al. Effect of human erythropoietin derived from recombinant DNA on the anaemia of patients maintained by chronic haemodialysis. *Lancet* 1986;2:1175–1178.
73. Eschbach JW, Egrie JC, Downing MR et al. Correction of the anemia of end-stage renal disease with recombinant human erythropoietin. Results of a combined phase I and II clinical trial. *N Engl J Med* 1987;316:73–78.
74. Adamson JW, Eschbach JW. Treatment of the anemia of chronic renal failure with recombinant human erythropoietin. *Annu Rev Med* 1990;41:349–360.
75. The US Recombinant Human Erythropoietin Predialysis Study Group. Double-blind, placebo-controlled study of the therapeutic use of recombinant human erythropoietin for anemia associated with chronic renal failure in predialysis patients. *Am J Kidney Dis* 1991;18:50–59.
76. Mohini R. Clinical efficacy of recombinant human erythropoietin in hemodialysis patients. *Semin Nephrol* 1991;9:16–21.
77. Satoh K, Masuda T, Ikeda Y et al. Hemodynamic changes by recombinant erythropoietin therapy in hemodialyzed patients. *Hypertension* 1990;15:262–266.
78. Brown WS, Marsh JT, Wolcott D et al. Cognitive function, mood and P3 latency: effects of the amelioration of anemia in dialysis patients. *Neuropsychologia* 1991;29:35–45.
79. Chmielewski CM. Case study of the anemic patient: epoetin alfa – focus on erythrokinetics. *ANNA J* 1995;22:420–423.
80. Nowrousian MR. *Definition, classification and characterization of anemia in cancer. Recombinant Human Erythropoietin (rhEPO) in Clinical Oncology*. Springer-Verlag/Wien, New York 2008;5:117–148.
81. Beguin Y. Erythropoietin and the anemia of cancer. *Acta Clin Belg* 1996;51:36–52.
82. Ludwig H, Fritz E. Anemia of cancer patients: patient selection and patient stratification for epoetin treatment. *Semin Oncol* 1998;25:35–38.
83. Ludwig H, Fritz E. Anemia in cancer patients. *Semin Oncol* 1998;25:2–6.
84. Bron D, Meuleman N, Mascaux C. Biological basis of anemia. *Semin Oncol* 2001;28:1–6.
85. Ludwig H, Pecherstorfer M, Leitgeb C et al. Recombinant human erythropoietin for the treatment of chronic anemia in multiple myeloma and squamous cell carcinoma. *Stem Cells* 1993;11:348–355.
86. DeGowin RL, Gibson DP. Erythropoietin and the anemia of mice bearing extramedullary tumor. *J Lab Clin Med* 1979;94:303–311.
87. Zarrabi MH, Lysik R, DiStefano J et al. The anaemia of chronic disorders: studies of iron reutilization in the anaemia of experimental malignancy and chronic inflammation. *Br J Haematol* 1977;35:647–658.
88. Leng HM, Albrecht CF, Kidson SH et al. Erythropoietin production in anemia associated with experimental cancer. *Exp Hematol* 1999;27:806–810.
89. Fisher JW, Roh BL. Influence of alkylating agents on kidney erythropoietin production. *Cancer Res* 1964;24:983–988.
90. Magid E, Hansen P. The erythropoietin content of plasma from sheep treated with nitrogen mustard. *Scand J Clin Lab Invest* 1966;18:347–352.
91. Barcelo AC, Bozzini CE. Erythropoietin formation during hypoxia in mice with impaired responsiveness to erythropoietin induced by irradiation or 5-fluorouracil injection. *Experientia* 1982;38:504–505.
92. Zucker S, Friedman S, Lysik RM. Bone marrow erythropoiesis in the anemia of infection, inflammation, and malignancy. *J Clin Invest* 1974;53:1132–1138.
93. Lockner D. Iron absorption and serum erythropoietin concentration in patients with cancer of the uterine cervix. *Acta Med Scand* 1966;180:651–655.
94. Schreuder WO, Ting WC, Smith S et al. Testosterone, erythropoietin and anaemia in patients with disseminated bronchial cancer. *Br J Haematol* 1984;57:521–526.
95. Miller CB, Jones RJ, Piantadosi S et al. Decreased erythropoietin response in patients with the anemia of cancer. *N Engl J Med* 1990;322:1689–1692.
96. Smith DH, Goldwasser E, Vokes EE. Serum immunoerythropoietin levels in patients with cancer receiving cisplatin-based chemotherapy. *Cancer* 1991;68:1101–1105.
97. Greendyke RM, Sharma K, Gifford FR. Serum levels of erythropoietin and selected other cytokines in patients with anemia of chronic disease. *Am J Clin Pathol* 1994;101:338–341.
98. Ponchio L, Beguin Y, Farina G et al. Evaluation of erythroid marrow response to recombinant human erythropoietin in patients with cancer anaemia. *Haematologica* 1992;77:494–501.

99. Ozguroglu M, Arun B, Demir G et al. Serum erythropoietin level in anemic cancer patients. *Med Oncol* 2000;17:29–34.
100. Groopman JE, Itri LM. Chemotherapy-induced anemia in adults: incidence and treatment. *J Natl Cancer Inst* 1999;91:1616–1634.
101. Corazza F, Beguin Y, Bergmann P et al. Anemia in children with cancer is associated with decreased erythropoietic activity and not with inadequate erythropoietin production. *Blood* 1998;92:1793–1798.
102. Itri LM. The use of epoetin alfa in chemotherapy patients: a consistent profile of efficacy and safety. *Semin Oncol* 2002;29:81–87.
103. Migliaccio AR, Migliaccio G, D'Andrea A et al. Response to erythropoietin in erythroid subclones of the factor-dependent cell line 32D is determined by translocation of the erythropoietin receptor to the cell surface. *Proc Natl Acad Sci USA* 1991;88:11086–11090.
104. Huang LJ, Constantinescu SN, Lodish HF. The N-terminal domain of Janus kinase 2 is required for Golgi processing and cell surface expression of erythropoietin receptor. *Mol Cells* 2001;8:1327–1338.
105. Nakamura Y, Komatsu N, Nakauchi H. A truncated erythropoietin receptor that fails to prevent programmed cell death of erythroid cells. *Science* 1992;257:1138–1141.
106. Sinclair AM, Rogers N, Busse L et al. Erythropoietin receptor transcription is neither elevated nor predictive of surface expression in human tumour cells. *Br J Cancer* 2008;98:1059–1067.

Erythropoietins, Erythropoietic Factors and Erythropoiesis
Edited by S.G. Elliott, M.A. Foote, and G. Molineux

Erythropoiesis – genetic abnormalities

Josef T. Prchal[1] and Xylina T. Gregg[2]

[1] *University of Utah, 30 North 1900 East, Rm 5C310 SOM, Salt Lake City, UT 84132-2408, USA*
[2] *Utah Cancer Specialists, 3838 South 700 East, Ste 100, Salt Lake City, UT 84106, USA*

Introduction

We review genetic mechanisms altering erythropoiesis that lead to either an increased red cell mass (polycythemia/erythrocytosis) or decreased red cell mass (anemia). Since polycythemia is often and inconsistently referred to by an alternative term, erythrocytosis, and, as no consensus on usage has been reached, we refer to the individual entities by the term used in the original description. Anemias are far more common than polycythemic states, and most are due to acquired nutritional, autoimmune, and toxic causes; or to inherited globin, cytoskeleton, and red cell enzyme mutations. These causes result in hemolytic, microcytic/hypochromic, hypoproliferative, or inefficient erythropoiesis anemia phenotypes. In this review, we concentrate on the few well-delineated germline or somatic mutations disturbing normal control of erythropoiesis that cause a disease phenotype either in humans or in mouse.

Erythropoiesis

Although regulation of erythropoiesis is described in other chapters of this book, for better understanding of this chapter we provide the pertinent details that are relevant for delineation of polycythemias and anemias. Erythrocyte production, i.e., erythropoiesis, is a tightly regulated system, but the details of its regulation are not fully elucidated. Much remains to be learned from uncovering the molecular basis of many congenital and acquired mutations that disrupt the control of erythropoiesis.

Erythropoiesis can be viewed as composed of three stages. In the initial stage, commitment of pluripotent hematopoietic progenitors to committed erythroid precursors takes place. The second stage is characterized by expansion of erythroid progenitors that is largely regulated by erythropoietin (Epo) and is made effective by the appearance of its receptor (EpoR) on surface of these progenitors. The expression of EpoR peaks in early erythroblasts and then declines. The terminal stage consists of enucleation and removal of remnants of organelles and nucleotides that may be toxic to mature circulating erythrocytes. Defined defects of erythropoiesis leading to polycythemia occur at the second stage of erythropoiesis and are generally those enhancing proliferation and or delaying apoptosis.

Prevention of apoptosis is a well-recognized mechanism assuring productive early erythropoiesis and preventing anemia; however, the importance of proapoptotic processes for productive terminal erythropoiesis is newly recognized. Whether a congenital imbalance of proapoptotic and antiapoptotic processes can contribute to polycythemia remains to be determined.

The role of microRNAS (miRNAs) in regulation of erythropoiesis is being defined. Some miRNAs are mainly expressed in early stages of erythropoiesis, others in late stages, and some have biphasic expression during erythroid differentiation; and some appear to have erythroid specific expression. miRNAs exact role in erythropoiesis and their molecular targets are currently being defined at many laboratories [1].

Polycythemias

Classification

Polycythemia can be classified as primary polycythemia, secondary polycythemia, or polycythemia due to abnormal hypoxia sensing, of which one such entity (Chuvash polycythemia) has features of both primary and secondary polycythemia. Primary polycythemias are caused by intrinsic defects in the erythroid precursors that result in an excessive inappropriate response to normal stimulators of erythropoiesis. In contrast, in secondary polycythemias, the cytokine responsiveness of erythroid progenitors is normal. Secondary polycythemias are driven by hormonal factors (predominantly Epo but also insulin-like growth factor-1, angiotensin 2, and cobalt) extrinsic to the erythroid progenitor cells, and the increased erythrocyte mass represents a physiologic response to tissue hypoxia or abnormal autonomous Epo production [2]. Secondary polycythemias may result from an appropriate physiologic response to tissue hypoxia; for example, tissue hypoxia due to pulmonary or cardiac disease of from high affinity hemoglobins, which release less oxygen per gram of hemoglobin, results in an increased compensatory erythropoiesis that is an appropriate physiologic response to accommodate for decreased oxygen delivery. Secondary polycythemias resulting from increased amounts of Epo secreted from tumors or from dysregulated control of Epo production are inappropriate physiologic responses.

Polycythemias due to abnormal hypoxia sensing may blur the distinction between primary and secondary polycythemias. Increased amounts of Epo (or inappropriately normal Epo in conjunction with concomitant increase of red cell mass) resulting from abnormal hypoxia sensing are characteristic of these polycythemias. Yet when erythroid progenitors responses to Epo were studied in Chuvash polycythemia, the erythroid progenitors were also hypersensitive to Epo.

Both primary and secondary polycythemias can be either acquired or congenital, but polycythemia due to abnormal hypoxia sensing is usually congenital. Acquired conditions that lead to increased Epo production, such as chronic hypoxia and a variety of tumors, are the most common causes of secondary polycythemias. Congenital cyanotic heart or lung disorders leading to tissue hypoxia and increased amounts of Epo are examples of secondary congenital polycythemias. These types of polycythemia are not addressed in this review, which focuses only on polycythemias due to known genetic mechanisms. Hypoxemia from pulmonary or heart disease or from exposure to extreme high altitude, however, leads in some but not all of the affected individuals to an appropriate increase of red cell mass. The individual variation of erythropoietic responses to a hypoxic stimulus suggests that there is a genetically determined pleomorphism underlying the erythropoietic response to hypoxia.

Primary polycythemias

Primary polycythemias result from intrinsic defects in red blood cell precursors that cause increased responsiveness to Epo. The most common type of primary polycythemia is polycythemia vera, which is due to an acquired somatic mutation. An inherited primary polycythemia is primary familial and congenital polycythemia (PFCP).

Polycythemia vera (PV), an acquired primary polycythemia, is characterized by clonal expansion of hematopoietic precursors, and is the most common type of primary polycythemia [3]. Hematopoiesis in PV is characterized by accumulation of phenotypically normal erythrocytes with variable overproduction of platelets and myeloid leukocytes. *In vitro*, PV erythroid progenitors (BFU–Es) develop in the absence of exogenously added Epo [4]; a hallmark of PV. A single nucleotide substitution (G1849T) generates *JAK2 V617F* [5] that is present in the bone marrow and peripheral blood myeloid cells of >95% of patients with PV. This mutation leads to loss of function of the inhibitory auto regulatory pseudokinase domain [5], which results in augmented or constitutively active JAK2 activity that explains many of the *in vitro* defined characteristic abnormalities of PV erythroid progenitors [5]. Most patients with PV have at least some progenitors homozygous for *JAK2 V617F* mutation as a consequence of uniparenteral disomy [6]. However, *JAK2 V617F* is not specific for PV as it is also found in approximately 50% of patients with essential thrombocythemia (ET) and primary myelofibrosis (PMF) [7, 8]. *JAK2 V617F* also has been identified in a small proportion of patients with chronic neutrophilic leukemia, acute megakaryocytic leukemia, and in most patients with refractory anemia with ringed sideroblasts and thrombocytosis (RARS-T) [7,

8]. The nonspecificity of the mutation, the absence of *JAK2 V617F* in some patients, the existence of familial PV in which affected members can be either *JAK2 V617F*-negative or positive, the fact that only a proportion of clonal PV cells are *JAK2 V617F*-positive, and the existence of Epo-independent BFU–Es that are either *JAK2 V617F*-negative or positive demonstrate that this somatic mutation of *JAK2* is not the sole pathogenic process in PV [9–12]. Moreover, several mutations in exon 12 of JAK2 have been described in *JAK2 V617F*-negative patients [13, 14].

Evidence exists that the acquired augmented erythropoiesis in PV, which is caused by somatic mutations, often has a familial predisposition [15, 16], which suggests the existence of germ-line mutation(s) that either facilitate the somatic mutation(s) or contribute to the acquisition of somatic acquired mutations to result in familial clustering of PV. The molecular basis of these interactions remains to be defined.

PFCP is characterized by an autosomal dominant mode of inheritance, and less frequently, by the occurrence of sporadic cases [17, 18]. The clinical features of PFCP include the presence of isolated erythrocytosis, absence of predisposition to development of acute leukemia or other myeloproliferative disorders, absence of splenomegaly, normal white blood cell and platelet counts, low plasma Epo values, normal hemoglobin-oxygen dissociation curve (indicated by a normal P50), and hypersensitivity of erythroid progenitors to exogenous Epo in *in vitro* studies [17–20]. PFCP is generally thought to be a benign condition, but it has been reported to be associated with predisposition to cardiovascular problems, such as hypertension, coronary artery disease, and cerebrovascular events, that are not clearly related to an increased hematocrit. Association with cardiovascular disease however, has not been described in all series [21–23].

The distal cytoplasmic region of EpoR, in association with SHP-1, is required for down-regulation of Epo-mediated activation of JAK2/STAT5 proteins [24–26]. To date, nine mutations of *EPOR* have been convincingly linked with PFCP [27]. All of these mutations result in truncation of the EpoR cytoplasmic carboxyl terminal leading to loss of its negative regulatory domain, resulting in a gain-of-function of EpoR. Three additional missense *EPOR* mutations have been described in families with PFCP, but they have not been linked to PFCP or any other disease phenotype [28]. The absence of a polycythemic phenotype in some patients with *EPOR* mutation is suggestive of a role played by gene modifiers or epigenetic factors in phenotypic penetrance [29]. On the other hand, mutations of the *EPOR* were found in only 12% of subjects with a PFCP phenotype, suggesting that in a majority of PFCP families, mutations in genes other than *EPOR* result in defective Epo signaling and accumulation of erythrocytes [27, 30, 31]. In those sporadic cases with a proven *EPOR* gain-of-function mutation, a *de novo* occurrence of this mutation is typically demonstrated [27, 32], underscoring the likely deleterious impact of this mutation on survival.

Secondary polycythemias

Secondary congenital polycythemia results from conditions that lead to increased amounts of endogenous Epo. Genetic causes include hemoglobin variants with high affinity for oxygen, congenitally lower amount of erythrocyte 2, 3 biphosphoglycerate (2,3-BPG), and inherited methemoglobinemias. All these conditions are characterized by a left shift in the hemoglobin dissociation curve, which in turn leads to tissue hypoxia and physiologically appropriate increased amounts of Epo. Although these entities represent the first polycythemias with a delineated pathophysiology and molecular cause, in our experience, they are frequently misdiagnosed as PV and PFCP.

High-oxygen affinity hemoglobin mutants

An autosomal dominant polycythemia with a normal-to-increased Epo concentration is suggestive of a mutant hemoglobin with high affinity for oxygen [33]. The affinity of hemoglobin with oxygen is expressed as the P50, which is the partial pressure of oxygen in blood at which 50% of the hemoglobin is saturated with oxygen. An high-affinity hemoglobin reflected in the laboratory by an abnormally low P50 reflects an increased affinity of hemoglobin for oxygen and is supportive of a high-oxygen affinity hemoglobin variant or decreased 2,3-BPG level. If cooximeter equipment is unavailable, the P50 can be conveniently calculated from venous gas parameters [34].

Amino acid substitutions involving the hemoglobin $\alpha 1/\beta 2$ interface, the *C*-terminal residues of the β chain, or the 2, 3-BPG binding sites can all affect the affinity of hemoglobin for oxygen. Most mutations affecting oxygen affinity result in high-affinity hemoglobin variants that produce relative tissue hypoxia resulting in compensatory secondary appropriate polycythemia [33, 34]. Almost 100 hemoglobin variants are known to be associated with high affinity for oxygen. Establishing a correct diagnosis of a high-affinity hemoglobin variant is important as these patients have normal life expectancy and do not require phlebotomy.

2, 3-BPG deficiency

Congenitally low amounts of erythrocyte 2, 3-BPG can occur because of deficiency of the red cell enzyme 2, 3-BPG mutase [35, 36]. This condition is an extremely rare autosomal recessive one but should be suspected in the case of isolated polycythemia (without any feature of myeloproliferative disorders such as progressive increase of red blood cell mass, high platelet and granulocyte count, and splenomegaly), absence of a family history, and low P50 (signifying high-oxygen affinity). A mutant hemoglobin needs to be ruled out first. In both 2, 3-BPG deficiency and high-affinity hemoglobins, the red cells will have high oxygen affinity; however, unlike high-affinity hemoglobin, the oxygen affinity of the dialyzed hemolysate is normal and the level of 2, 3-BPG is very low.

Congenital methemoglobinemias

Three types of hereditary methemoglobinemias have been identified (reviewed in [36]). Two are inherited as autosomal recessive traits: cytochrome b5R deficiency and cytochrome b5 deficiency. The third type is an autosomal dominant disorder, hemoglobin M disease in which there is a mutation of one of the globin genes. All congenital methemoglobinemias are associated with suboptimal delivery of oxygen per red cell and this may result in compensatory secondary appropriate polycythemia. The resulting polycythemia is typically mild and therapy is not necessary as it would only decrease tissue oxygen delivery and lead to tissue hypoxia.

Polycythemia due to abnormal hypoxia sensing

Chuvash polycythemia

Chuvash polycythemia was the first recognized hereditary condition of augmented hypoxia-sensing. Chuvash polycythemia (CP), the only known endemic polycythemia, is an autosomal recessive hereditary polycythemia. The Chuvash people reside in the mid-Volga River region in Russia wherein CP affects hundreds of people, making it the most common congenital polycythemia [37]. In a study of five multiplex Chuvash families with CP, a homozygous mutation of the von Hippel Lindau (*VHL)* gene was found in the affected individuals [38, 39]. This mutation impairs the interaction of pVHL with both hypoxia inducible factor (HIF)-1α and HIF-2α, reducing the rate of ubiquitin-mediated destruction of HIF-1α. As a result, the amount of the HIF-1 and HIF-2 heterodimers increase and lead to increased expression of target genes including *EPO*, *VEGF*, and plasminogen activator inhibitor (*PAI*) among others [39, 40]. CP is associated with a predisposition to development of thrombosis, bleeding, cerebral vascular events, and increased mortality. It is characterized by an intact response to hypoxia despite increased basal expression of a broad range of hypoxia-regulated genes in normoxia. Unexpectedly, CP erythroid progenitors are hypersensitive to Epo [38, 39], thus CP shares the features of both primary as well as secondary polycythemia. Despite increased expression of HIF-1α and HIF-2α and VEGF in normoxia, patients with CP do not display predisposition to tumor formation. Imaging studies of 33 patients with CP revealed unsuspected cerebral ischemic lesions in 45% but no tumors characteristic of VHL syndrome [40].

Homozygosity for the *VHL 598C>T* has been reported to occur sporadically also in whites in the United States and Europe and in people of Southeast Asian (Indian subcontinent) ancestry [41–43]. A high prevalence of this disorder in the Italian island of Ischia has been reported [44]. To address the question of whether the *VHL 598C>T* substitution occurred in a single founder or resulted from recurrent mutational events, haplotype analysis was performed on subjects bearing the *VHL 598C>T* mutation and normal unrelated individuals from Chuvash, Asian, white, Hispanic, and African-

American ethnic groups [45]. These studies indicated that in most individuals, the *VHL 598C>T* mutation arose in a single ancestor between 12,000 and 51,000 years ago. However, a Turkish polycythemic family had a different haplotype indicating that the *VHL 598C>T* mutation in this family occurred independently [46].

In contrast, autosomal dominant mutations of the *VHL* gene cause VHL syndrome [47]. Patients heterozygotic for dominant *VHL* mutations are at increased risk of developing hemangioblastomas, renal cell carcinoma, pheochromocytoma, pancreatic endocrine tumors, and endolymphatic sac tumors when they acquire a somatic mutation in the normal *VHL* allele [47–50]. Some patients with VHL syndrome also develop acquired polycythemia [47]. The development of hemangioblastoma and renal cell carcinoma associated with VHL tumor predisposition syndrome has been proposed to be related to increased expression of HIF and possibly VEGF [51, 52]. The absence of a predisposition to tumorigenesis in patients with CP implies that deregulation of HIF-1 and VEGF may not be sufficient to cause predisposition towards tumor formation in VHL syndrome.

Non-Chuvash germline *VHL* mutations also cause polycythemia. Some patients with congenital polycythemia have proven to be compound heterozygotes for the Chuvash mutation, *VHL 598C>T*, and other *VHL* mutations including *562C>G*, *574C>T*, *388C>G*, and *311G>T* [27, 41, 42]. A Croatian boy was homozygous for *VHL 571C>G*, the first example of a homozygous *VHL* germline mutation other than *VHL 598C>T* causing polycythemia [27, 42]. Additionally, a polycythemic Portuguese girl was a compound heterozygote for *VHL 562C>G* and *VHL 253C>T* [27, 53].

A few cases of congenital polycythemia, known to have mutations of only one *VHL* allele, confound an obvious pathophysiological explanation. Two Ukrainian children with polycythemia were heterozygotes for *VHL 376G>T*, but the father with the same mutation was not polycythemic [41]. Peripheral blood erythroid progenitors from the children and father were hyper-responsive to recombinant human Epo in *in vitro* clonogenic assays in a way similar to what is seen in patients with CP. An English patient was a heterozygote for *VHL 598C>T* [43], although the inheritance of a deletion of a *VHL* allele or a *null VHL* allele in a trans position was not excluded in this patient. Subsequently, two patients with polycythemia who were heterozygous for VHL were described in whom a *null VHL* allele was more rigorously excluded [27, 46]; one of these patients also had ataxia-telangiectasia [53].

Proline hydroxylase and HIF-2α *mutations*

Two families, one with a proline hydroxylase (*PHD2)* mutation [54] and other with a *HIF-2α* mutation [55], were described. In both cases, heterozygotes for this mutation had a mild or borderline polycythemia that the authors referred to as erythrocytosis [54, 55]. Testing for Epo-hypersensitivity of erythroid progenitors was not reported. Because of the small family size, the possibility of a nonerythroid etiology for the erythrocytosis could not be excluded, but this work further supports the concept that abnormalities in factors involved in the regulation of HIF signaling can result in polycythemia.

Based on their study of the *HIF-2α* mutation [55], Percy and colleagues concluded that HIF-2α may be sole and central to regulating amounts of Epo. While it is clear that HIF2 is the transcription factor that regulates hepatic Epo production, the kidney is the major site of Epo synthesis, HIF1α was identified by its binding to a 3' hypoxia responsive element of kidney *EPO* [56], and severe anemia develops after nephrectomy. In contrast, liver *EPO* is regulated by upstream nucleotide sequences [57] and only about 10% to 20% of Epo is produced by the liver (reviewed in [55]).

A potential alternative explanation for the phenotype described is that it may result predominantly from autonomous liver-generated Epo (possibly with suppressed renal Epo production) [58]. These investigators created a mouse with a conditional deletion of *Vhl* in epidermal keratinocytes, which caused cutaneous vasodilatation and increased expression of *Hif-1α* and *Hif-2α*. Although keratinocytes do not make Epo, the amount of Epo in these mice was increased, and they became polycythemic. Further studies of this epidermal *Vhl* knockout mouse revealed that the increased amounts of HIF-1 caused upregulation of inducible nitric oxide synthase, which in turn led to increased cutaneous nitric oxide (NO), a potent vasodilator. This NO-induced skin vasodilatation resulted in decreased perfusion of other organs, most notably the liver, with subsequent hypoxia-induced increased expression of hepatic *Hif-2α*, which in turn caused increased expression of the *Epo* gene. In follow-up experiments using mice with wild-type *Vhl*, the authors deleted the cutaneous genes for either *HIF-1α* or *HIF-2α* and crossbred these mice with those with the conditional skin *vhl* deletion. Unexpectedly, under conditions of normoxia, the loss of *HIF-1α* and *HIF-2α* had no effect on Epo levels. Under hypoxic conditions, however, the *HIF-1α* epidermal knockout mice did not display an appropriate increase in renal *EPO* gene transcription and were unable to mount an appropriate renal Epo response. These experiments show the importance of epidermal HIF in sensing environmental oxygen levels and regulating systemic hypoxic responses in mice, with physiologic regulation mediated primarily by HIF-1α while the pathologic loss of *Vhl* is mediated primarily by HIF-2α [58]. Whether this intriguing mechanism of regulation of the mouse *EPO* gene by a skin/NO pathway is present only in mice, or also operates in human *EPO* regulation remains to be proven.

Unexplained congenital polycythemias with increased or inappropriately normal amounts of erythropoietin

In our experience, most patients with congenital polycythemias with normal or increased amounts of endogenous Epo do not have *VHL* mutations, hemoglobinopathies, or 2,3-BPG deficiency, and the molecular basis of polycythemia in these cases remains to be elucidated. Proline hydroxylase and *HIF-2α* mutations have been excluded in some, but not in all of these patients. It is not clear why in some families, the polycythemia is dominantly inherited

[59], in others recessively, and in some it is sporadic, or why in families with the same mutation the phenotype differs. Lesions in genes linked to hypoxia independent regulation of HIF as well as oxygen-dependent gene regulation pathways are leading candidates for mutation screening in polycythemic patients with normal or increased concentrations of Epo without *VHL* or proline hydroxylase mutations.

Anemias

Classification

A review of the myriad of acquired and congenital causes of anemia is beyond the scope of this chapter. Anemia is caused by variety of genetic conditions, including red cell enzyme deficiencies, hemoglobinopathies, thalassemias, red cell membrane defects, iron disorders, B12 and folate deficiency, and sideroblastic anemias. This chapter focuses solely on a few examples of acquired and congenital, and experimental genetic lesions that result in dysregulated erythropoiesis. Some of the congenital molecular lesions leading to dysregulated erythropoiesis have unanticipated phenotypes and underscore the complexity of erythroid regulation.

Mutations leading to anemia, tend to be more clinically symptomatic and appear to be more numerous than those causing polycythemia. Mutations leading to anemias occur in all stages of erythropoiesis, including its early stage. Some mutations disturb both erythroid and nonerythroid regulatory events, as exemplified by the motheaten mouse. In other conditions, such as myelodysplastic syndrome, the exact molecular basis of the disease is still largely unknown.

Myelodysplastic syndromes

Myelodysplastic syndromes (MDS) is a group of acquired clonal hematopoietic disorders due to mostly uncharacterized somatic mutations associated with one or more cytopenias, of which anemia is most common. While some subtypes of MDS are associated with well-defined chromosomal abnormalities, the exact molecular consequences of these lesions largely remain to be elucidated [60]. One exception is in MDS with the 5q-chromosomal abnormality, in which dysregulation of ribosomal- and translation-related genes was described [61]. Somatic mutations of MDS are being defined using increasingly comprehensive single nucleotide polymorphisms (SNP) chip technologies [62, 63]. It is anticipated that uncovering the molecular basis of these mutations and their function will lead to better and more complete understanding of regulation of erythropoiesis.

Gas6 and erythropoiesis
Growth arrest-specific 6 (Gas6) protein is a secreted vitamin K-dependent protein that interacts with cell membranes and leads to intracellular signaling (through its receptor, tyrosine kinases), including activation of the PI3K and Akt pathways. Gas6 receptors Tyro3, Axl, and Mertk are expressed in hematopoietic tissue, megakaryocytes, myelomonocytic precursors, and bone marrow stromal cells. Gas6 has been shown to amplify the erythropoietic response to Epo using a mouse model of Gas6 knockout. These findings warrant further investigation of the therapeutic potential of Gas6 for the treatment of patients who are hyporesponsive or resistant to Epo [64]. Microarray data has shown that the *Gas6* gene is a HIF-1 target gene [65]. Moreover, Gas6 is known to downregulate the expression of inflammatory cytokines such as tumor necrosis factor (TNF)-α by macrophages [66]. Because apoptotic depletion of bone marrow progenitors by TNF-α has been implicated in the development of anemia [67], Gas6 might also play a beneficial role in erythropoiesis by reducing TNF-α production by macrophages.

Motheaten mouse
Deficiency of hematopoietic phosphatase (SHP-1) would be predicted to result in expansion of erythroid and other hematopoietic progenitors, and this phenotype has indeed been observed in motheaten mice. While viable motheaten mice that are deficient in SHP-1 have hyperproliferation of myeloid cells, widespread inflammatory lesions, and their myeloid and erythroid progenitors are hypersensitive to cytokines, they are nevertheless anemic [68]. The anemia in motheaten mice is independent of the presence of anti-erythrocyte antibodies, and is at least in part due to increased erythrocyte fragility and augmented oxidant stress by an incompletely defined mechanism [69].

GATA-1 mutation with anemia and other phenotypes
A hypochromic, microcytic anemia with elevated fetal hemoglobin, thalassemia, porphyria, and thrombocytopenia was described in a boy, and proved to be due to *GATA-1* mutation (*R216W*) that was also present in one maternal allele [70]; however, since the *GATA-1* gene is an X-chromosome encoded gene, the heterozygous mother was not affected. A bone marrow allograft corrected the anemia, thrombocytopenia, porphyria, and the thalassemia. This report is the first of association of anemia, porphyria with thalassemia, and thrombocytopenia due to a mutation of a trans-acting factor.

Bnip3L$^{-/-}$ mouse
Prevention of apoptosis is a well-recognized mechanism assuring productive early erythropoiesis and preventing anemia, but the importance of pro-apoptotic processes for productive terminal erythropoiesis is a more recent finding. Bnip3L is a proapoptotic Bcl-2 family member. Bnip3L$^{-/-}$ mice predictably developed expansion of erythroid precursors and splenomegaly. The mice unexpectedly develop a hemolytic anemia due, at least in part, to failure to tar-

get the mitochondria into autophagosomes for degradation during erythroid maturation, which results in increased oxidative erythrocyte stress and shortened red cell survival [71]. Thus, this mouse model demonstrates that both pro- and anti-apoptotic mechanisms at appropriate stages of erythropoiesis are essential for proper formation of a normal erythrocyte.

Other causes of dysregulated erythropoiesis
Many other examples exist of dysregulated erythropoiesis by well-defined mechanisms including podocalyxin [72, 73], DMT-1 defect in which the human mutation causes iron overload but also a hypochromic microcytic Epo-responsive anemia [74], as well as others, but is beyond the scope of this chapter.

Conclusion

Regulation of erythropoiesis is a complex and incompletely understood process. We conclude that elucidation of molecular basis of anemic and polycythemic phenotypes will lead to a better understanding of the regulation of erythropoiesis and eventually to the development of novel targeted therapies of human diseases.

References

1. Bruchova H, Yoon D, Agarwal AM et al. Regulated expression of microRNAs in normal and polycythemia vera erythropoiesis. *Exp Hematol* 2007;35:1657–1667.
2. Gregg XT, Prchal JT. Recent advances in the molecular biology of congenital polycythemias and polycythemia vera. *Curr Hematol Rep* 2005;4:238–242.
3. Spivak J. Polycythemia vera: myths, mechanisms, and management. *Blood* 2002;100:4272–4290.
4. Prchal JF, Axelrad AA. Bone-marrow responses in polycythemia vera. *N Engl J Med* 1974;290:1382 (letter).
5. James C, Ugo V, Le Couedic JP et al. A unique clonal JAK2 mutation leading to constitutive signalling causes polycythaemia vera. *Nature* 2005;434:1144–1148.
6. Kralovics R, Guan Y, Prchal JT. Acquired uniparental disomy of chromosome 9p is a frequent stem cell defect in polycythemia vera. *Exp Hematol* 2002;30:229–236.
7. Jelinek J, Oki Y, Gharibyan V et al. JAK2 mutation 1849G>T is rare in acute leukemias but can be found in CMML, Philadelphia-chromosome negative CML and megakaryocytic leukemia. *Blood* 2005;106:3370–3373.
8. Levine RL, Gilliland DG. JAK-2 mutations and their relevance to myeloproliferative disease. *Curr Opin Hematol* 2007;14:43–47.
9. Kralovics R, Teo SS, Li S et al. Acquisition of the V617F mutation of JAK2 is a late genetic event in a subset of patients with myeloproliferative disorders. *Blood* 2006;108:1377–1380.
10. Nussenzveig RH, Swierczek SI, Jelinek J et al. Polycythemia vera is not initiated by JAK2V617F mutation. *Exp Hematol* 2007;35:32–38.
11. Bellanné-Chantelot C, Chaumarel I, Labopin M et al. Genetic and clinical implications of the Val617Phe JAK2 mutation in 72 families. *Blood* 2006;108:346–352.
12. Rumi E, Passamonti F, Pietra D et al. JAK2 (V617F) as an acquired somatic mutation and a secondary genetic event associated with disease progression in familial myeloproliferative disorders. *Cancer* 2006;107:2206–2211.
13. Scott LM, Tong W, Levine RL et al. *JAK2* exon 12 mutations in polycythemia vera and idiopathic erythrocytosis. *N Engl J Med* 2007;356:459–468.

14. Pietra D, Li S, Brisci A et al. Somatic mutations of JAK2 exon 12 in patients with JAK2 (V617F)-negative myeloproliferative disorders. *Blood* 2008;111:1686–1689.
15. Kralovics R, Stockton DW, Prchal JT. Clonal hematopoiesis in familial polycythemia vera suggests the involvement of multiple mutational events in the early pathogenesis of the disease. *Blood* 2003;102:3793–3796.
16. Kralovics R, Skoda RC. Molecular pathogenesis of Philadelphia chromosome negative myeloproliferative disorders. *Blood Rev* 2005;19:1–13.
17. Prchal J, Crist W, Goldwasser E et al. Autosomal dominant polycythemia. *Blood* 1985;66:1208–1214.
18. Juvonen E, Ikkala E, Fyhrquist F et al. Autosomal dominant erythrocytosis causes by increased sensitivity to erythropoietin. *Blood* 1991;78:3066–3069.
19. Emanuel P, Eaves C, Broudy V et al. Familial and congenital polycythemia in three unrelated families. *Blood* 1992;79:3019–3030.
20. Forget B, Degar BA, Arcasoy MO. Familial polycythemia due to truncations of the erythropoietin receptor. *Trans Am Clin Climatol Assoc* 2000;111:38–44.
21. Queisser W, Heim ME, Schmitz JM et al. Idiopathic familial erythrocytosis. Report on a family with autosomal dominant inheritance. *Dtsch Med Wochenschr* 1988;113:851–856.
22. Prchal J, Semenza GL, Prchal J et al. Familial polycythemia. *Science* 1995;268:1831–1832.
23. Sokol L, Kralovics R, Hubbell GL. A novel erythropoietin receptor mutation associated with primary familial polycythemia and severe cardiovascular and peripheral vascular disease. *Blood* 2001;98:[abstract 937].
24. Winkelmann J, Penny L, Deaven L et al. The gene for the human erythropoietin receptor: Analysis of the coding sequence and assignment to chromosome 19q. *Blood* 1990;76:24.
25. D'Andrea A, Yoshimura A, Youssoufian H et al. The cytoplasmic region of the erythropoietin receptor contains non-overlapping positive and negative growth-regulatory domains. *Mol Cell Biol* 1991;11:1980–1987.
26. Yoon D, Watowich SS. Hematopoietic cell survival signals are elicited through non-tyrosine-containing sequences in the membrane-proximal region of the erythropoietin receptor (EPOR) by a Stat5-dependent pathway. *Exp Hematol* 2003;31:1310–1316.
27. Gordeuk VR, Stockton DW, Prchal JT. Congenital polycythemias/erythrocytoses. *Haematologica* 2005;90:109–116.
28. Prchal JT. Pathogenetic mechanisms of polycythemia vera and congenital polycythemic disorders. *Semin Hematol* 2001;38:10–20.
29. Kralovics R, Sokol L, Prchal JT. Absence of polycythemia in a child with a unique erythropoietin receptor mutation in a family with autosomal dominant primary polycythemia. *J Clin Invest* 1998;102:124–129.
30. Kralovics R, Prchal JT. Genetic heterogeneity of primary familial and congenital polycythemia. *Am J Hematol* 2001;68:115–121.
31. Jedlickova K, Stockton DW, Prchal JT. Possible primary familial and congenital polycythemia locus at 7q22.1–7q22.2. *Blood Cells Mol Dis* 2003;31:327–331.
32. Percy MJ, McMullin MF, Roques AW et al. Erythrocytosis due to a mutation in the erythropoietin receptor gene. *Br J Haematol* 1998;100:407–410.
33. Wajcman H, Galacteros F. Hemoglobins with high oxygen affinity leading to erythrocytosis. New variants and new concepts. *Hemoglobin* 2005;29:91–106.
34. Agarwal N, Mojica-Henshaw MP, Simmons ED et al. Familial polycythemia caused by a novel mutation in the beta globin gene: essential role of P50 in evaluation of familial polycythemia. *Int J Med Sci* 2007;4:232–236.
35. Labie D, Leroux JP, Najman A et al. Familial diphosphoglyceratemutase deficiency. Influence on the oxygen affinity curves of hemoglobin. *FEBS Lett* 1970;9:37–40.
36. Gregg X, Prchal JT. Red cell enzymopathies. In: Hoffman RB, ed. Hematology: Basic Principles and Practice. 5th ed; 2007.
37. Sergeyeva A, Gordeuk VR, Tokarev YN et al. Congenital polycythemia in chuvashia. *Blood* 1997;89:2148–2154.
38. Ang S, Chen H, Gordeuk VR et al. Endemic polycythemia in Russia: mutation in the VHL gene. *Blood Cells Mol Dis* 2002;28:57–62.
39. Ang S, Chen H, Hirota K et al. Disruption of oxygen homeostasis underlies congenital Chuvash polycythemia. *Nat Genet* 2002; 32:614–621.
40. Gordeuk VR, Sergueeva AI, Miasnikova GY et al. Congenital disorder of oxygen sensing: associ-

ation of the homozygous Chuvash polycythemia VHL mutation with thrombosis and vascular abnormalities but not tumors. *Blood* 2004;103:3924–3932.
41. Pastore Y, Jelinek J, Ang S et al. Mutations in the VHL gene in sporadic apparently congenital polycythemia. *Blood* 2003;101:1591–1595.
42. Pastore Y, Jedlickova K, Guan Y et al. Mutations of von Hippel-Lindau tumor-suppressor gene and congenital polycythemia. *Am J Hum Genet* 2003;73:412–419.
43. Percy M, McMullin MF, Jowitt SN et al. Chuvash-type congenital polycythemia in four families of Asian and Western European ancestry. *Blood* 2003;102:1097–1099.
44. Perrotta S, Nobili B, Ferraro M et al. Von Hippel-Lindau-dependent polycythemia is endemic on the island of Ischia: identification of a novel cluster. *Blood* 2006;107:514–519.
45. Liu E, Percy MJ, Amos CI et al. The worldwide distribution of the VHL 598C>T mutation indicates a single founding event. *Blood* 2004;103:1937–1940.
46. Cario H, Schwarz K, Jorch N et al. Mutations in the von Hippel-Lindau (VHL) tumor suppressor gene and VHL haplotype analysis in patients with presumable congenital erythrocytosis. *Haematologica* 2005;90:19–24.
47. Friedrich C. Genotype-phenotype correlation in von Hippel-Lindau syndrome. *Hum Mol Genet* 2001;10:763–767.
48. Maher E, Webster AR, Richards FM et al. Phenotypic expression in von Hippel-Lindau disease: correlations with germline VHL gene mutations. *J Med Genet* 1996;33:328–332.
49. Maher E. Von Hippel-Lindau disease. Curr. *Mol Med* 2004;4:833–842.
50. Clifford S, Cockman ME, Smallwood AC et al. Contrasting effects on HIF-1alpha regulation by disease-causing pVHL mutations correlate with patterns of tumourigenesis in von Hippel-Lindau disease. *Hum Mol Genet* 2001;10:1029–1038.
51. Kondo K, Klco J, Nakamura E et al. Inhibition of HIF is necessary for tumor suppression by the von Hippel-Lindau protein. *Cancer Cell* 2002;1:237–246.
52. Turner K, Moore JW, Jones A et al. Expression of hypoxia inducible factors in human renal cancer: relationship to angiogenesis and to the von Hippel-Lindau gene mutation. *Cancer Res* 2002;62:2957–2961.
53. Bento M, Chang KT, Guan Y et al. Congenital polycythemia with homozygous and heterozygous mutations of von Hippel-Lindau gene: five new Caucasian patients. *Haematologica* 2005;90:128–129.
54. Percy MJ, Zhao Q, Flores A et al. A family with erythrocytosis establishes a role for prolyl hydroxylase domain protein 2 in oxygen homeostasis. *Proc Natl Acad Sci USA* 2006;103:654–659.
55. Percy MJ, Furlow PW, Lucas GS et al. A gain-of-function mutation in the HIF2A gene in familial erythrocytosis. *N Engl J Med* 2008;358:162–168.
56. Hirota K, Semenza GL. Regulation of angiogenesis by hypoxia-inducible factor 1. *Crit Rev Oncol Hematol* 2006;59:15–26.
57. Semenza GL, Koury ST, Nejfelt MK et al. Cell-type-specific and hypoxia-inducible expression of the human erythropoietin gene in transgenic mice. *Proc Natl Acad Sci USA* 1991;88:8725–8729.
58. Boutin AT, Weidemann A, Fu Z et al. Epidermal sensing of oxygen is essential for systemic hypoxic response. *Cell* 2008;133:223–234.
59. Maran J, Jedlickova K, Stockton D et al. Finding the novel molecular defect in a family with high erythropoietin autosomal dominant polycythemia. *Blood* 2003;102:162b.
60. Haferlach T, Bacher U, Haferlach C et al. Insight into the molecular pathogenesis of myeloid malignancies. *Curr Opin Hematol* 2007;14:90–97.
61. Pellagatti A, Hellström-Lindberg E, Giagounidis et al. Haploinsufficiency of RPS14 in 5q- syndrome is associated with deregulation of ribosomal- and translation-related genes. *Br J Haematol* 2008;142:57–64.
62. Mohamedali A, Gäken J, Twine NA et al. Prevalence and prognostic significance of allelic imbalance by single-nucleotide polymorphism analysis in low-risk myelodysplastic syndromes. *Blood* 2007;110:3365–3373.
63. Gondek LP, Dunbar AJ, Szpurka H et al. SNP uniparental disomy and cryptic chromosomal abnormalities in MDS/MPD-U and MPD. PLoS ONE 2007;2:e1225.
64. Angelillo-Scherrer A, Burnier L, Lambrechts D et al. Role of Gas6 in erythropoiesis and anemia in mice. *J Clin Invest* 2008;118:583–596.
65. Gregg Semenza, personal communication, May 2008.
66. Lemke G, Lu Q. Macrophage regulation by Tyro 3 family receptors. *Curr Opin Immunol* 2003;15:31–36

67. Papadaki HA, Kritikos HD, Gemetzi C *et al.* Bone marrow progenitor cell reserve and function and stromal cell function are defective in rheumatoid arthritis: evidence for a tumor necrosis factor alpha-mediated effect. *Blood* 2002;99:1610–1619.
68. Kasturi KN, Mayer R, Bona CA et al. Germline V genes encode viable motheaten mouse autoantibodies against thymocytes and red blood cells. *J Immunol* 1990;145:2304–2311.
69. Lyons BL, Lynes MA, Burzenski L et al. Mechanisms of anemia in SHP-1 protein tyrosine phosphatase-deficient "viable motheaten" mice. *Exp Hematol* 2003;31:234–243.
70. Phillips JD, Steensma DP, Pulsipher MA et al. Congenital erythropoietic porphyria due to a mutation in GATA1: the first trans-acting mutation causative for a human porphyria. *Blood* 2007;109:2618–2621.
71. Sandoval H, Thiagarajan P, Dasgupta SK et al. Essential role for Nix in autophagic maturation of erythroid cells. *N. Nature* 2008; [Epub ahead of print], PMID:18454133.
72. Sathyanarayana P, Menon MP, Bogacheva O et al. Erythropoietin modulation of podocalyxin and a proposed erythroblast niche. *Blood* 2007;110:509–518.
73. Doyonnas R, Kershaw DB, Duhme C et al. Anuria, omphalocele, and perinatal lethality in mice lacking the CD34-related protein podocalyxin. *J Exp Med* 2001;194:13–27.
74. Mims MP, Guan Y, Priwitzerova M et al. Identification of a human mutation of DMT1 in a patient with microcytic anemia and iron overload. *Blood* 2005;105:1337–1342.

Erythropoiesis-stimulating agents

Erythropoietins, Erythropoietic Factors and Erythropoiesis
Edited by S.G. Elliott, M.A. Foote, and G. Molineux

Studies of erythropoiesis and the discovery and cloning of recombinant human erythropoietin

MaryAnn Foote

MA Foote Associates, Westlake Village, CA 91362, USA

Introduction

Patients who are anemic because of chronic kidney disease, cancer, arthritis, or chemotherapy or radiation therapy often report fatigue [1–6]. Before the introduction of recombinant human erythropoietin (rHuEPO), anemia and its sequelae fatigue were treated with red blood cell transfusion, androgen stimulation of red blood cell production, and/or iron supplementation, among other treatments [7]. While effective in increasing red blood cell counts, both transfusions and androgen therapy have inherent risks [8]. Transfusions of red blood cells can be complicated by blood-borne pathogens, iron overload, immunologic consequences, and lack of or delayed hemoglobin response. Transfusions often improve but do not correct anemia and usually must be given frequently, and androgen therapy can cause viralization or abnormal liver function. rHuEPO is an ideal therapy because it mimics the action of the endogenous hormone by stimulating the production of red blood cells. Patients with chronic kidney disease are unable to produce adequate amounts of endogenous erythropoietin (EPO) to stimulate red blood cell production. Patients with cancer often have damaged bone marrow, with or without the insult of chemotherapy, that does not completely respond to the endogenous hormone.

The cloning of the human *EPO* gene by Fu Kuen Lin and colleagues was a difficult and frustrating endeavor. This milestone and the subsequent creation and production of rHuEPO as a therapeutic option was a breakthrough that has enabled physicians to ameliorate anemia and its sequelae. Patients treated with rHuEPO report a return to more normal lives. rHuEPO is the standard of care for treatment of anemia in patients with chronic kidney failure or receiving chemotherapy, and in other disease settings.

This chapter is a literature review of the history of the early work in erythropoiesis and the discovery and cloning of EPO. Other chapters discuss its commercial production and the clinical uses of rHuEPO.

Early studies in erythropoiesis

Bright [9] is credited with being the first scientist to recognize that anemia was a complication of kidney disease, but Jourdanet [10] has been credited as the first scientist to observe the relationship between altitude and blood viscosity. Jourdanet noted the similarity of symptoms reported by patients with altitude sickness and the symptoms reported by patients who had experienced severe blood loss. Several years later, Viault [11, 12] expanded knowledge of red blood cells and the effect of altitude on them, and quantified the change in red cell counts as altitude increased. On a train trip from the city of Lima, Peru to the high-altitude tin mines of that country, he repeatedly sampled his blood, blood of willing fellow travelers, and blood of a dog, rooster, and llama. Viault noted an increase in his red blood cell count from $5 \times 10^6/mm^3$ to $8 \times 10^6/mm^3$ during the ascent to higher altitude.

Other early scientists continued studies in an attempt to understand the mechanism of erythropoiesis in rabbits [13] and immigrants to the high Alps [14, 15]. One theory proposed at the time to explain the polycythemia seen at high altitudes was that low oxygen pressure directly stimulated bone marrow to increase red blood cell production. This theory held for nearly 50 years.

In 1906, Carnot developed the concept of humoral regulation of erythropoiesis [16–18]. Serum from anemic rabbits was injected into normal rabbits, and caused an increase in the red blood cell counts of the normal rabbits. Carnot suggested that "hemopoietine" present in the serum of anemic rabbits was responsible for the increase in cell numbers. Many other investigators repeated these experiments in anemic rabbits or rabbits raised at high altitudes [19–23].

For almost 30 years, researchers continued to repeat Carnot's work. Because some investigators were successful and others were not, controversy continued about the mechanism of erythropoiesis. Finally, Erslev [24] modified Carnot's original study: He injected large amounts of plasma from anemic rabbits into normal rabbits and found that the number of nucleated red blood cells in the bone marrow, the number of peripheral reticulocytes, and the hematocrit of the normal rabbits increased. This study suggested that red blood cell production is mediated by a humoral factor in rabbits. Four years later, Jacobson et al. [25] demonstrated that this factor, EPO, was produced by the kidney. Progress was being made in understanding the relationship between oxygen supply and demand of the body and EPO and erythropoiesis; however, despite more than 100 years of research, nothing was known about the structure of EPO or the *EPO* gene. Some debate continued whether EPO was produced as an inactive precursor in the kidney that was activated in some other tissue or organ.

The role of kidney and bone marrow in erythropoiesis

In the fetus, the liver is the primary site of endogenous EPO production [26]. In the adult, EPO is produced primarily (i.e., >90%) in the adult kidney [25, 27]. The liver [28, 29] and the brain [30] both synthesize some EPO, but the amount produced by these tissues alone is insufficient to maintain adequate erythropoiesis. Thus, kidney disease causes anemia due to loss of the main source of EPO production.

An oxygen sensor within renal cells detects the oxygen content of the blood and the kidney regulates the amount of EPO released into the blood (Fig. 1). The hormone acts on red blood cell precursor cells in the bone marrow to stimulate their proliferation and maturation and to increase the number of red blood cells in the peripheral circulation. The feedback loop is completed when the kidney cells recognize the change in oxygen delivery secondary to the

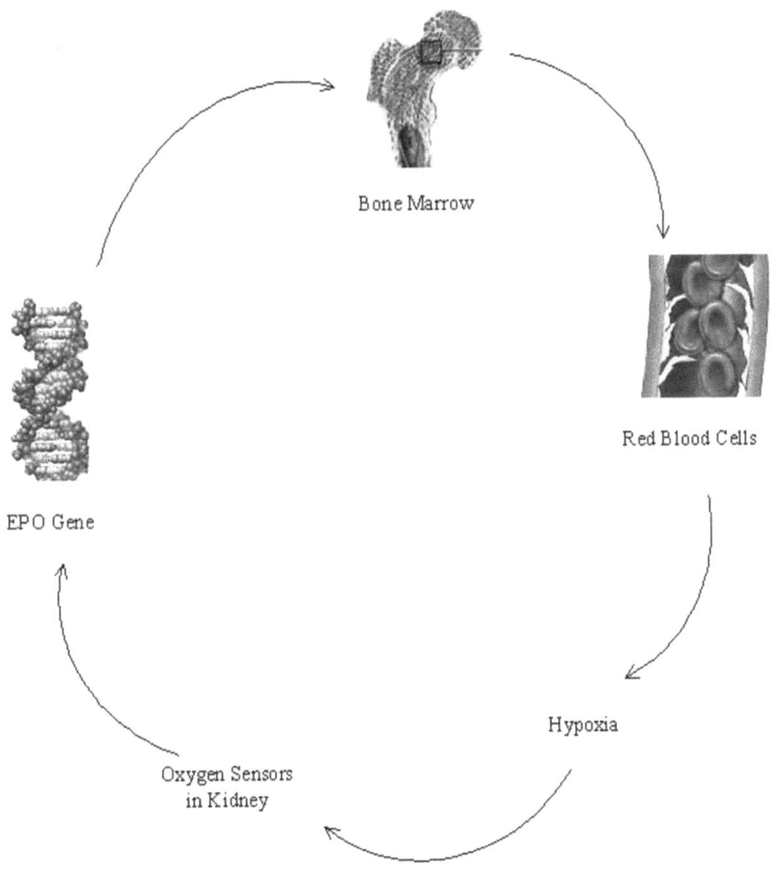

Figure 1. Relationship of kidney and bone marrow in production of red blood cells. (Figure courtesy of Amgen.)

change in the number of circulating red blood cells. Production of new red blood cell slows until the sensor cells recognize a need for increased erythropoiesis. Some EPO, albeit in trace amounts, is always detectable in the circulating blood, even in patients with total kidney failure, suggesting that a subset of cells provides a continuous output of EPO even when oxygen delivery is normal.

Isolation of EPO from urine

Even though the amounts of endogenous EPO increase under conditions of hypoxia or anemia, EPO represents a minor fraction of the total protein in the blood. A significant obstacle to the development of EPO as a therapeutic agent was the difficulty in isolating and purifying adequate amounts of the hormone to allow for its characterization. Several groups attempted to purify human [31, 32] and sheep [33, 34] EPO. The results were inconclusive, with the purity of the product questionable or produced in insufficient amounts to allow chemical characterization.

In 1977, Miyake, Kung, and Goldwasser succeeded in isolating and purifying milligram amounts of EPO from 1500 L of urine from patients with aplastic anemia [35]. Patients with aplastic anemia characteristically overproduce EPO, however the amounts present are still small. Large volumes of urine were essential for the recovery of a sufficient amount to purify EPO.

Miyake et al. [35], using a seven-step process that included ion exchange chromatography, ethanol precipitation, gel filtration, and adsorption chromatography, produced a preparation with a potency of 70,400 units/mg protein, 21% yield, and a purification factor of 930. This method allowed the production of enough material to partially characterize the hormone. With a source of EPO now available, a strategy for the cloning and expression of the human gene was devised.

Cloning of the *EPO* gene

In the 1980s, the nucleic acid sequences for cynomalogous monkey and human EPO were finally isolated and characterized, each by a different method. Several hurdles needed to be overcome, including the limited amount of data about the primary structure of human or monkey EPO, lack of a known source of mRNA, no information about the genomic structure of the gene, lack of simple tests to confirm that the cloned sequence encoded the *EPO* gene, and controversy about the induction mechanism for EPO production due to hypoxia. The difficulty in cloning the gene was further complicated by an inability to determine which step in the process was responsible for the failure.

In 1981, Goldwasser presented the first 26 amino acids of the protein at a meeting. Sue and Sytkowski [36] published the sequence in a paper that described the development of polyclonal antibodies to EPO. The Sue and Sytkowski paper was subsequently shown to have two errors, unknown at the time, in the amino acid sequence. Another group [37] published a putative amino acid sequence for the first 31 amino acids of EPO, but it was proved to be erroneous, as it had the same two errors reported by Sue and Sytkowski as well as three additional errors in the amino acid sequence, again unknown at the time of publication. Goldwasser had provided Lin and colleagues at Amgen Inc. with the sequence he and his colleagues had obtained for the first 26 amino acids of human EPO, but of course, none were aware at the time of the errors. They were aware of the possibility that the available peptide sequence may overlap the intron-exon boundary of the *EPO* gene thereby preventing successful cloning using oligonucleotide sequences based on the peptide sequence. Earlier, Goeddel et al. [38] had successfully sequenced, cloned, and produced recombinant human insulin. This protein has only 51 amino acids, compared with EPO's 165 amino acids, and required nearly 10 years of work to sequence. With newer biotechnology techniques, it was naively thought that the sequencing of EPO would be easy. Lin needed not only to isolate the gene with no knowledge of its structure or of a simple way of confirming that the gene was in hand, but also to express the gene in a suitable host cell to provide a product with the proper structure, including the carbohydrate and polypeptide components of the molecule.

Lin used many approaches, including the standard gene-cloning routes known at the time, all of which failed. He persisted and eventually succeeded only because he used a technique far more complex than any technique tried earlier. This novel approach involved the use of multiple sets of fully degenerate oligonucleotide probes to screen a human genomic library. Two small pools of oligonucleotides corresponding to short fragmented samples of EPO amino acid sequences were used. Both pools, one of 20 nucleotides and the other of 17 nucleotides, had low codon degeneracy. Because of the degeneracy of the genetic code, the same amino acid can be encoded by more than one codon; Lin and colleagues accounted for every possible codon that encoded these putative amino acid sequences necessitating 128 different probes in each pool. The probes were labeled with radioactive phosphorus to identify any matches of a single probe with the human genome. The gene library on which the probes were tested consisted of the total human genome, fragmented into pieces 10,000 to 20,000 nucleotides long. Lin and colleagues found that probes in both mixtures hybridized with four of the 1.5 million clones in a human fetal liver genomic library [39]. Analysis of these clones showed that at least one contained the entire coding region of the human gene for EPO and it was the basis for developing the expression system using transfected Chinese hamster ovary (CHO) cells. In parallel with this effort, cDNA from the kidneys of anemic monkeys was prepared using mixed probes based on the human EPO peptide sequences [40] and the monkey gene also was cloned.

After Lin and his colleagues successfully cloned the gene, Jacobs and colleagues [41] also cloned the human *EPO* gene using degenerate oligonucleotides and a peptide sequence derived from urinary EPO supplied by Miyake. Cloning of additional *EPO* genes from other species used the sequence information from the human and monkey genes and proceeded rapidly. The mouse *EPO* gene was cloned using monkey and human EPO DNA segments as hybridization probes [42, 43]. Subsequently, *EPO* genes from other species were cloned by hybridization or by polymerase chain reaction using probes or primers based on the known EPO sequences, including the genes from the rat [44], pig [45, 46], sheep [45, 47], and cow [48], among other animals.

Expression of the *EPO* gene

Clones of the gene for EPO were inserted into CHO cells, which synthesized the 193 amino acid precursor protein, removed the signal peptide and carboxy-terminal arginine, added *N*- and *O*-linked carbohydrate to glycosylation sites, and released the mature protein into the culture medium. Immunologic, biologic, and biochemical assays showed that the recombinant hormone had the *in vivo* biologic activity and was immunologically equivalent to human EPO, as revealed within the limits of the available assays [39]. EPO expressed by CHO cells has a molecular weight of 30.4 Kd and contains 40% carbohydrate [49].

Discussion

The isolation and expression of the gene for human EPO was a major achievement that capped almost 155 years of exploration into the nature of anemia and the production of red blood cells. EPO is present in minute quantities in the blood, and difficulty in isolating and purifying the hormone in amounts that would allow investigation of its properties posed a significant obstacle to the development of EPO as a therapeutic agent. After Miyake, Kung, and Goldwasser developed a technique to purify urinary EPO, a strategy for the cloning and expression of the human gene could be devised. The innovative approach of Lin et al. allowed the successful isolation, cloning, sequencing, and development of the recombinant protein. Large-scale production began and clinical trials started. The recombinant protein, epoetin alfa, produced dose-dependent increases in erythropoiesis that paralleled the expected response to endogenous EPO. The commercial production of epoetin alfa has been translated into benefits for millions of patients (Tab. 1). Epoetin alfa ameliorates the debilitating symptoms of anemia and allows these patients to have more normal lives. Other chapters in this volume will elaborate on the clinical use, production, formulation, and other important and evolving aspects in the study of rHuEPO.

Table 1. Some benefits of epoetin alfa therapy

• Increased exercise tolerance
• Improved central nervous system function
• Reduced heart enlargement
• Reduced extreme fatigue
• Increased ability to perform daily functions of life
• Reduced risk of alloimmunization in transplant recipients
• Improved coagulation

References

1 Eschbach JW, Abdulhadi MH, Browne JK, Delano BG, Downing MR, Egrie JC, Evans RW, Friedman EA, Graber SE, Haley NR et al. (1989) Recombinant human erythropoietin in anemic patients with end-stage renal disease. Results of a phase III multicenter clinical trial. *Ann Intern Med* 111: 992–1000

2 Volgelzang NJ, Brietbart W, Cella D, Curt GA, Groopman JE, Horning SJ, Itri LM, Johnson DH, Scherr SL, Portenoy RK (1997) Patient, caregiver, and oncologist perceptions of cancer-related fatigue: Results of tripart assessment survey. The Fatigue Coalition. *Semin Hematol* 34: 4–12

3 Demetri GD, Kris M, Wade J, Degos L, Cella D (1998) Quality-of-life benefit in chemotherapy patients treated with epoetin alfa is independent of disease response of tumor type: Results from a prospective community oncology study. Procrit Study Group. *J Clin Oncol* 16: 3412–3425

4 Glaspy J, Bukowski R, Steinberg D, Taylor C, Tchekmedyian S, Vadhan-Raj S (1997) Impact of therapy with epoetin alfa on clinical outcomes in patients with nonmyeloid malignancies during cancer chemotherapy in community oncology practice. Procrit Study Group. *J Clin Oncol* 15: 1218–1234

5 Peeters HR, Jongen-Lavrencic M, Bakker CH, Vreugdenhil G, Breedveld FC, Swaak AJ (1999) Recombinant human erythropoietin improves health-related quality of life in patients with rheumatoid arthritis and anaemia of chronic disease; utility measures correlate strongly with disease activity measures. *Rheumatol Int* 18: 201–206

6 Gabrilove JL, Cleeland CS, Livingston RB, Sarokhan B, Winer E, Einhorn LH (2001) Clinical evaluation of once-weekly dosing of epoetin alfa in chemotherapy patients: Improvements in hemoglobin and quality of life are similar to three-times-weekly dosing. *J Clin Oncol* 19: 2875–2882

7 Foote MA, Colowick A, Goodkin DA (2002) Basics of anatomy and physiology: Red blood cells and anemia. *AMWA J* 17: 13–20

8 Watson AJ (1989) Adverse effects of therapy for the correction of anemia in hemodialysis patients. *Sem Nephrol* 9: 30–34

9 Bright R (1836) Cases and observations illustrative of renal disease accompanied with the secretion of albuminous urine. *Guy's Hospital Reports* 1: 338–379

10 Jourdanet D (1863) De l'anemie des altitudes et de l'anemie en general dans ses rapports avec la pression del l'atmopshere. *Balliere* (Paris) 44

11 Viault F (1891) Sur l'augmentation considerable du nombre des globules rouges dans le sange chez les habitants des haut plateaux de l'Amerique du Sud. *CR Acad Press* (Paris) 111: 917–918

12 Viault F (1891) Sur la quantité d'oxygen contenue dans le sang des animaux des hauts pleateaux de L'Amerique du Sud. *CR Acad Sci* (Paris) 112: 295–298

13 Muntz A (1891) De l'enrichissement du sang en hemoglobine, suivant les conditions d'existence. *CR Acad Sci* (Paris) 112: 298–301

14 Meischer F (1893) Ueber die Beziehungen zwischen Meereshöhe und Beschaffenheit des Blutes. *Korresp-Bl Schweiz Arz* 23: 809–830

15 Meischer F (1893) Bemerkungen über eine verbesserte Form der Mischipette und ihren Einfluss auf die Genauigkeit der Blutkörperzählung. *Korresp-Bl Schweiz Arz* 23: 830–832

16 Carnot P (1906) Sur le mecanisme de l'hyperglobulie provoquée par le serum d'animaux en renovation sanguine. *CR Acad Sci* (Paris) 111: 344–346
17 Carnot P (1906) Sur l'activite cytopoietique du sang et des organs regeneres au cours des regenerations viserales. *CR Biol* (Paris) 111: 463–465
18 Carnot P, Deflandre C (1906) Sur l'activité cytopoietique du sang et des organs regeneres au cours des regeneration du sang. *CR Acad Sci* (Paris) 143: 432–435
19 Gibelli C (1911) Uber den wert des serums anamisch gemachten tiere bei der regernation des blutes. *Arch Exp Path Pharmark* 65: 284–302
20 Giribaldi G (1920) Sul potere emopoietico del siero di sangue di animali resi anemici con iniezioni sottocutanee di glicerina. *Biochim Terap Sper* 7: 52–65
21 Forster J (1924) Luftverdünnung und Blutregeneration durch "Hemopoietine". *Biochem Z* 145: 309–317
22 Sandor G (1932) Über die blutbildene Wirkung des Serums von Tieren, die in verdünnter Luft gehalten wurden. *Z Ges Exp Med* 82: 633–646
23 Thorling EB (1969) The history on the early theories of humoral regulation of the erythropoiesis. *Dan Med Bull* 16: 159–164
24 Erslev A (1952) Humoral regulation of red cell production. *Blood* 8: 349–357
25 Jacobson LO, Goldwasser E, Fried W, Plzak L (1957) Role of the kidney in erythropoiesis. *Nature* 179: 633–634
26 Congote LF (1977) Regulation of fetal liver erythropoiesis. *J Steroid Biochem* 8: 423–428
27 Mirand EA, Prentice TC (1957) Presence of plasma erythropoietin in hypoxic rats with and without kidneys or spleen. *Proc Soc Exp Biol Med* 96: 49–51
28 Koury MJ, Bondurant MC, Graber SE, Sawyer ST (1988) Erythropoietin messenger RNA levels in developing mice and transfer of ^{125}I-erythropietin by the placenta. *J Clin Invest* 82: 154–159
29 Erslev AJ, Caro J, Kanusu E, Silver R (1980) Renal and extrarenal erythropoietin production in anemic rats. *Br J Haematol* 45: 65–72
30 Marti HH, Wenger RH, Rivas LA, Straumann U, Digicaylioglu M, Henn V, Yonekawa Y, Bauer C, Gassmann M (1996) Erythropoietin gene expression in human, monkey, and murine brain. *Eur J Neurosci* 8: 666–676
31 Espada J, Gutnisky A (1970) A new method for concentration of erythropoietin from human urine. *Biochem Med* 3: 475–484
32 Espada J, Langton AA, Dorado M (1972) Human erythropoietin: Studies on purity and partial characterization. *Biochim Biophys Acta* 285: 427–435
33 Goldwasser E, Kung CKH (1971) Purification of erythropoietin. *Proc Natl Acad Sci USA* 68: 697–698
34 Goldwasser E, Gross M (1975) Erythropoietin: Assay and study of its mode of action. *Methods Enzymol* 37: 109–121
35 Miyake T, Kung CK, Goldwasser E (1977) Purification of human erythropoietin. *J Biol Chem* 252: 5558–5564
36 Sue JM, Sytkowski AJ (1983) Site-specific antibodies to human erythropoietin directed towards the NH_2-terminal region. *Proc Natl Acad Sci USA* 80: 3651–3655
37 Yanagawa S, Hirade K, Ohnota H, Saski R, Chiba H, Ueda M, Goto M (1984) Isolation of human erythropoietin with monoclonal antibodies. *J Biol Chem* 259: 2707–2710
38 Goeddel DV, Kleid DG, Bolivar F, Heyneker HL, Yansura DG, Crea R, Hirose T, Kraszewski A, Itakura K, Riggs AD (1979) Expression in *Escherichia coli* of chemically synthesized genes for human insulin. *Proc Natl Acad Sci USA* 76: 106–110
39 Lin FK, Suggs S, Lin CH, Browne JK, Smalling R, Egrie JC, Chen KK, Fox GM, Martin F, Stabinsky Z et al. (1985) Cloning and expression of the human erythropoietin gene. *Proc Natl Acad Sci USA* 82: 7580–7582
40 Lin FK, Lin CH, Lai PH, Browne JK, Egrie JC, Smalling R, Fox GM, Chen KK, Castro M, Suggs S (1986) Monekey erythropoietin gene: Cloning, expression and comparison with the human erythropoietin gene. *Gene* 44: 201–209
41 Jacobs K, Shoemaker C, Rudersdorf R, Neill SD, Kaufman RJ, Mufson A, Seehra J, Jones SS, Hewick R, Fritsch EF et al. (1985) Isolation and characterization of genomic and cDNA clones of human erythropoietin. *Nature* 313: 806–810
42 McDonald JD, Lin FK, Goldwasser E (1986) Cloning, sequencing, and evolutionary analysis of the mouse erythropoietin gene. *Mol Cell Biol* 6: 842–848
43 Shoemaker CB, Mitsock LD (1986) Murine erythropoietin gene: cloning, expression, and human

gene homology. *Mol Cell Biol* 6: 849–858
44 Nagao M, Suga H, Okano M, Masuda S, Narita H, Ikura K, Sasaki R (1992) Nucleotide sequence of rat erythropoietin. *Biochim Biophys Acta* 1171: 99–102
45 Wen D, Boissel JP, Tracy TE, Gruninger RH, Mulcahy LS, Czelusniak J, Goodman M, Bunn HF (1993) Erythropoietin structure-function relationships: high degree of sequence homology among mammals. *Blood* 82: 1507–1516
46 Pearson PL, Smith TP, Sonstegard TS, Klemcke HG, Christenson RK, Vallet JL (2000) Porcine erythropoietin receptor: molecular cloning and expression in embryonic and fetal liver. *Domest Anim Endocrinol* 19: 25–38
47 Fu P, Evans B, Lim GB, Mortiz K, Wintour EM (1993) The sheep erythropoietin gene: Molecular cloning and effect of hemorrhage on plasma erythropoietin and renal/liver messenger RNA in adult sheep. *Mol Cell Endocrinol* 93: 107–116
48 Suliman HB, Majiwa PA, Feldman BF, Mertens B, Logan-Henfrey L (1996) Cloning of a cDNA encoding bovine erythropoietin and analysis of its transcription in selected tissues. *Gene* 171: 275–280
49 Davis JM, Arakawa T, Strickland TW, Yphantis DA (1987) Characterization of recombinant human erythropoietin produced in Chinese hamster ovary cells. *Biochemistry* 26: 2633–2638

Erythropoietins, Erythropoietic Factors and Erythropoiesis
Edited by S.G. Elliott, M.A. Foote, and G. Molineux

Commercial production of recombinant erythropoietins

Alice S. Chuck[1], Adrian Distler[2], Shane A. Sander[2], Rohini R. Deshpande[2], Gregg Nyberg[3] and James E. Seely[3]

[1] *Seattle Genetics, Bothell, Washington, USA*
[2] *Amgen Inc., Thousand Oaks, California, USA*
[3] *Amgen Inc., Longmont, CO, USA*

Introduction

For the production of recombinant products from genetically engineered cells, a number of cell hosts may be used, but they generally belong to one of five categories: plant, bacterial, yeast, insect, or mammalian. Depending on the type of product that is desired, the features of that product, the intended use of the product, and the preferred method of manufacture, an appropriate host cell can be selected. With the appropriate genetic engineering, the gene of interest such as erythropoietin (EPO) can be produced by a host cell. In the case of EPO production, the sequence of amino acids, as well as the amount of glycosylation, must be correct to achieve the desired efficacy *in vivo*. This chapter describes one method of EPO production using cells genetically engineered to secrete recombinant human erythropoietin (rHuEPO). In this method of production, mammalian cells, which are capable of producing glycosylation forms with the desired efficacy in humans, are typically selected as hosts. The host mammalian cells secrete the rHuEPO product into the medium environment in which they are cultured, making the remainder of production a matter of separating the rHuEPO product from the cells and other components in the cell culture broth. This chapter summarizes the generation of rHuEPO-producing cell lines, the production of rHuEPO, the separation of rHuEPO from components of the cell culture broth, and the packaging of the final rHuEPO drug product.

Cell line development

The development of a cell line generates a consistent source of cells that is capable of satisfying commercial demand and that adheres to regulatory guidelines for genetically engineered cells. Development begins with the creation of genetically engineered cells that produce the desired product, and ends with a cell bank, a homogeneous population of genetically engineered cells frozen at the same population doubling level.

Cell line selection and screening

After the initial genetic modification of a host cell population to produce the product of interest, an appropriate production cell line must be isolated and selected. Generally, populations of cells may be screened for desirable properties, such as productivity and the ability to secrete product of the desired quality. Products, such as rHuEPO, may be posttranslationally modified by the host cells (e.g., glycosylation), and these modifications can impact *in vivo* efficacy. Different populations of genetically modified cells may produce product with varying degrees of posttranslational modifications. Therefore, product quality can be an important endpoint when screening for an appropriate production cell line. Selecting the final production cell line may involve several stages of isolating (i.e., cloning) and screening candidate cell lines. To ensure a homogeneous production cell line, the final cell line will have typically been single-cell cloned.

Single-cell cloning

The industry standard for generating a homogenous population is single-cell cloning. Single-cell cloning is accomplished by placing one cell in a microculture environment, and then expanding that cell through increasingly larger cultures to generate a greater number of cells (e.g., 1×10^9 cells) (Fig. 1).

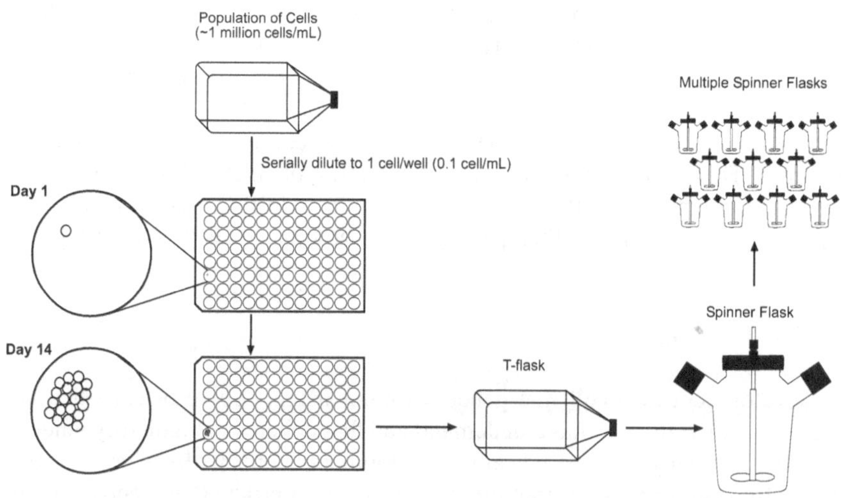

Figure 1. Single-cell cloning technique

Doubling of a single cell should produce two identical cells. Doubling again results in four cells that are assumed to be identical. Within a limited number of generations, expansion of the cells in this manner results in a homogenous culture of cells with identical genomes. When sufficient cells are amassed, cell banks may be created.

Cell bank

Once a cell line has been identified as the manufacturing line, a cell bank must be generated. The cell bank provides the starting material for all lots of product manufactured. For this reason, it is carefully scrutinized from both a business and regulatory perspective.

The master cell bank is the base starting material for the life cycle of the product. Cell banks are created in a two-tier system (Fig. 2). Each vial of master cell bank material can be used to create a working cell bank. Likewise, a single vial provides the starting material for a commercial cell culture lot.

Two-tier system: creation of master and working cell banks

A two-tiered approach is commonly used in industry to generate a supply of cell bank vials for the lifetime of a product. The master cell bank is generated first and typically consists of 100–400 vials. The master cell bank is not used by the manufacturing facility directly. Instead, the second-tier cell bank, the working cell bank, is used by the manufacturing facility. The working cell bank is generated by expanding one vial of material from the master cell bank. The working cell bank typically contains 300–1,000 vials.

Safety assessment

Once cell banks are generated, they must be characterized before they can be used to manufacture a drug for clinical use, as per regulatory guidelines [1–3]. Characterization is critical to ensure that the correct protein is being produced by the cells and that the cell line is not contaminated by adventitious agents.

The cell bank is the primary source for the recombinant protein produced, so verification that the correct molecule is being produced must be done at the inception of the bank. Confirmation of the DNA and/or RNA coding sequence ensures that the cells encode the required genetic sequence for the protein. Verification of the genetic sequence within the cell must be followed with verification of the amino acid sequence of the purified protein. Additionally, the cell bank must be examined to confirm that it is not producing any altered forms of the protein. For example, during gene integration in mammalian cells (after transfection or amplification), the cell's genome is rearranged, and

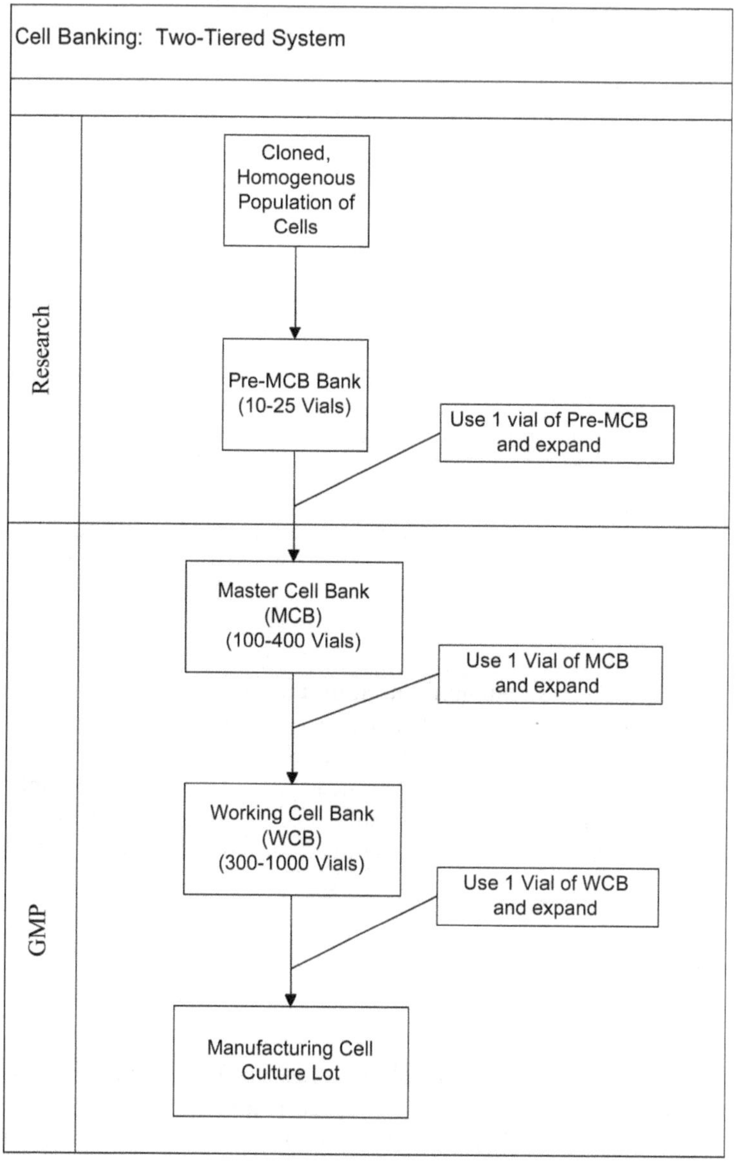

Figure 2. Cell banking is a two-tiered system. GMP = Good Manufacturing Processes; MCB = master cell bank; WCB = working cell bank

rearrangement possibly may involve the coding sequence of the protein of interest. One concern is that the rearrangement could occur at a site that results in a molecule containing a portion of the correct protein and a portion of another protein. In this event, the rearranged molecule could retain some of the correct protein's characteristics and be carried through a purification process;

however, the net result may be either the incorrect protein or a mixture of correct and incorrect proteins. Therefore, the cell bank is examined in great detail to ensure the absence of rearrangements that could lead to unwanted proteins being carried through manufacturing.

Ensuring that adventitious agents are not present is crucial. When a cell bank comprises mammalian cells, which theoretically can act as hosts to viruses, the cell bank is examined for mycoplasma, bacteria, fungi, and viruses. Numerous *in vitro* and *in vivo* assays are used to assess viral contamination. In the end, all of the aforementioned tests for adventitious agents must be negative. If they are not, the cell bank cannot used to make material intended for clinical use.

Cell bank stability assessment

The master cell bank is designed to last the lifetime of the product, so monitoring it and the working cell bank for storage stability is important. Cell banks are generally stored at –130 °C or below to ensure cell stability. The cell banks are monitored by testing cells for viability upon thawing. If the cell bank's viability is stable, then no other testing is necessary. If viability is not stable, the bank is re-examined, and another master cell bank may need to be generated.

Cell culture process

One method of EPO manufacture uses cell lines genetically engineered to produce rHuEPO. In this chapter, an example of rHuEPO production in a cell culture process is presented, where cells from a mammalian cell bank are grown and product is secreted into the cell culture medium. The ability of the cell culture to produce product is affected by the nutrient environment and other physical parameters, such as temperature, pH, osmolality, and concentrations of dissolved gases. The cell culture process is designed to support the production of rHuEPO of a consistent quality, tailored to enable the product specifications and efficacy targeted by the manufacturer.

Raw materials

The raw materials for the cell culture process consist of the cell bank and the nutrients used to sustain and expand these cells. Nutrients in the cell culture medium generally are amino acids, additional carbon sources (e.g., glucose), vitamins, trace elements, growth factors, hormones, and salts. A review of media composition is given in [4]. Historically, cell culture medium also has contained serum from bovine sources, which may vary in composition depending on its source or the processing. Guidance for the industry for use of raw materials for the manufacture of biologicals is provided by CBER/CDER [1].

Process options

Many process options are available to produce a recombinant protein. One of the simplest cell culture options is batch mode, where the cells and the nutrient-containing media are added to a production vessel, and no further additions are made. Physical parameters (e.g., temperature, pH, and dissolved oxygen concentration) may be controlled within set ranges. The cells grow to the extent that can be supported by the initial nutrient concentrations; cells will stop growing and begin dying when nutrients become insufficient and/or if the amounts of waste products (the by-products of nutrient consumption by the cells) increase.

Fed-batch processes differ from batch processes by providing for the addition of nutrients into the culture as required. This method allows a much greater number of cells to be supported than batch processes. More cells typically translate to greater productivity due to the constitutive production feature engineered into most commercial cell lines. With constitutive production, as long as a cell is present and viable, it will continue to make product. Thus, the viable cell number correlates closely with the amount of recombinant protein produced.

In addition to supplementing a culture with nutrients as required, spent medium in the culture may be removed to reduce the amount of waste products. The process of feeding fresh nutrients and concurrently removing spent media is called perfusion. The spent medium, called the harvest, contains the protein product; the device used to separate the spent medium from the cells is called a cell retention device.

Operationally, batch and fed-batch processes are relatively easy to execute and perform reliably; perfusion processes are relatively more complex to exe-

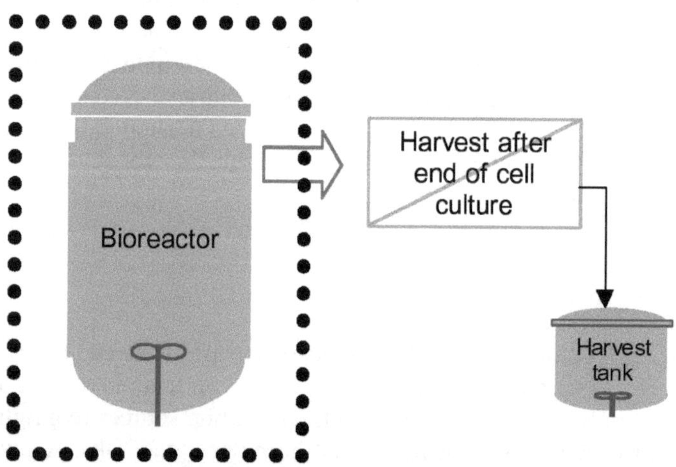

Figure 3. Schematic of a bioreactor operating in batch mode. The sterile envelope, depicted by the dotted line, encompasses only the bioreactor. The harvest operation, which includes cell separation from the conditioned medium and concentration of the conditioned medium, is not performed under sterile conditions

cute and in general use longer production times. The numbers of pieces of equipment that must be operated in a sterile manner within the sterile envelope are depicted in Figure 3, Figure 4, and Figure 5. The number of pieces and complexity of equipment in the sterile envelope correlates with the complexity in executing the three types of processes. A schema of the equipment (bioreactor, media feed tank, and harvest tank) used for each of these process options and the relative scale of tanks and production vessels is depicted. For batch and fed-batch processes, the production vessel is typically the largest piece of

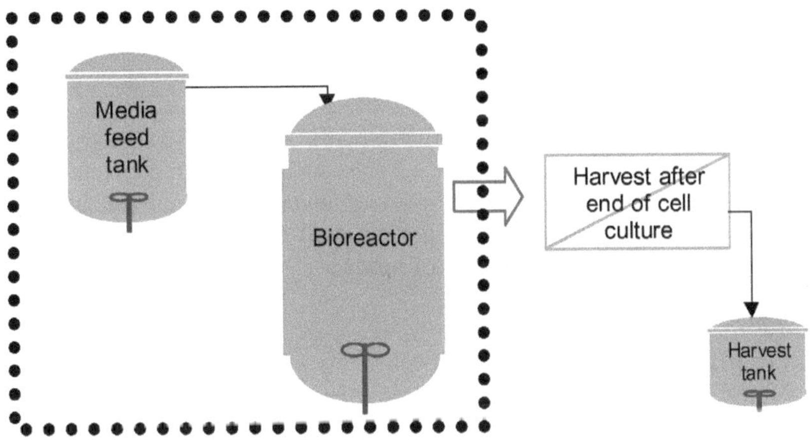

Figure 4. Schematic of a bioreactor operating in fed-batch mode. The sterile envelope, depicted by the dotted line, includes the media tank and the bioreactor. The harvest operation, which includes cell separation from the conditioned medium and concentration of the conditioned medium, is not performed under sterile conditions

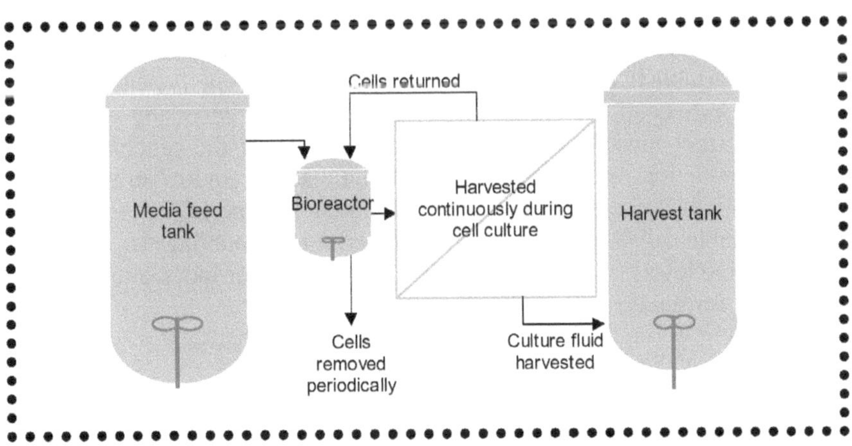

Figure 5. Schematic of a bioreactor operating in perfusion mode. The sterile envelope, depicted by the dotted line, includes the media tank, bioreactor, cell separation device (harvest unit), and harvest tank. The harvest operation here does not include any concentration of the conditioned medium.

equipment used in the cell culture process; for perfusion processes, the feed and harvest tanks are typically the largest equipment used.

The size of the equipment and the time over which each piece of equipment needs to be occupied for a given production run will influence the design of a manufacturing facility and the number of production runs a manufacturing plant can execute. The manufacturing facility is designed to comply with regulatory guidelines for GMP manufacture (i.e., CFR parts 210 and 211 are used in the US; guidelines differ slightly in other countries). Additional information on process options is available [5, 6].

Commercial manufacture

A commercial cell culture manufacturing process consists of a seed train followed by a production step. The purpose of the seed train is to generate a sufficient number of cells with which to initiate production in the production vessel. The required process conditions are imposed on a production vessel to generate optimal quantity and quality of product.

Seed train

A seed train describes the expansion of cells from a frozen cell bank vial to the inoculum of the production reactor. The volumes involved may range from 1 mL to 10,000 L. The vessels involved may include t-flasks, shaker flasks, roller bottles, disposable bag bioreactors and stirred tank reactors (Fig. 6). The reactors may be operated in batch, fed-batch, or perfusion modes. The purpose of each step of the seed train is to generate a sufficient number of cells of known quality to enable inoculation of the subsequent step. The culmination of the seed train occurs at inoculation of the production reactor.

Given a cell bank vial containing 1×10^7 viable cells, the initial seed train time to the production reactor may be estimated. First, the doubling time for the cell line must be established. For mammalian cells, doubling time typically ranges from 16–48 h. Second, the requirements for cell number generated at the end of the seed train must be established. For example, if the process uses a 10,000 L production reactor and the production reactor requires an inoculum density of 1×10^6 viable cells/mL and an initial volume of 7,500 L, then 7.5×10^{12} viable cells are required. The number of cell doublings (n, or population doubling level) from vial thaw to production reactor inoculation can be given by the formula:

$$n = LN(XS / X0) / LN(2)$$
$$n = LN(7.5 \times 10^{12}/1 \times 10^7) / LN(2)$$
$$n = 19.5$$
$$n \sim 20$$

where LN = natural logarithm and XS final cell number = 7.5×10^{12} and X0 = initial cell number = 1×10^7

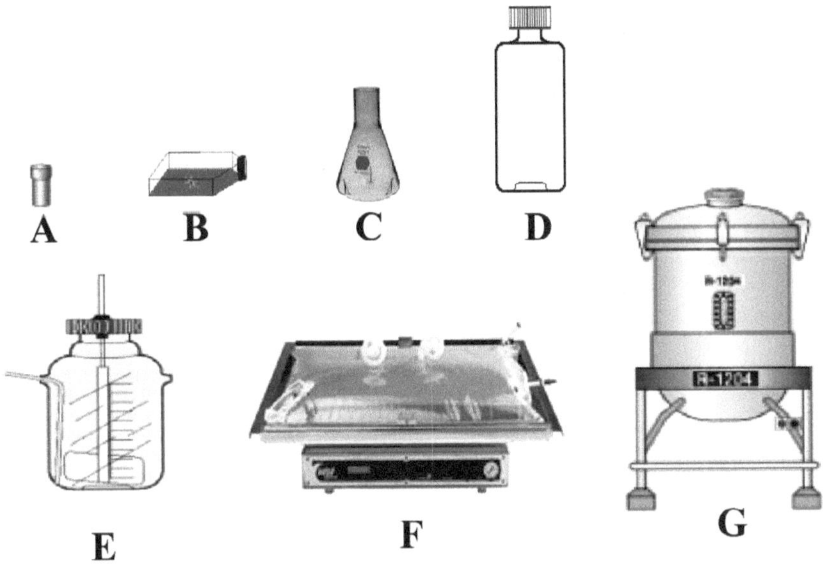

Figure 6. Exemplary vessels used in the preparation of a seed train: (A) freezer vial, (B) T-flask, (C) shaker flask, (D) roller bottle, (E) spinner flask, (F) Cellbag, (G) stirred tank reactor

Using a conservative estimate of 2 days for the doubling time (td), the overall seed train time is n × td = 20 × 2 days = 40 days, or approximately 6 weeks.

Multiple seed trains using culture taken from an existing seed train may be initiated if cell numbers allow and if the cell line's productivity remains stable. Thus, multiple production reactors stemming from a single cell bank vial may be considered to shorten the overall time to inoculate production reactors. Alternatively each production reactor may be derived from a single vial thaw.

Production

Once the seed train has generated a sufficient number of cells from the cell bank, these cells can produce rHuEPO. Cells can adhere to a solid substrate and secrete rHuEPO into the liquid cell culture medium. Examples of solid substrates include tissue culture plastic such as roller bottles or microcarrier beads that may be made of ceramic or other porous materials. Alternatively, cells can be freely suspended in liquid medium where they secrete rHuEPO. Exemplary cultures in roller bottles and tank vessels are shown in Figure 7 and Figure 8, respectively.

Commercial cell lines are designed and selected for constitutive production; therefore, the amount of rHuEPO produced correlates with the viable cell number (Fig. 9). For adherent culture, the number of cells is limited by surface area. To increase production of rHuEPO, the surface area is increased, typically by increasing the number of roller bottles or microcarrier beads. Since nutrients are supplied in the liquid phase and products are secreted (by the cells) from the

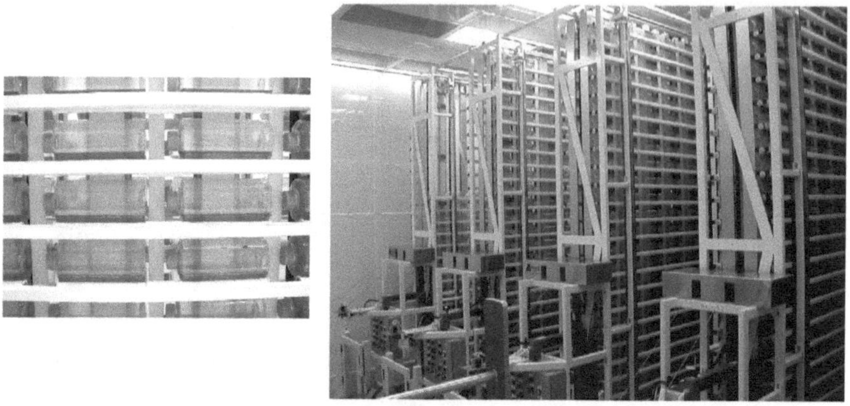

Figure 7. Production of rHuEPO in roller bottles (left panel). The right panel shows large racks containing many of the same type of roller bottle cultures.

Figure 8. Production of rHuEPO in tank vessels. A medium containing rHuEPO-producing cells is shown in the spinners on the left. The stirred tank reactor shown on the right also contains a culture of rHuEPO-producing cells.

solid phase into the liquid phase, the successful mass transfer of nutrients to the cells, and rHuEPO and metabolic waste products from the cells must be carefully considered in the design of the process. A commercial production lot from adherent cultures with roller bottles typically entails using many small production vessels; the challenge of such a process for commercial scale production is to demonstrate equivalent control (i.e., monitoring) of each of the multiple production vessels. The degree of control required is imposed to achieve comparable growth and productivity performance from each production vessel. After product is produced, it is separated from the cells in the harvest process. Since cells are attached to the solid substrate, harvesting the rHuEPO product in the

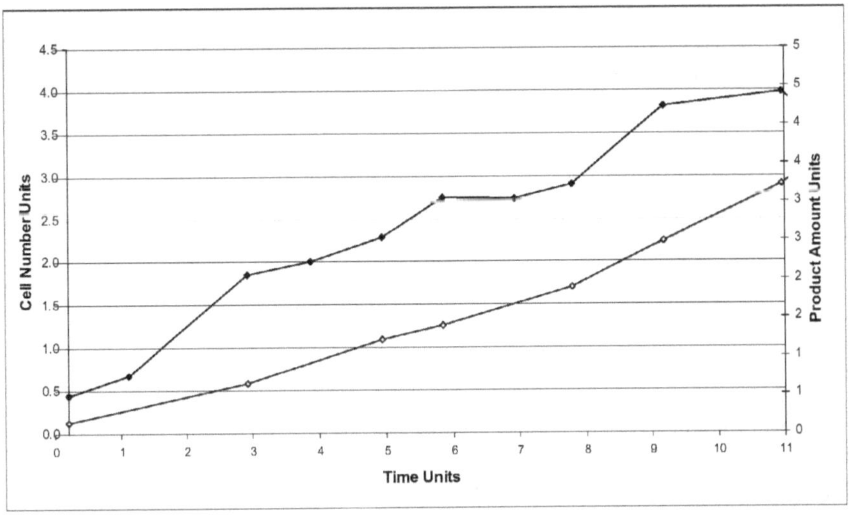

Figure 9. Correlation between number of cells and amount of product produced. Solid diamonds = cell number profile; hollow diamonds = product amount profile.

liquid conditioned medium is operationally straightforward and involves pouring off or decanting the liquid before volume reduction and purification.

For suspension cultures, scale-up is more straightforward. To increase production, the number of cells per unit volume (i.e., cell density) may be maximized, and/or the volume of culture may be maximized. To maximize the cell density, limiting nutrients within the culture environment are identified and supplements are added to the culture. The medium is sufficiently enriched to support the maximal number of cells. To maximize the culture volume, a larger scale of production vessel may be used. Both cells and product are distributed homogeneously in well-stirred reactors so that the growth and production performance within each volume of liquid is consistent. The volume of liquid may be increased as long as homogeneity is achieved (i.e., mixing is adequate). Stirred tank reactors for suspension cultures of mammalian cells typically range between 1,000 L and 20,000 L. To harvest the product, the cells must be separated from the liquid medium. Separation is achieved by microfiltration or centrifugation. The clarified liquid-containing product may then be concentrated for volume reduction and purification.

Within any of the process modes described, a cell line that has been engineered to produce rHuEPO may grow and secrete product encoded by the *EPO* gene. Not all of the secreted product may qualify as the rHuEPO sold by a manufacturer. The EPO molecule has 3 *N*-glycosylation sites and 1 *O*-glycosylation site, leading to the possibility of differently charged molecules with different isoelectric points (i.e., isoforms). Those molecules with the lowest isoelectric points, typically corresponding to those having the greatest amount of sialylation and glycosylation branching, have the greatest *in vivo* efficacy.

Molecules with higher isoelectric points, or less sialylation and less glycosylation, have less *in vivo* efficacy, and can be removed in purification. Additionally, even if the cell produces product of the desired charge profile, this product may be degraded in culture by proteases. Further, the product may form undesired complexes such as aggregates, which must be separated from the desired monomer. The cell may be influenced to produce increasing amounts of the desired charge profile by controlling process conditions, such as temperature, pH, osmolality, and nutrient and waste product concentrations. Similarly, undesired forms of the product may also be minimized.

Recovery and purification of erythropoietin

A number of purity criteria must be met for rHuEPO to be a viable human pharmaceutical. Host cell and other contaminants, such as DNA, host cell proteins, and endotoxin, must be removed to appropriate levels. The removal or inactivation of any theoretical endogenous or adventitious viruses must be shown. Amounts of product-related contaminants, such as aggregates, proteolytically degraded, underglycosylated, or oxidized forms, are minimized. A targeted purity is specified by the manufacturer, so each product lot must meet the criteria described in the manufacturer's product specification. Product specifications may differ among manufacturers.

The clearance rate of rHuEPO in the human body is affected by the number of sialic acid residues on the carbohydrate portion of the molecule. Under-sialylated rHuEPO is cleared in the liver by the asialoglycoprotein receptor [7]. Selecting for molecules with the greatest degree of sialylation gives a product with the desired pharmacokinetic properties, which affects the biological potency of rHuEPO. Different manufacturers may have different selection criteria (and therefore different product quality characteristics) for the EPO molecules that are retained in the recovery and purification processes.

In addition to having a well-understood and reproducible recovery process, a number of regulatory requirements exist for the manufacture of biopharmaceuticals. The product must be produced under Good Manufacturing Practices (GMP), which encompasses a range of topics from plant design, which affects air and water quality, solvent usage, and waste disposal, to day-to-day operations, which include operator training, batch record writing and review, quality assurance, and cleaning validation. Many of these guidelines can be found in Q7A 'Good Manufacturing Practice Guidance for Active Pharmaceutical Ingredients' [8].

Concentration and diafiltration

After the clarified cell culture media is harvested, it must be concentrated, and the buffer salts and other high conductivity/low-molecular-weight components must be removed. Concentration is an important step, since it is difficult to store and handle large pool volumes. The expense of storing large volumes of harvest media in a frozen state can be prohibitive. In addition, the time required to load a large volume of media on to a column decreases plant productivity. Prolonged column load times can also lead to proteolytic degradation of product in the harvest media as it awaits loading onto the column. Removal of buffer salts and other high-conductivity, low-molecular-weight components is important because these may interfere with binding to the purification columns. For example, ion-exchange chromatography depends on ionic interactions between charged groups on the protein and oppositely charged groups on the chromatography media. The high conductivity of the harvest media would prevent protein from binding to the ion-exchange chromatography media.

Concentration and diafiltration are done using ultrafiltration membranes. This operation allows for the passage of low-molecular-weight solutes and water through membrane pores and the retention of larger-molecular-weight solutes, such as rHuEPO and other proteins. The driving force for the passage of these solutes and water through the pores is the pressure difference across the membrane. The membranes can be made of cellulose, polyether sulfone, or other polymers. Membranes are available that have different 'nominal molecular-weight cut-offs', ranging in molecular weight from 5,000 to 500,000. Proteins and other high-molecular-weight materials that do not pass through the membranes are referred to as the retentate; the material that passes through the membranes is called permeate. The rate at which the water and solutes go through the membrane is called the permeate flux rate and is usually measured in litres per minute.

During the concentration phase (Fig. 10a), both water and solutes are forced through the pores, leading to a volume reduction in the harvest medium. The

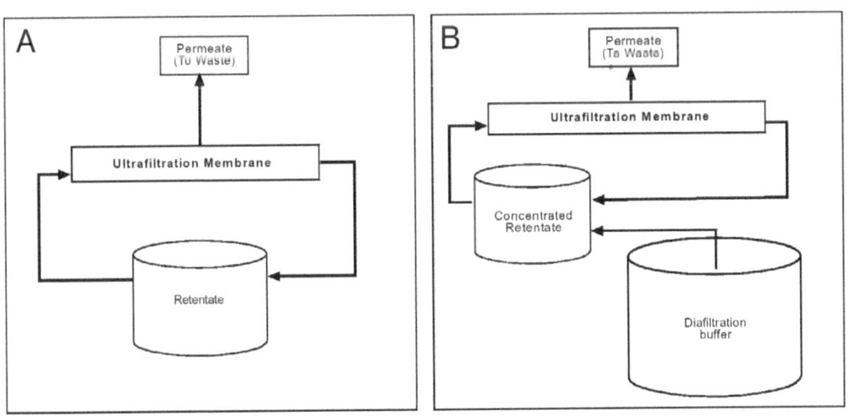

Figure 10. Panel A – Concentration mode. Clarified cell culture medium volume is reduced in retentate to make concentrated retentate. Panel B – Diafiltration (DFM) mode. Removal of high-conductivity components in concentrated retentate to make DFM.

range of this volume reduction may be 5- to 100-fold depending on what is required for a given protein. During the diafiltration phase, the concentrated harvest continues to be fed through the membrane; at the same time, however, a lower conductivity buffer (diafiltration buffer) is pumped into the harvest media retentate at approximately the same rate as the permeate flux rate (Fig. 10b). This process causes a buffer exchange of the original high conductivity buffer for the lower one. The greater the number of retentate volumes pumped though the membrane during the diafiltration step, the more complete the buffer exchange. This product pool is referred to as diafiltered media, which either can be immediately processed through the subsequent purification steps or stored frozen and processed later.

Chromatography: isoform selection

An anion-exchange media can be used to separate molecules on the basis of molecular charge. This chromatography may be used to separate rHuEPO from host cell proteins, nucleic acids, and endotoxin.

Chromatography: removal of contaminant proteins

A reversed-phase chromatography step can be used to remove host cell proteins and additional nucleic acids. The interactions of the column matrix with proteins are mostly hydrophobic in nature. Once bound to the column, a protein can be eluted with another hydrophobic, low-dielectric organic solvent. The protein portion of rHuEPO is quite hydrophobic because of the large proportion of hydrophobic residues and its solubility characteristics after removal of the polysaccharide portion of the molecule. Gel filtration chromatography, also known as size-exclusion chromatography or molecular sieving, may also be used to separate a protein mixture based on molecular weight. The media consists of agarose or acrylic beads of defined pore sizes. Proteins too large to penetrate the pores are excluded from the interior volume of the beads and have a shorter elution path through the column (hence they elute sooner). Smaller proteins either partially or completely penetrate the pores, resulting in a longer elution path and longer elution times.

Viral clearance

If rHuEPO is made in mammalian cells, the ability to remove theoretical endogenous or adventitious viruses from the product must be shown. Because virus titers are well below the level of detection, bench-scale studies are done in which different types of model viruses are deliberately added to process streams to demonstrate the ability of the process to remove viruses. Usually several types of viruses are used, including enveloped and nonenveloped viruses.

Dosage form (drug product) manufacturing

In general, dosage form manufacturing of rHuEPO, as with all recombinant protein products, is governed by national governmental regulatory agencies to ensure that biologic production facilities adhere to sound quality control and current GMP. Facilities that produce rHuEPO are routinely audited and inspected to ensure compliance and patient safety. In the United States, biologic production for parenteral applications is governed by the Food and Drug Administration (FDA CBER/CDER); regulations are covered under section 21 of the Code of Federal Regulations Parts 210 and 211. These regulations differ slightly from country to country. All manufacturing plants and processes are validated and filed with the FDA (or regional authority) before marketing approval is given. All operations are conducted in an aseptic, temperature-controlled, highly monitored environment, with final filling operations done under conditions that typically allow only 100 particle counts/cubic foot of air in the filling operation suite. For dosage form/drug product manufacturing, maintaining validated processes, quality control, and aseptic conditions are critical to patient safety and product integrity.

In addition to regulatory compliance, a primary concern for recombinant proteins, and in particular rHuEPO, is the maintenance of product integrity by minimizing physical or chemical degradation. Product integrity is maintained from the bulk stage through the final dosage form by adjusting process parameters and the composition of the final formulation. To mitigate process impacts, rHuEPO is formulated in solutions that often contain human serum albumin or polysorbate that acts as a protector against surface adsorption, surface denaturization, shear forces, chemical degradants, and other deleterious factors. Because of the relatively low concentrations of rHuEPO used clinically, additives are important to protect the product during processing. Formulations are optimized for stability and parenteral delivery. Ingredients such as citrates, chlorides, and phosphates at relatively neutral pH are often used. Nonoptimized formulations can result in degradants that arise before the end of shelf life. The formation of degradants can lead to loss of activity or by-products that may potentially be antigenic.

rHuEPO is marketed in a variety of formulations and concentrations, providing the practitioner with a wide selection of dosages for achieving optimal hematocrit values for patients. The final dosage form available in clinics is typically an aqueous formulation in a vial. Prefilled syringes, lyophilization, frozen liquid, or other delivery forms also can be used. This variety of formats can complicate production because of the need to formulate at different strengths, different fill volumes, and different delivery vehicles.

Process

Typical processing of rHuEPO consists of several unit operations: buffer preparation, formulation of the purified bulk, filtration of the formulated product, filling into a delivery vehicle, lyophylization (i.e., freeze drying if applicable), unit inspection, and packaging. Figure 11 depicts a generic process flow from bulk though packaging operations. A formulation-and-fill process may comprise several vessels, a fill line, and inspection packaging lines. One vessel may be used for buffer preparation, another for active addition to the buffer, and a third for holding the final formulated product before filling. The set up can vary depending on batch size and the formulation being produced.

General processing conditions must be closely examined to ensure product compatibility. Typically, the following items must be examined: materials in the processing vessels and transfer lines; duration of product exposure to various temperature and sterility conditions; and shear forces associated with mixing, filtration, and filling. During the formulation-and-filling operation, rHuEPO primarily comes in contact with stainless steel. Exposure to silicone tubing, various filtration membranes, ceramics, and glass can occur, however. Generally, process conditions are tested, and the product is placed on a stability testing protocol to ensure that no deleterious effects result from processing.

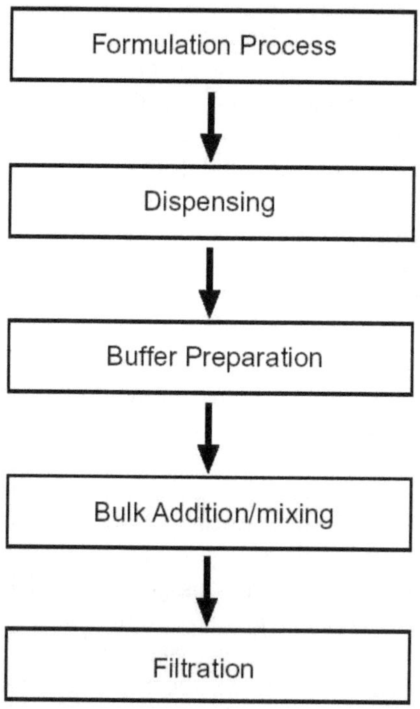

Figure 11. Generic process flow from bulk to packaging operations

The initial step of the dosage form manufacturing is bulk dispensing and addition of materials associated with the final formulated product. Product amounts are adjusted according to batch size and concentration (potency). Additional excipient raw materials are added before production. Before use, each material is examined to ensure that safety, quality standards, and product-formula requirements are met. Buffer preparation is completed by the addition of the excipients to purified Water For Injection in a processing vessel. The buffer can be filtered to clarify and to remove particulates. After buffer preparation, the bulk rHuEPO is added. Additional components that protect or stabilize the rHuEPO can also be added. The formulations or stockkeeping units that are used in multidose settings will contain an antibacterial agent to minimize microbial growth after initial breech of the stopper septum. The bulk rHuEPO product together with the excipients added to make up the formulation is referred to as the formulated bulk.

The formulated bulk is sterile filtered into a holding vessel and stored until filled. Dosage-form components such as vials or syringes are filled using automated equipment in a clean-room environment. The filled vials and syringes are immediately sealed with a stopper, and capped (in the case of vials). Product contact components such as stoppers, vials, and syringes are washed and depyrogenated at high temperature or sterilized in the presence of steam. Some components can be purchased from vendors pre-sterilized and ready to use.

Each filled vial or syringe should be inspected for particulate matter in solution or cosmetic defects, such as marks or scratches. Additionally, each unit is inspected to ensure that proper lot number and expiration date have been applied to the label.

Release

Before the release of any rHuEPO lot, the final material is thoroughly tested to ensure sterility, concentration, lack of endotoxin and bioburden, and product integrity according to the manufacturer's quality control requirements. Additionally, batch records from the manufacturing run are reviewed to ensure that no deviation was made during the production run. Finally, as each lot is released, the product is shipped to wholesalers according to validated procedures.

Summary

An example of commercial production of rHuEPO is discussed in this chapter, where the manufacture starts with a cell line that is engineered to produce product in commercial quantities of required quality. The cell line is banked, and the cell bank is used as the starting material for the manufacturing process. In this example, rHuEPO is secreted from genetically engineered mammalian cells in a cell culture process and then recovered and purified as rHuEPO bulk in a purification process, to achieve the desired product characteristics specified by the manufacturer. Product from different manufacturers may have different quality characteristics. The rHuEPO bulk drug substance is formulated to achieve the required dosage forms for the final drug product. The manufacturing process is performed in a cGMP facility and monitored for consistent performance. Regulatory and other safety requirements are followed to reproducibly produce a safe and efficacious product.

References

1. Center for Biologics Evaluation and Research (CBER) and Center for Drug Evaluation and Research (CDER) Guidance for industry for the submission of chemistry, manufacturing, and controls information for a therapeutic recombinant DNA-derived product or a monoclonal antibody product for *in vivo* use. 1996.
2. International Conference on Harmonisation (ICH) of Technical Requirements for Registration of Pharmaceuticals for Human Use. Q5D. 1997
3. International Conference on Harmonisation (ICH) of Technical Requirements for Registration of Pharmaceuticals for Human Use Q5A(R1) Viral Safety Evaluation of Biotechnology Products Derived From Cell Lines of Human or Animal Origin. 1999.
4. Freshney RI. *Culture of animal cells.* Alan R Liss Inc, New York; 1983.
5. Ozturk SS. Engineering challenges in high density culture systems. *Cytotechnol* 1996;22:3–16.
6. Chu L, Robinson DK. Industrial choices for protein production by large-scale cell culture. *Curr Opin Biotechnol* 2001;12:180–187
7. Monroe RS, Huber BE. The major form of murine asialoglycoprotein receptor – cDNA sequence and expression in liver, testis and epididymus. *Gene* 1994;148:237–244
8. Center for Drug Evaluation and Research. Q7A Good manufacturing practice guidance for active pharmaceutical ingredients. 2001.

Erythropoietins, Erythropoietic Factors and Erythropoiesis
Edited by S.G. Elliott, M.A. Foote, and G. Molineux

Biosimilar epoetins

Huub Schellekens

Departments of Pharmaceutical Sciences and Innovation Studies, Utrecht University, Utrecht, The Netherlands

Introduction

Recombinant DNA technology and hybridoma techniques enabled the development of therapeutic proteins such as the interferons (IFN), erythropoietin (EPO), insulin, growth hormones, cytokines, and monoclonal antibodies [1]. These proteins have revolutionised the treatment of many diseases, including, cancer, hepatitis, and multiple sclerosis. The patents of the first generation of biopharmaceuticals have expired or are about to expire, which enables marketing of noninnovator versions of these products, named 'biosimilars' in the European Union (EU), and 'follow-on protein products' in the USA. The EU has established a legal and regulatory framework for acquiring a marketing authorisation for biosimilars to market [2–9], while in the United States the regulatory pathway is still being developed [10].

Biosimilars are defined as biologic products similar, but not identical, to reference products that are submitted for separate marketing approval after patent expiration of the reference products [3, 11–13]. Biosimilars are not considered to be generic versions. Conventional generics of classical chemical drugs are allowed on the market if pharmaceutical equivalence (i.e., identical active substances) and bioequivalence (i.e., comparable pharmacokinetics) compared with the innovator drug have been established. Thus, clinical efficacy and safety studies are not needed. The term 'biogeneric' should be avoided because this implies the active substance of a biosimilar can be shown to be identical to the active substance of the reference product. This generic paradigm cannot be applied to therapeutic proteins that, in general, are a collection of isoforms and not a single molecule as usually the situation with conventional small-molecule drugs.

The market approval is more complicated for biosimilar compared with classical generic drugs. To establish therapeutic equivalence of biosimilars with reference products, clinical trials are essential [12, 14, 15]. Safety of biosimilars is also an important issue. The main difference concerning safety between biopharmaceuticals and conventional drugs is the potential to induce an immune response (immunogenicity) [16, 17]. The European Medicines Evaluation Agency (EMEA) outlined the requirements for market approval of biosimilars [18]. The EMEA also published public assessment reports for the approved biosimilar products that discuss product characteristics and the clinical data supporting approval. The EMEA has approved biosimilar versions of recombinant somatropin [19, 20] and recombinant human EPO (rHuEPO) [21–25]. One follow-on somatropin product has been approved in the United States [10].

Immunogenicity of proteins

Nearly all therapeutic proteins are immunogenic, and may induce antibodies by two different mechanisms. In the case of products of plant or microbial origin, a classical reaction as to any foreign substance is induced. Human homologues, such as IFNs and rHuEPO, induce antibodies by breaking immune tolerance existing normally to self-antigens. The mechanisms by which tolerance is induced or broken are not completely understood. An important factor to break tolerance is to present the self antigens in a repetitive way. A periodicity of these antigens as present in aggregates of proteins is apparently very efficient in activating ignorant or anenergetic B cells that are responsible for tolerance.

The clinical manifestations of an immune response are highly dependent on the mechanism involved. The classical activation of the immune system by foreign proteins and the breaking of B-cell tolerance by human proteins differ in time of onset and response level. Also the way these types of response are activated differ and, therefore, also the characteristics of the products.

Immune tolerance exists to products that are copies of endogenous proteins such as IFNs, colony stimulating factors, and rHuEPO. Breaking B-cell tolerance starts with the activation of autoreactive B-cells with aggregates as the most important risk factor.

These aggregates may appear during storage making stability and solubility important in predicting immunogenicity. Factors important for the formation of aggregates are oxidation and reduced solubility by the lack of glycosylation, e.g., if glycoprotein is expressed in a prokaryotic host.

Different methods can identify the presence of aggregates and include size exclusion chromatography, sodium dodecyl sulfate-polyacrilamide gel electrophoresis (SDS-PAGE), and Western blotting. These methods, however, may miss immunogenic aggregates. Other methods, such as analytical ultracentrifugation, field-flow fractionation, and light scattering techniques, have shown to be able to detect aggregates that go undetected by standard assays.

The only biologic test to study the capacity of a protein product to break B-cell tolerance are mice that have been made transgenic for the specific protein. These mice are immune tolerant and there is a good correlation of an immune response between these mice and patients. Although these models have helped to identify the factors important for breaking B-cell tolerance and also have been useful in improving the formulation of products, not enough experience allows them to be used as absolute predictors of immunogenicity of human proteins. The only way to test the immunogenicity of any human product, including biosimilars, is through clinical trials.

When designing clinical trials for testing the immunogenicity of human proteins, many factors must be taken into account. Breaking tolerance by human protein takes, in general, more than 6 months of chronic treatment. The route of administration influences the likelihood of an antibody response independent of the mechanism of induction. The probability of an immune response is the highest with subcutaneous administration, less probable after intramuscular administration, and intravenous administration is the least immunogenic route.

The biologic activities of the product also influences its immunogenicity. An immune-stimulating therapeutic protein is more likely to induce antibodies than a immune-suppressive protein. Impurities may influence immunogenicity. Oxidised proteins were shown to induce antibodies that crossreacted with the unmodified product, and host cell-derived endotoxins have been reported to act as adjuvants. The probability of an immune response, therefore, increases with the level of impurities.

Sex, age, and ethnic background have been reported to influence the incidence of antibody response to specific therapeutic proteins. Type of disease has been identified as the main risk factor of patients. Patients with cancer are less likely to produce antibodies to therapeutic protein than other patients for reasons that are not clear.

Concomitant therapy should be considered. Immune suppressive therapy reduces the probability to develop an immune response to proteins. In addition, immune suppressive drugs, such as methotrexate, are used in conjunction with monoclonal antibodies and other protein drugs to reduce the immune reactions. Thus, immunogenicity data obtained in one population cannot always be extrapolated to another population.

Consequences of antibodies

There may be no clinical consequence for developing an immune response to a biopharmaceutical. The patient may develop binding antibodies that do not significantly affect the activity of the biopharmaceutical or the endogenous protein. On the other hand, anti-product antibodies can bind to and influence the pharmacokinetics of a biopharmaceutical.

A severe consequence of immunogenicity is the loss of efficacy of the product, which can be a severe problem if there is no alternative treatment and the product has a clinically important effect. An example is the case with antibodies to factor VIII that inhibit its activity in patients with haemophilia. Side effects, such as anaphylaxis-like reactions and serum sickness, are mainly caused by the formation of immune complexes seen when the protein is administered in high doses. Major clinical impact can occur if the endogenous protein with essential biologic activity is also neutralised. For example, neutralising endogenous EPO can result in a rare condition known as antibody-mediated pure red cell aplasia (PRCA).

Between 1998 and 2003 a large increase in the incidence of antibody-mediated PRCA occurring in patients with chronic kidney disease and anaemia who had been treated with Eprex, a formulation of epoetin alfa marketed by Johnson & Johnson [17, 25–27]. The PRCA cases were associated with a breakdown of immune tolerance to treatment with rHuEPO, particularly with subcutaneous administration, resulting in neutralising antibody formation against both recombinant and endogenous EPO [28]. Previously, PRCA caused by production of neutralising anti-EPO antibodies occurred very rarely with treatment. The apparent increase in immunogenicity coincided with a relatively minor formulation change for Eprex – replacement of human serum albumin as a stabiliser with glycine and polysorbate 80. The withdrawal of this formulation of epoetin alfa for subcutaneous administration led to a subsequent decrease in the incidence of PRCA. The reason why Eprex induced PRCA is still not fully understood. A number of possible causes have been proposed [28–30], including micelle formation from polysorbate 80 and epoetin alfa [29, 31] or leachates from rubber stoppers breaking B-cell tolerance through an adjuvant effect [30, 32, 33]. Although many factors are reported to influence the immunogenicity of therapeutic proteins, aggregates play a role in most cases. The Eprex preparation showed an increase in the amounts of aggregates during storage, although the amount was reported never to have exceeded specifications. Product specifications, however, are not defined on the basis of biologic effects, and aggregates may still be relevant for PRCA induction.

The Eprex case highlights concerns regarding the unpredictability and seriousness of immunogenicity of biopharmaceuticals and the potential clinical consequences of their extensive use. Only a small change in the manufacturing process appears to have altered the product's characteristics with a drastic impact on clinical outcome.

The biosimilars assessed by the EMEA

The EMEA provides information on the approval process for human medicines (the European Public Assessment Report [EPAR]), including a scientific discussion on the clinical data submitted for approval. Generally, the EPAR for biosimilars have stated that the biosimilar received approval because it was shown to have a comparable quality, safety, and efficacy profile to the reference product [34–37]. Despite the comparability of these biosimilars to the reference products, clinicians should be aware of some of the issues that emerged during the development and approval of these products that highlight the challenges of biosimilars.

Two biosimilar somatropins, Omnitrope and Valtropin, have been approved by the EMEA. Omnitrope is a biosimilar version of the reference product,

Genotropin. Like Genotropin, Omnitrope is a recombinant form of human somatropin that is manufactured with recombinant DNA technology synthesised in *Escherichia coli*. During the development of Omnitrope, an immunogenicity issue emerged with an early version of the product. As many as 60% of patients enrolled in a study developed anti-growth hormone antibodies, which did not appear to affect growth rate. The cause of immunogenicity was linked to excess host cell protein contamination, which was resolved by the manufacturer with additional purification steps [41].

The first biosimilar epoetin alfa received marketing authorisation in August 2007, which was followed by the publication of the EPAR of Binocrit (Sandoz GmbH), Epoetin alfa Hexal (Hexal Biotech Forschungs GmbH), and Abseamed (Medice Arzneimittel Pütter GMBH & Co). Although the biosimilars are marketed by three companies, they are the product of the same manufacturing plant and the applications for marketing authorisations were identical.

The second biosimilar products of epoetin alfa were approved for marketing in October 2007, and the EPAR of Silapo (Stada) and Retacrit (Hospira) were published in January 2008.

'Biosimilar' is a regulatory term and not a scientific term. Only the evaluations by the regulatory agencies will ultimately define how similar a 'similar' must be. The introduction of biosimilars will reveal how the unavoidable differences between the products translate in biologic and clinical differences. EPAR, however, only provide restricted information, especially information concerning the physical-chemical characteristics of the biosimilar.

As expected, the biggest differences in the structure are in the glycosylation profiles. Compared with the reference epoetin alpha (Eprex), Binocrit contains more high mannose, while Silapo has a reduced amount of *O*-glycan, sialic acid, *N*-glycolyl neuraminic acid, and *O*-acetyl neuraminic acid, which the EMEA considers to be 'undesired variants'.

In the case of Binocrit, this difference has no influence on the *in vitro* behaviour. In the case of Stada, the *in vitro* data were incomplete, showed significant differences between batches, and did not always included direct comparison with the reference product. It is apparent that Stada has a slightly reduced specific activity that translates into reduced pharmacodynamic and clinical activity when dosing patients on the basis of mass (µg).

The reduced specific activity may be related to the possible differences between the product and the standard provided by the European Pharmacopoeia Commission. The suggested different dose-response *in vitro* and *in vivo* of the Stada makes the use of this standard a problem. The use of the murine model to test the relative potency is intrinsically not very precise and allows a relative high variation.

Differences in specific activity are not unusual between comparable products. The two IFN-β products produced in Chinese hamster ovary (CHO) cells (Avonex and Rebif) show a considerable difference in specific activity but are clinically comparable.

For the biosimilar epoetin products, the Committee for Medicinal Products for Human Use (CHMP) has issued a product-specific guideline that describes that two phase III clinical trials are necessary. In the case of Binocrit, however, the European regulators have accepted the data from a single trial, because the subcutaneous route of the comparator Eprex was contraindicated at the time of the studies.

Stada was evaluated in two pivotal trials, a correction phase study and a maintenance study. Both studies with the Stada met their primary endpoint with respect to mean haemoglobin concentrations. In the correction phase study, however, the prespecified equivalence margins of the treatment difference in the dosage were not met. In both studies, Stada showed a reduced clinical activity, which is probably based on the difference in specific activity that was considered clinically not relevant by the CHMP.

Both the Hexal and Stada have exploratory clinical studies assessing efficacy and safety of the biosimilar in the treatment of chemotherapy-associated anaemia. Overall, no significant difference was seen between the treatment groups for the incidence or type of adverse events. Both biosimilar epoetins had a safety profile similar to that of the innovator comparator Eprex.

The possible immunogenicity of the biosimilars and the possible development of PRCA has been an important issue in the evaluation. The Eprex-associated PRCA was the result of a very specific change in the formulation. It is unlikely, therefore, that other immunogenicity problems with an erythropoiesis-stimulating agent (ESA) will result in PRCA.

An antibody response induced by another epoetin may have a different character and may lead to different clinical consequences. With both biosimilars, a radioimmune precipitation assay (RIA) used to monitor the immunogenicity. This assay has been used as the gold standard in the diagnosis of antibody induced PRCA. How well this assay is capable to determine early immune responses in a low-risk population, however, is far from clear. With this assay, a transient binding antibody response was seen before treatment in a number of patients with renal disease who were treated intravenously in the pivotal trials with the innovator and/or biosimilar products. The questions remain whether the specificity of RIA is sufficient to screen for immunogenicity in low-risk populations.

In the risk management programs of both biosimilars, a cohort study is described to assess the incidence of PRCA after marketing authorisation. The number of patients to be involved or the length of the study is not disclosed, but considering the rarity of PRCA, these studies will need to enrol tens of thousands of patients to assess the rate of the event.

The CHMP has clearly taken no position in the discussion about the naming of the biosimilars. The innovative producers of the original products have argued for each biosimilar to have a specific International Nonproperity Name (INN). The CHMP has accepted both the INN name to be identical to the original epoetin alfa in the case of Hexal. In the case of Stada, however, it has accepted epoetin zeta as the INN name.

The approval of these biosimilar products does not substantiate interchangeability with reference products [42]. Furthermore, the EMEA has not approved all biosimilar applications. Alpheon, a biosimilar version of Roferon-A (IFN alfa-2a), was rejected by the EMEA. The manufacturer of Alpheon had submitted nonclinical data (protein structure, composition, and purity) on the biosimilar and conducted a randomised controlled trial study in patients with hepatitis C to demonstrate comparable efficacy and safety between the biosimilar and reference product. The reasons for the rejection by the EMEA included quality and clinical differences between Alpheon and the reference product, inadequate data on the stability of the active substance, inadequate validation of the process for the finished process, and insufficient validation of immunogenicity testing [43].

Conclusions

Biosimilar products are very complex molecules and, therefore, cannot be treated the same as conventional generic drugs. Although a variety of assays are available, they may not be adequate to reliably predict the safety and efficacy of a biosimilar product. The regulatory approval of biosimilars requires much more than the demonstration of pharmaceutical equivalence and pharmacokinetic bioequivalence associated with conventional generics.

The immunogenicity of recombinant therapeutic proteins has become a significant safety concern. Ultimately, only clinical studies and post-authorisation pharmacovigilance to monitor potential immunogenicity will provide definitive evidence for product comparability to the innovator product with respect to safety and efficacy [13, 15].

Outstanding issues will need to be resolved, including substitution, naming, and labelling [44]. It is unclear whether biosimilars and innovator biopharmaceuticals should be considered interchangeable. The labels of the approved biosimilars are nearly identical or are very similar to those of the reference product. A more transparent label that included relevant clinical data for the biosimilar, i.e., the data included in the EPAR, would help clinicians make informed treatment decisions.

Epoetin products from manufacturers outside Europe and the USA have been available in areas of the world without patent protection. These products were shown to differ widely in composition [45]. Although this is no proof that these products are clinically inferior to the innovator product, some of the products failed to meet their own specifications. Some of the manufacturers do not have adequate control over their production process. The content of the vials often did not meet the labelled activity. This mislabelling may lead to undesired clinical effects. Many biopharmaceuticals, including epoetin, are heterogeneous with respect to isoform distribution. As the contribution of different isoforms to the overall activity and toxicity is unknown, consistency between production runs is essential. With that in mind, the most alarming aspect of this comparison was the variation observed between different lots of the same product (Fig. 1). This inconsistency between batches exposes major flaws in the control of the production process, or worse yet, may even hint to different sources of what is claimed to be an identical product.

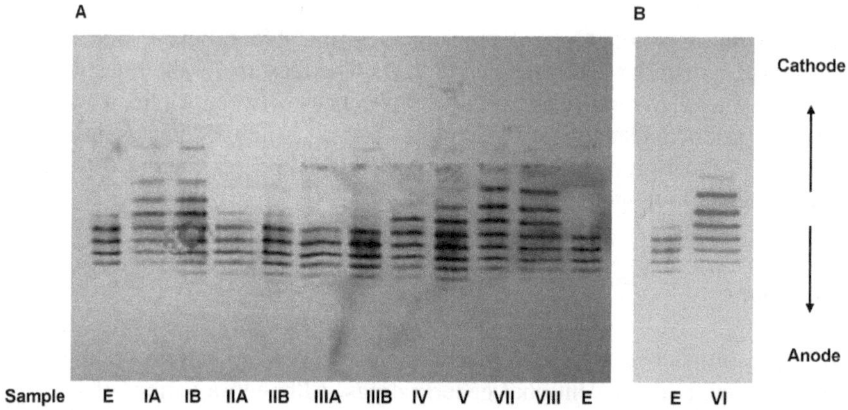

Figure 1. Isoelectric Focusing/Western Blot. Isoform distribution of each sample is shown. For comparison, the Eprex (E) control is shown in the first and last lanes of Figure A and in the first lane of Figure B. I-VIII are epoetin alfa formulations from different manufacturers. A and B means different batches from the same manufacturer.

Acknowledgement
The author contributed to meetings and publications sponsored by Amgen, Roche, Hospira, Novartis, Johnson and Johnson, and Shire.

References

1. Avidor Y, Mabjeesh NJ, Matzkin H. Biotechnology and drug discovery: from bench to bedside. *South Med J* 2003; 96:1174–1186.
2. European Parliament and Council. Directive 2004/27/EC of the European Parliament and of the Council of 31 March 2004. *Offical Journal of the European Union* Vol 47; 2004:34–57.
3. European Medicines Agency. Guideline on similar biological medicinal products 2005. Available at: http://www.emea.europa.eu/pdfs/human/biosimilar/043704en.pdf. (Accessed 4 March 2008)
4. European Medicines Agency. Guideline on similar biological medicinal products containing biotechnology-derived proteins as active substance: Quality issues. 2006. Available at: http://www.emea.europa.eu/pdfs/human/biosimilar/4934805en.pdf. (Accessed 4 March 2008)
5. European Medicines Agency. Annex to guideline on similar biological medicinal products containing biotechnology-derived proteins as active substance: Non-clinical and clinical issues. Guidance on similar medicinal products containing recombinant erythropoie. *Trends Neurosci* 2006. Available at: http://www.emea.europa.eu/pdfs/human/biosimilar/9452605en.pdf. (Accessed 4 March 2008)
6. European Medicines Agency. Annex to guideline on similar biological medicinal products containing biotechnology-derived proteins as active substance: Non-clinical and clinical issues. Guidance on similar medicinal products containing somatropin. 2006. Available at: http://www.emea.europa.eu/pdfs/human/biosimilar/9452805en.pdf. (Accessed 4 March 2008)

7. European Medicines Agency. Annex to guideline on similar biological medicinal products containing biotechnology-derived proteins as active substance: Non-clinical and clinical issues. Guidance on similar medicinal products containing recombinant human insulin. 2006. Available at: http://www.emea.europa.eu/pdfs/human/biosimilar/3277505en.pdf. (Accessed 4 March 2008)
8. European Medicines Agency. Annex to guideline on similar biological medicinal products containing biotechnology-derived proteins as active substance: Non-clinical and clinical issues. Guidance on similar medicinal products containing recombinant granulocyte-colony stimulating factor (G-CSF). 2006. Available at: http://www.emea.europa.eu/pdfs/human/biosimilar/3132905en.pdf. (Accessed 4 March2008)
9. European Medicines Agency. Guideline on similar biological medicinal products containing biotechnology-derived proteins as active substance: non-clinical and clinical issues 2006. Available at: http://www.emea.eu.int/pdfs/human/biosimilar/4283205en.pdf. (Accessed 1 May 2006)
10. US FDA. Omnitrope (somatropin [rDNA origin]) questions and answers 2006. Available at: http://www.fda.gov/cder/drug/infopage/somatropin/qa.htm. (Accessed 3 March 2008)
11. Crommelin DJ, Bermejo T, Bissig M et al. Pharmaceutical evaluation of biosimilars: important differences from generic low-molecular weight pharm. *Eur J Hosp Pharm Sci* 2005;1:11–17.
12. Roger SD. Biosimilars: how similar or dissimilar are they? *Nephrology (Carlton)* 2006;11:341–346.
13. Schellekens H. Follow-on biologics: challenges of the 'next generation'. *Nephrol Dial Transplant* 2005;20: v31–36.
14. Chirino AJ, Mire-Sluis A. Characterizing biological products and assessing comparability following manufacturing changes. *Nat Biotechnol* 2004;22:1383–1391.
15. Locatelli F, Roger S. Comparative testing and pharmacovigilance of biosimilars. *Nephrol Dial Transplant* 2006;21 Suppl 5:v13–16.
16. Kessler M, Goldsmith D, Schellekens H. Immunogenicity of biopharmaceuticals. *Nephrol Dial Transplant* 2006;21 Suppl 5:v9–12.
17. Schellekens H. Factors influencing the immunogenicity of therapeutic proteins. *Nephrol Dial Transplant* 2005;20:vi3–9.
18. Wiecek A, Mikhail A. European regulatory guidelines for biosimilars. *Nephrol Dial Transplant* 2006;21 Suppl 5:v17–20
19. Omnitrope. Summary of product characteristics. Available at: http://www.emea.europa.eu/humandocs/PDFs/EPAR/Omnitrope/H-607-PI-en.pdf. (Accessed 4 March 2008)
20. Valtropin. Summary of product characteristics. Available at: http://www.emea.europa.eu/humandocs/PDFs/EPAR/valtropin/H-602-PI-en.pdf. (Accessed 4 March 2008)
21. Abseamed. Summary of product characteristics. Available at: http://www.emea.europa.eu/humandocs/PDFs/EPAR/abseamed/H-727-PI-en.pdf. (Accessed 4 March 2008)
22. Binocrit. Summary of product characteristics. Available at: http://www.emea.europa.eu/humandocs/PDFs/EPAR/binocrit/H-725-PI-en.pdf. (Accessed 4 March2008)
23. Epoetin alfa Hexal. Summary of product characteristics. Available at: http://www.emea.europa.eu/humandocs/PDFs/EPAR/epoetinalfahexal/H-726-PI-en.pdf. (Accessed 4 March 2008)
24. Silapo. Summary of product characteristics. Available at: http://www.emea.europa.eu/humandocs/PDFs/EPAR/silapo/H-760-PI-en.pdf. (Accessed 4 March 2008)
25. Bennett CL, Luminari S, Nissenson AR et al. Pure red-cell aplasia and epoetin therapy. *N Engl J Med* 2004;351:1403–1408.
26. Casadevall N, Eckardt KU, Rossert J. Epoetin-induced autoimmune pure red cell aplasia. *J Am Soc Nephrol* 2005;16 Suppl 1:S67–69.
27. Casadevall N, Rossert J. Importance of biologic follow-ons: experience with EPO. *Best Pract Res Clin Haematol* 2005;18:381–387.
28. Schellekens H. Immunologic mechanisms of EPO-associated pure red cell aplasia. *Best Pract Res Clin Haematol* 2005;18:473–480.
29. Hermeling S, Schellekens H, Crommelin DJ et al. Micelle-associated protein in epoetin formulations: a risk factor for immunogenicity? *Pharmacol Res* 2003;20:1903–1907.
30. Schellekens H, Jiskoot W. Eprex-associated pure red cell aplasia and leachates. *Nat Biotechnol* 2006;24:613–614.
31. Louet S. Lessons from Eprex for biogeneric firms. *Nat Biotechnol* 2003;21:956–957.
32. Villalobos AP, Gunturi SR, Heavner GA. Interaction of polysorbate 80 with erythropoietin: a case

study in protein-surfactant interactions. *Pharmacol Res* 2005;22: 1186–1194.
33. Boven K, Knight J, Bader F et al. Epoetin-associated pure red cell aplasia in patients with chronic kidney disease: solving the mystery. *Nephrol Dial Transplant* 2005;20 Suppl 3: iii33–40.
34. European Medicines Agency. Omnitrope: EPAR summary for the public 2007. Available at: http://www.emea.europa.eu/humandocs/PDFs/EPAR/Omnitrope/060706en1.pdf. (Accessed 3 March 2008)
35. European Medicines Agency. Valtropin: EPAR summary for the public 2006. Available at: http://www.emea.europa.eu/humandocs/PDFs/EPAR/valtropin/H-602-en1.pdf. (Accessed 4 March 2008)
36. European Medicines Agency. Retacrit: EPAR summary for the public 2008. Available at: http://www.emea.europa.eu/humandocs/PDFs/EPAR/retacrit/H-872-en1.pdf. (Accessed 4 March 2008)
37. European Medicines Agency. Silapo: EPAR summary for the public 2007. Available at: http://www.emea.europa.eu/humandocs/PDFs/EPAR/silapo/H-760-en1.pdf. (Accessed 4 March 2008)
38. European Medicines Agency. Epoetin alfa Hexal: EPAR summary for the public 2007. Available at: http://www.emea.europa.eu/humandocs/PDFs/EPAR/epoetinalfahexal/H-726-en1.pdf.
39. European Medicines Agency. Binocrit: EPAR summary for the public 2007. Available at: http://www.emea.europa.eu/humandocs/PDFs/EPAR/binocrit/H-725-en1.pdf.
40. European Medicines Agency. Abseamed: EPAR summary for the public 2007. Available at: http://www.emea.europa.eu/humandocs/PDFs/EPAR/abseamed/H-727-en1.pdf
41. European Medicines Agency. Omnitrope: scientific discussion 2006. Available at: http://www.emea.europa.eu/humandocs/PDFs/EPAR/Omnitrope/060706en6.pdf. (Accessed 4 March 2008)
42. Smith N. EMEA "will not guarantee" that biosimilars are interchangeable with originator. APM Health Europe 2006. Available at: http://health.apmnews.com/story.php?mots=BIOSIMILAR &searchScope=1&searchType=0&depsPage=2&numero=L3250 (Accessed 1 May 2006)
43. European Medicines Agency. Questions and answers on recommendation for refusal of marketing application for Alpheon 2006. Available at: http://www.emea.europa.eu/pdfs/human/opinion/19089606en.pdf. (Accessed 3 March 2008)
44. Mellstedt H, Niederwieser D, Ludwig H. The challenge of biosimilars. *Ann Oncol* 2008;19:411–419.
45. Schellekens H. Biosimilar epoetins: how similar are they? *EJHP* 2004;3;243–247.

Erythropoietins, Erythropoietic Factors and Erythropoiesis
Edited by S.G. Elliott, M.A. Foote, and G. Molineux

New molecules and formulations

Steven G. Elliott

Amgen Inc., One Amgen Center Drive, Thousand Oaks, CA 91320, USA

Introduction

Recombinant human erythropoietin (rHuEPO; epoetin alfa), is a glycosylated hormone commonly used for the treatment of anemia associated with chronic kidney disease [1–4]. It is also indicated for the treatment of certain anemias associated with cancer, HIV infection, and for use in surgical situations to reduce allogeneic blood transfusion requirements. A number of studies have demonstrated that rHuEPO is well tolerated and effective at ameliorating anemia, restoring energy levels, and improving patient quality of life in these indications [5–7]. It has also been shown to reduce transfusions, a procedure that can carry inherent risks including transmission of infectious agents and iron overload. In addition, the blood supply is limited and transfusion-induced immune reactions can complicate organ transplantation, for example in patients with kidney transplants.

A more effective anemia treatment option was needed and the discovery and development of rHuEPO via cloning of the human *EPO* gene began in the early 1980s (*see* chapter by M.A. Foote). While rHuEPO has a defined amino acid sequence, differences in its production result in changes in post-translational modifications including glycosylation, conformation, and impurities. Currently it is not possible for another manufacturer to duplicate exactly the product profile of the innovator [8]. Thus, the term ‘generic’ is not used to describe rHuEPO molecules made by different manufacturers. Instead, the descriptors ‘follow-on biologics’ (FOB), ‘generic biosimilars’, or ‘generic biopharmaceuticals’ are used [9].

The first commercial rHuEPO product introduced in the United State and European Union was epoetin alfa (Epogen, Amgen Inc; Procrit, Ortho Biotech Eprex, Ortho Biotech). It was followed by the introduction of a second product, epoetin beta (e.g., NeoRecormon, Roche). Epoetins alfa and beta have minor structural differences but similar physiological effects [10, 11]. Both epoetins are produced by Chinese hamster ovary (CHO) cells. Other FOBs have also been developed. Their approval and acceptance in the EU has been slow, primarily because of incomplete information regarding their efficacy and safety profiles (e.g., likelihood that the agent will promote pure red-cell aplasia [PRCA]).

As patents for epoetins alfa and beta expire, follow-on epoetins, or biosimilar epoetins, have been approved by the European Medicines Agency (EMEA), including three epoetins alfa: Binocrit (Sandoz), HEXAL (Hexal Biotech Forschungs), and Abseamed (Medice Arzneimittel Puetter). An epoetin produced in baby hamster kidney cells, epoetin omega, differs somewhat from epoetin alfa and epoetin beta in its glycosylation profile. While it was available in Eastern European countries and parts of Latin America, it is no longer marketed. Epoetin delta (Dynepo, Shire), which is produced from an engineered human fibrosarcoma cell line HT1080, also has been marketed [12].

In addition to epoetin alfa, epoetin beta, and FOBs, other classes of molecules that may stimulate erythropoiesis have been explored. Together with the epoetins, these types of agents have been given the name 'erythropoiesis-stimulating agents' or ESAs. Most of these agents are peptidic in nature; however small molecule drugs also have been explored. ESAs can also be categorized according to their mechanism of action and where in the erythropoiesis process they may intervene (Fig. 1).

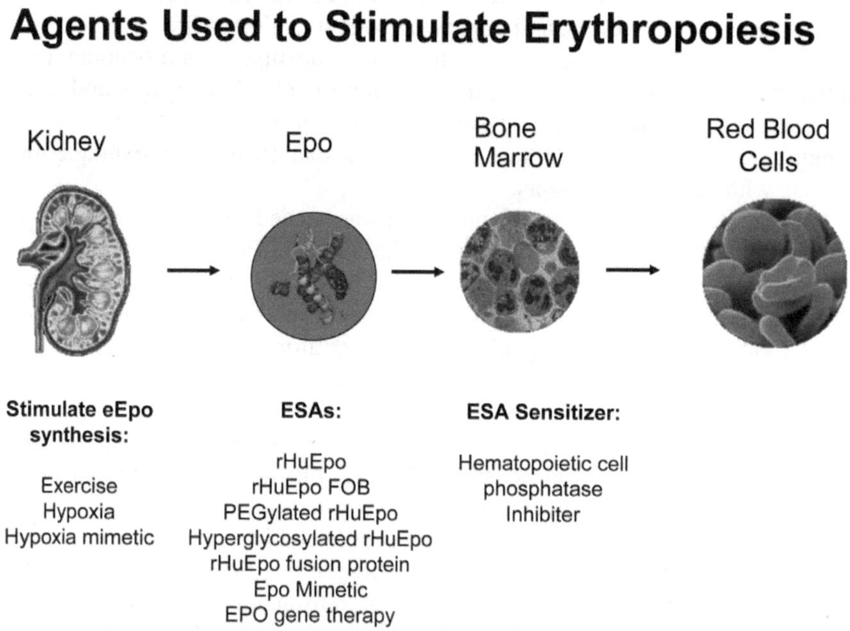

Figure 1. ESAs and points of intervention ESAs can induce EPO synthesis by hypoxia or through use of hypoxia mimetics. Endogenous EPO, rHuEPO, or mimetics of EPO can bind and activate EPOR stimulating erythropoiesis. EPOR activation results in signal transduction involving phosphorylation of EPOR and phosphorylation of JAK2 followed by downstream events. Signal transduction is halted by dephosphorylation of EPOR by hematopoietic cell phosphatase (HCP) or induction of suppressors of cytokine signaling (SOCS). Constitutive phosphorylation of EPOR, constitutive activation of JAK (or JAK2 agonists), loss of HCP activity or inhibition of SOCS activity can induce erythrocytosis.

Considerations of new formulations and drug entities

The purpose of any therapeutic intervention is to treat the patient with an appropriate risk-benefit profile. rHuEPO has been particularly successful in this regard. The molecule is effective at stimulating erythropoiesis with a well characterized safety profile. This risk-benefit profile has created a high standard against which any new erythropoiesis-stimulating molecule will be measured. Thus, the preferred properties of any new drug or formulation should retain or increase efficacy and avoid new or unwanted side effects or toxicities.

One particular concern that has arisen with some rHuEPO molecules is anti-EPO antibodies [13, 14]. Antibody formation to drug entities is of concern not only because the drug may lose efficacy, but also because such antibodies might cross-react with endogenously-produced EPO resulting in pure red cell aplasia (PRCA), a very serious and severe form of anemia [14]. Potential causes for antibody formation not only include the structure of the molecule itself, but also the breakdown products or aggregates generated during manufacturing or storage. Accordingly, new drugs should be designed and manufacturing procedures put in place to minimize this risk.

The production and clearance rates of endogenous EPO protein are naturally optimized for maintenance of hemostasis in the body. This natural regulation, however, is difficult to recreate using a therapeutic. For example, rHuEPO is indicated for three times per week administration in humans. Needlesticks and the inconvenience to the patient and healthcare provider burdens both groups. The clinical setting and other practical concerns may also influence the choice of ESA: areas that lack of refrigeration may desire ESAs that can be stored at room temperature. Infrequent subcutaneous administration of rHuEPO or orally available molecular entities may be useful for home administration. Patients who have frequent visits to the clinic, such as is common for hemodialysis, who already have intravenous access, may prefer frequent administration of intravenous formulations so their hematocrit and drug dose can be monitored frequently. These and other factors are considered in design and development of new molecular entities and formulations.

Approaches to enhancing or modifying a drug's properties have included new formulations and delivery systems of the existing drugs. Pumps or slow-release formulations can be used to maintain the circulating concentration of the drug for extended periods of time (sustained delivery). Modification of the drug by fusion with polymers or other proteins or peptides can be used to extend the half-life (sustained duration of action). Attempts to modify the protein itself by *in vitro* mutagenesis also have been explored. Glycoengineering has been successfully applied to rHuEPO. With this process, new carbohydrate attachment points were introduced into the protein by *in vitro* mutagenesis of the *EPO* gene and followed by expression in mammalian cells. Some of the glycoengineered forms had an increased number of attached carbohydrates, increased *in vivo* activity and serum half-life. The new molecule may be administered with extended dosing intervals with no loss of efficacy.

Some limitations of rHuEPO due to its peptidic composition can potentially be bypassed with new molecular entities such as small molecules or antibodies (EPO mimetics). These compounds may have advantageous properties not present in rHuEPO, such as an oral delivery (small molecule) or structural conformations that hold little in common with EPO, thus exploiting different routes of absorption or elimination. Finally, stimulation of erythropoiesis by mechanisms different from those of rHuEPO, such as those steps upstream or downstream from the receptor activation step have been explored. This chapter discusses new ESAs, their formulations and delivery systems being considered or developed and their limitations, if any.

Erythropoiesis-stimulating agents

EPO molecules with altered activity

One approach to increase activity of EPO is to alter the interaction with the EPO receptor (EPOR). EPO activates erythroid precursor cells by binding and activating EPOR on the surface of erythroid progenitor cells [15]. Receptor activation occurs as a result of a conformational change in the EPOR triggered by EPO binding [16, 17]. The conformational change is induced when the two binding sites on a single EPO molecule crosslink 2 EPOR [18–21]. The two binding sites on rHuEPO have different affinities: high (approximately 1 nM) and low (approximately 10 μM) [22, 23]. Initially, one receptor binds to the high-affinity binding site and is followed by binding of a second receptor to the low-affinity binding site.

Molecules that have increased affinity at either of the two sites can have increased *in vitro* potency. Yet despite this property, no ESAs suitable for clinical development have been reported that have increased biologic activity *in vivo* as a function of increased receptor affinity. The reasons are several-fold. First is the theoretical concern that antibodies, were they to develop against such a modified ESA, could negatively affect patient outcome. Although there are numerous mechanisms to prevent the generation of antibodies against endogenous proteins, administration of a slightly modified form of an endogenous protein may trigger antibody generation. In general, these antibodies recognize the regions of the protein that have been modified. If an ESA was modified in the region that contacts the EPOR (the EPO:EPOR interface), it is likely that antibodies developed against this region would disrupt the EPO:EPOR interaction, thereby inhibiting ESA-stimulated erythropoiesis. If in addition to binding the modified ESA, the antibodies also neutralized endogenous EPO, PRCA would likely result.

A second reason ESAs with increased EPOR affinity haven't been developed is that increased affinity does not always translate into increased *in vivo* potency [24]. Recombinant HuEPO binding to EPOR can trigger rapid internalization of the EPO:EPOR complex with subsequent degradation [25, 26].

Because erythropoiesis requires continuous stimulation of multiple EPOR through multiple binding events, serum half-life is a more important determinant of *in vivo* potency than is receptor binding activity. The EpoR-signaling pathway is down modulated shortly after activation [26, 27]. As a consequence, molecules cleared quickly have low *in vivo* activity [24], which has been observed with carbohydrate-modified EPO analogs that demonstrate increased EPOR binding *in vitro* and decreased *in vivo* efficacy [28–30]. Increased dose can partially compensate for the increased clearance; however, these compounds may need be administered more frequently.

In contrast to molecules with increased affinity for EPOR, some ESAs approved for clinical use have reduced receptor affinity, e.g., darbepoetin alfa [24, 31] (Tab. 1) and pegylated epoetin beta [32]. Usually reduced EPOR binding activity results in both decreased *in vitro* and *in vivo* activity; however, some molecules that demonstrate reduced receptor-binding activity can have increased *in vivo* potency because the increased serum half-lives of these ESAs can more than compensate for reduced receptor binding activity [24, 31]. This activity is highlighted by results shown in Table 1 where rHuEPO analogs with mutations in the receptor-binding sites (either the high-affinity site, site 1; or the low-affinity site, site 2) had reduced EPOR-binding activity that correlated with both reduced *in vitro* and *in vivo* activity. However, darbepoetin alfa had increased *in vivo* activity despite its reduced receptor binding activity because of its increased serum half-life.

Table 1. Activity of rHuEpo analogs

Analog[a]	Description[b]	Receptor binding[c] % of rHuEpo	*In vitro* activity[d] % of rHuEpo	*In vivo* activity[e] % of rHuEpo
rHuEPO	Control	100	100	100
darbepoetin alfa	Increased carbo hydrate content	20	11	171
NM311(R14/Q)	Site 2 mutation, Same carbohydrate content	10	2.0	5.9
NM335 (N147/K)	Site 1 mutation, Same carbohydrate content	0.8	0.5	1.7

Activity was normalized to that of rHuEPO, arbitrarily set at 100.

[a] Mutations were introduced into the coding region of an rHuEPO cDNA by *in vitro* mutagenesis. rHuEPO and analogs were expressed in CHO cells and purified as described previously [35]. Aranesp has two additional *N*-linked carbohydrate chains compared with rHuEPO.

[b] Site 1 mutation: high-affinity-binding site, site 2 mutation: low-affinity-binding site. Carbohydrate content of NM311 and NM335 similar to that of rHuEPO.

[c] Receptor binding activity determined using competitive binding to OCIM1 cells [20].

[d] *In vitro* activity measured using EPO dependent 32D cells according to ref 20.

[e] *In vivo* activity determined using a mouse exhypoxic 59Fe uptake assay [24].

While ESAs are typically chosen because they have high-receptor binding activity, there are conditions where certain analogs with low EPOR-binding activity may have clinical utility, such as in patients with polycythemia. Compounds that bind through the high-affinity site but do not dimerize and activate EPOR because of reduced binding at the low-affinity site can function as antagonists [23]. These molecules may bind to EPOR, are inactive themselves, but inhibit binding of endogenous EPO. Such molecules may be monovalent Fab (fragments of antibodies, small molecules with only one binding site) or rHuEPO analogs where the low-affinity binding site is removed [33].

Other changes in rHuEPO that alter physical properties have been considered. Changes in rHuEPO amino-acid sequence can result in increased stability. These changes can include removal of amino acids that are unstable (e.g., Trp) [34]; or are subject to oxidation (e.g., Met), deamidation (e.g., Asn), or changes that confer increased conformational stability, such as those that stabilize alpha helices or connecting loops. Such molecules may be more amenable to long-term storage or suitable for formulations where more stable EPO molecules may be useful, such as in slow-release formulations or automated delivery systems. Removal of proteolytic cleavage sites by *in vitro* mutagenesis can enhance *in vivo* stability. It may be possible to remove antigenic sites, thereby reducing immunogenicity.

Molecules with increased serum half-life

A longer duration of action can allow for reduced frequency of administration and increased *in vivo* potency. One approach that has been successfully applied to rHuEPO is glycoengineering [35]. Glycoengineered molecules bind and activate the EPOR in the same manner as rHuEPO, resulting in similar biologic responses while at the same time reducing clearance and enhancing *in vivo* activity [31]. Other strategies to increase duration of action of EPO included chemical modifications, such as the addition of a polyethylene glycol molecule (PEGylation) or gene fusions between EPO and other proteins. In these cases, the goal is to reduce clearance rate by increasing hydrodynamic size.

Glycoengineering

Recombinant human EPO is a glycoprotein hormone consisting of approximately 40% carbohydrate [36]. The carbohydrate component consists of three *N*-linked carbohydrates attached to Asp at amino-acid positions 24, 38, and 83; and an *O*-linked carbohydrate attached to Ser at amino acid position 126 [36] (Fig. 2). Unlike the invariant protein sequence, the carbohydrate is variable in structure, resulting in glycoforms with modest differences in sizes, structures, and sugar content [37, 38]. A typical *N*-linked carbohydrate made by mammalian cells is branched with 2–4 arms. The end of each arm is typically capped

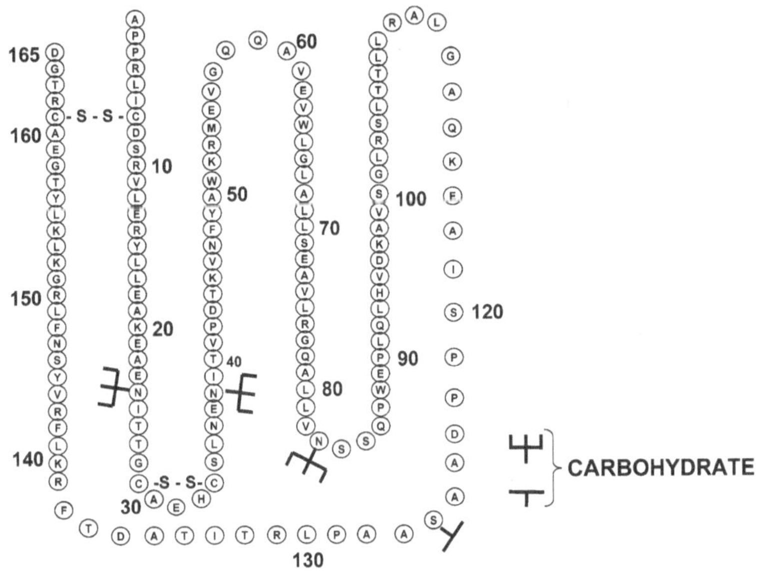

Figure 2. Amino Acid Sequence of Human EPO. Recombinant human erythropoietin (rHuEPO) is 165 amino acids in length. Disulfide bonds (-S–S-) join Cys 7 to Cys 161 and Cys 29 to Cys 33. The 3 *N*-linked glycosylation attachment points are at Asn 24, Asn 38, and Asn 83 and the *O*-linked carbohydrate is attached to Ser 126. Forked structures depict the attached carbohydrates.

by a sialic acid; sialic acid content exhibits microheterogeneity in the different glycoforms. The sialic acid is of importance because it is the only negatively charged sugar on the carbohydrate. Variations in the amount of sialic acid can affect the electrostatic properties of the molecules to which it is attached.

The carbohydrate is essential for *in vivo* but not for *in vitro* biologic activity [24, 28, 29]. The sialic acid component of carbohydrate, in particular, plays a critical role in the *in vivo* biologic activity of rHuEPO. Removal of sialic acid from the carbohydrate of EPO results in increased *in vitro* activity but almost complete loss of *in vivo* activity [24, 29]. Studies on glycoforms of rHuEPO containing different sialic acid contents demonstrated a direct relationship between increased sialic acid content and increased *in vivo* activity [39]. The increased *in vivo* activity was due to an increased serum half-life of the molecule and not increased affinity for the EPOR. The theoretical maximum number of sialic acids on rHuEPO is 14 (up to four sialic acids for each of the three *N*-linked carbohydrates and up to two sialic acids for the *O*-linked carbohydrate) [38]. It was hypothesized that *in vivo* activity may be increased beyond that observed with rHuEPO by adding new sialic acid containing *N*-linked carbohydrates. Each new *N*-linked chain could add up to four additional sialic acids.

To add more *N*-linked carbohydrate, *N*-linked glycosylation sites were introduced into the amino acid sequence of EPO. *N*-linked carbohydrate is attached to an Asn present in the consensus sequence Asn-Xxx-Ser/Thr (where Xxx can be any amino acid except proline) [40]. This sequence is necessary but not sufficient for *N*-linked carbohydrate addition [41]. During synthesis of a glycoprotein, appropriate consensus sequences are recognized by oligotransferases in the cell, resulting in attachment of carbohydrate. This carbohydrate is subsequently modified by the action of additional intracellular enzymes. The glycosylated protein is then secreted from the cell into the circulation [42].

For the purpose of potentially developing a new drug with properties superior to then available products, it became apparent that simply adding an *N*-linked consensus sequence to rHuEPO would not be sufficient. The changes needed to be introduced in such a way that the resultant molecule was efficiently glycosylated and retained activity, conformation, and stability. To increase the likelihood of success, the amino-acid changes were introduced into regions of the molecule distal to the receptor-binding site to ensure that the molecule would efficiently bind to and activate the EPOR. This effort was aided by structure-function studies that defined the active sites of rHuEPO and determination of amino acids important for maintenance of structure [20, 43, 44].

EPO glycosylation analogs with introduced *N*-linked glycosylation consensus sequences were constructed and tested [35]. Several candidates containing additional carbohydrate had acceptable activity and conformation characteristics. The consensus sequences successfully introduced into two of these candidates (Asn30Thr32Val87; and Asn88Thr90) were combined to generate a new molecule with two additional *N*-linked carbohydrates. This molecule (darbepoetin alfa, Aranesp) had near-normal *in vitro* activity, was glycosylated efficiently, and had a similar conformation and stability to rHuEPO. The carbohydrate content was increased from 40% to 51%, the size from approximately 30,400 Da to approximately 37,100 Da, and the maximum number of sialic acids was increased from 14 to 22.

Studies in mice with darbepoetin alfa revealed that more rHuEPO was required to obtain a response similar to that of darbepoetin alfa [31, 35, 39]. In nonclinical studies, animals were administered rHuEPO or darbepoetin alfa at various dose intervals. The relative *in vivo* activity difference of rHuEPO and darbepoetin alfa increased as the dosing interval increased [45]. Three-fold more rHuEPO than darbepoetin alfa was required to elicit a similar response when the drugs were administered three times per week. This difference increased to 13-fold when the molecules were administered at weekly intervals. The increased *in vivo* activity and the differing potencies with changes in dose interval could be explained by an observed three-fold increase in serum half-life of darbepoetin alfa over rHuEPO [24, 31]. The observation that the serum half-life increased in proportion to the number of added carbohydrate chains indicated that the carbohydrate directly affected clearance.

Testing of darbepoetin alfa in humans mirrored the results found in animals. Compared to epoetin alfa, the serum half-life of darbepoetin administered intravenously was increased approximately three-fold [46]. In clinical trials, patients were converted from rHuEPO administered two to three times per week to darbepoetin alfa administered once per week or from weekly rHuEPO to once-every-other-week darbepoetin alfa. The hemoglobin concentrations were successfully maintained with the less frequent dosing schedule [47, 48]. Additional clinical results demonstrated that hemoglobin concentrations can be successfully maintained when darbepoetin alfa is administered at once every 3 to 4 week dosing intervals [49–51].

One theoretical concern with any alteration in a protein's amino-acid sequence or structure is immunogenicity. Several characteristics of darbepoetin alfa and its methods of manufacture minimized the potential for antibody formation. The particular amino-acid substitutions in darbepoetin alfa had a minimal effect on structure and stability. The carbohydrate and sialic acid content of the material selected during the purification process was maximized for several reasons: first, the higher carbohydrate content enhances the *in vivo* activity; secondly, carbohydrate can increase solubility and stability of proteins thereby inhibiting formation of aggregates and other byproducts that may trigger antibody formation [52–54]. Finally, *N*-linked carbohydrate is large relative to the peptide backbone giving the carbohydrate a 'shielding' effect potentially inhibiting antibody formation. Antibody formation was monitored during clinical trials with darbepoetin alfa, and there was no enhanced formation of neutralizing antibodies compared to that of epoetin alfa made in the United States [14].

It was found that carbohydrate content could be increased even further than that contained on darbepoetin alfa by introducing additional *N*-linked glycosylation sites. AMG205, a rHuEPO glycosylation analog with four additional carbohydrates compared with epoetin alfa, had low 9G8a immunoreactivity, indicating a similar conformation to epoetin alfa [35]. 9G8a is a monoclonal antibody whose binding was increased to rHuEPO analogs with altered conformation or stability and assays with this antibody were used to assess conformation of rHuEPO analogs.

AMG205 *in vivo* potency and serum half-life were increased compared with darbepoetin alfa. In rats, AMG205 had nearly twice the half-life of darbepoetin alfa (intravenous, 31 *versus* 17.2 h; subcutaneous, 28.2 *versus* 13.2 h), respectively [55]. However, AMG205 had an *O*-linked glycosylation deficit. Another candidate, AMG114, exhibited increased efficiency of *O*-linked carbohydrate addition. It also had low 9G8a immunoreactivity. The serum half-life in rats was similar to that of AMG205 (16.7 h). In dogs, the serum half life of AMG114 administered intravenously was 68.9 h compared with 20.4 h for darbepoetin alfa.

Clinical testing in subjects with chemotherapy-induced anemia administered AMG114 revealed an approximately two-fold increase in serum half-life compared with darbepoetin alfa (131 *versus* 73.7 h) and a dose-dependent increase in hemoglobin concentrations when administered at every-3-weeks dosing intervals [56]. These results demonstrate that serum half-life could be increased in humans even further than that observed with darbepoetin alfa and that such molecules could maintain hemoglobin concentrations in humans.

The carbohydrates on darbepoetin alfa, AMG114 (and AMG205) are positioned distant from the recepotor binding sites (Fig. 3). The new chains are opposite to the two chains found on darbepoetin alfa. The areas of the molecule changed are distal to the region of rHuEPO that is involved in the interaction with the EPOR explaining the retention of *in vitro* activity of both darbepoetin alfa and AMG114.

AMG114 has not been developed for clinical use primarily because existing molecules including epoetin alfa and darbepoetin alfa can meet clinical needs.

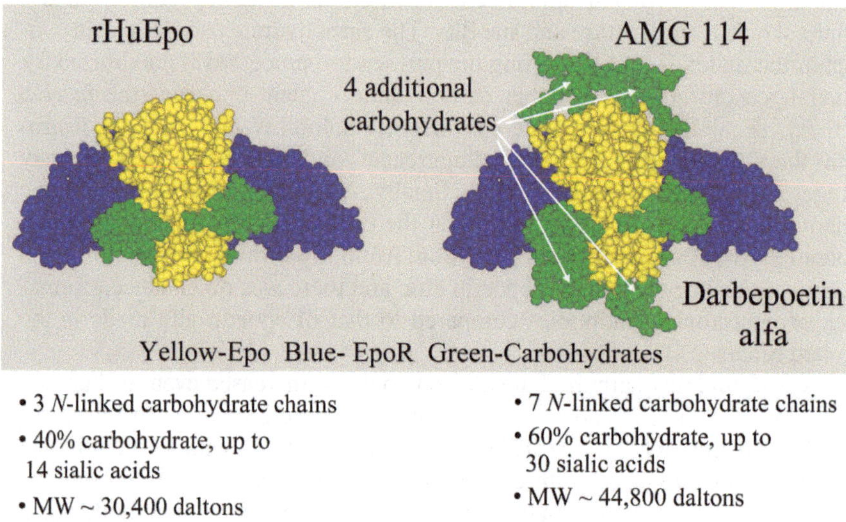

Figure 3. Darbepoetin alfa has two and AMG 114 has four additional carbohydrates distal to the receptor binding regions. A three-dimensional model of rHuEpo (yellow) was based on the crystal structure of EPO bound to the extracellular domains of 2 EPOR (blue)[18]. *N*-linked carbohydrates addition sites are green. The AMG114 glycosylation sites are at the end opposite the from the naturally occurring and added glycosylation sites in darbepoetin alfa. All the new *N*-linked carbohydrates in AMG114, including those in darbepoetin alfa, are distal from the EPOR- binding sites.

In addition, there was concern that molecules with very long half-lives might be more difficult to use clinically because of a potential increase in swings in hemoglobin concentrations.

PEGylation

Mixing activated polyethylene polymers with proteins under appropriate chemical reaction conditions produces PEGylated proteins. PEGylation involves chemical attachment of the polymer, polyethylene glycol (PEG), to reactive regions of proteins or carbohydrates. The attachment of PEG is thought to improve solubility and possibly reduce immunogenicity due to shielding by the conjugate. PEGylated molecules can have an increased hydrodynamic size because they create a 'water shell' around the molecule. In addition, the increased hydrodynamic size can result in reduced clearance and thus increased *in vivo* activity. Clearance studies with polyethylene glycol polymers showed a direct relationship between size of PEG and its rate of clearance. PEGylation has been used successfully to increase serum half life of many classes of proteins [57]. The attachment of PEG to rHuEPO would increase its serum half-life in proportion to the increased hydrodynamic size of the PEGylated ESA [58]. Polyethylene glycol (PEG) is thought to be relatively inert and nonimmunogenic by itself so it is a suitable starting material for protein conjugate therapeutics.

One issue with drugs made by solution or solid-phase chemistry can be poor specificity of conjugation in the chemical reaction or generation of undesirable byproducts. Many PEGylation chemistries have been tried in an attempt to reduce undesirable byproducts, improve the specificity and efficacy of polyethylene glycol attachments, and minimize immunogenicity risk of the protein conjugate while maximizing the *in vitro* and *in vivo* activity of the resultant molecule [59]. The current chemistries typically target the reactive amino groups on Lys or the amino terminal amine. rHuEPO has 8 Lys, some of which are part of the active site [18, 20]. Therefore, some PEGylated EPO molecules contained PEG conjugated to Lys may have low activity because the PEG interferes with receptor binding and activation [32]. Other PEGylated EPO molecules may have low activity because the attached polymer introduces structural alterations that interfere with receptor binding.

Amino acids can be introduced to increase the specificity of PEG attachment. For example, PEG moieties can be specifically conjugated to the Cys sulfhydryl group. Therefore Cys substitutions at specific regions in the protein may allow targeted PEGylation [60–62]. Another strategy is to make PEGylated EPO synthetically: during synthesis, a PEG-conjugated amino acid could be introduced instead of the unconjugated amino acid. This approach allows specific targeting of particular amino acid positions for PEG attachment, such as the glycosylation sites, and reduces heterogeneity and the potential for loss of *in vitro* activity [63, 64]. It is not clear that these molecules will retain the same stability, *in vivo* activity, and lack of immunogenicity as their glycosylated counterparts, however.

Fusion proteins

Several EPO gene fusion proteins have been reported, including EPO/interleukin (IL-3) [65] and EPO/granulocyte-macrophage colony-stimulating factor (GM-CSF) [66]. EPO dimer has also been generated as a potential therapeutic [67–69]. EPO dimer may have increased *in vitro* activity due to altered avidity to the receptor. However, the clinical relevance of such molecules is unclear. The two protein partners are typically joined by a linker peptide that includes Gly, Ala, and Ser. These three amino acids are thought to result in linkers that are flexible and relatively inert, allowing independent folding of the two proteins into their appropriate three-dimensional structures.

The promise of such molecules is that they might impart to the fusion protein biologic properties of both molecules. One can imagine that co-administration of a growth factor such as IL-3 that stimulates the early stages of erythropoiesis with rHuEPO can increase efficacy of rHuEPO. Because the fusion protein is larger and may have increased serum half-life, it may impart increased potency for both partners. The fusion protein also ensures that both molecular entities are simultaneously present. Simultaneous administration by fusing two drugs can simplify administration, especially when the two proteins have different pharmacokinetic parameters. The ability to independently adjust dosing of the fusion partners is lost, however. Full *in vitro* activity of both proteins in fusion proteins does not always occur [65]. Furthermore, the difficulty in retaining a nonimmunogenic structure has been a challenge [70] and no ESAs with protein fusions have been approved for use in humans.

EPO mimetics

rHuEPO is currently administered by either subcutaneous or intravenous injection. Because of its large size and peptidic nature, delivery by other routes such as oral, transdermal, or by inhalation can be quite challenging. One possible solution is a small molecule, an EPO mimetic, capable of stimulating the EPOR. EPO mimetics are compounds that mimic the activity of EPO but bear no structural homology. EPO mimetics can have new biologic or biophysical properties not present in EPO. Designed appropriately, such a compound has the potential to be delivered by routes that may be more convenient than those currently in use for rHuEPO, e.g. oral.

Significant challenges are associated with the identification and development of a useful small molecule EPO mimetic. First, the mimetic needs to be small to be delivered orally, yet it needs to be large enough to have sufficient affinity for the EPOR and to ensure it persists long enough in the serum to be efficacious. Finally, the compound should not have unwanted side effects due to either toxicities associated with off-target effects of the compound or from its breakdown products. In spite of these challenges, work has proceeded and progress has been made in attempt to identify compounds that may be amenable to oral delivery.

Several strategies have been used to identify EPO mimetics. The first is to screen peptide and small molecule libraries for those that can stimulate erythropoiesis using *in vitro* bioassays as screens. According to this strategy, the target of the drug may not be known, and compounds active in the assay may stimulate erythropoiesis downstream from the EPOR. Another strategy is to identify molecules that directly bind to and agonize the EPOR in a manner similar to that of rHuEPO. The latter strategy may be performed in two steps:

the first step is to identify compounds that bind the EPOR and the second step is to covalently link the compounds into bivalent dimers that can agonize the receptor by EPOR dimerization.

The latter strategy takes advantage of the observation that an EPOR mutant containing an Arg129 to Cys129 mutation was constitutively active [21, 71]. A disulfide bond formed between the Cys129 residues on the receptors resulted in dimerization and receptor activation demonstrating one instance where EPO was not essential for receptor activation. X-ray crystallography results demonstrated that the EPOR forms a 2:1 complex with EPO [18]. Each receptor uses the same region on its surface to bind to two surfaces on EPO, resulting in receptor dimerization. Further confirmation of the dimerization mechanism was the discovery of agonist monoclonal IgG antibodies that could dimerize and activate EPOR [19]. The anti-EPOR antibodies activated EPOR because they were bivalent, had two binding sites, and could simultaneously bind and cross-link 2 EPOR (Fig. 4). Monovalent Fab fragments could bind but did not agonize the receptor, confirming a requirement for bivalent binding.

Peptide phage libraries were screened and one peptide was identified that could bind and agonize EPOR [72]. This peptide, AF11154, had no homology

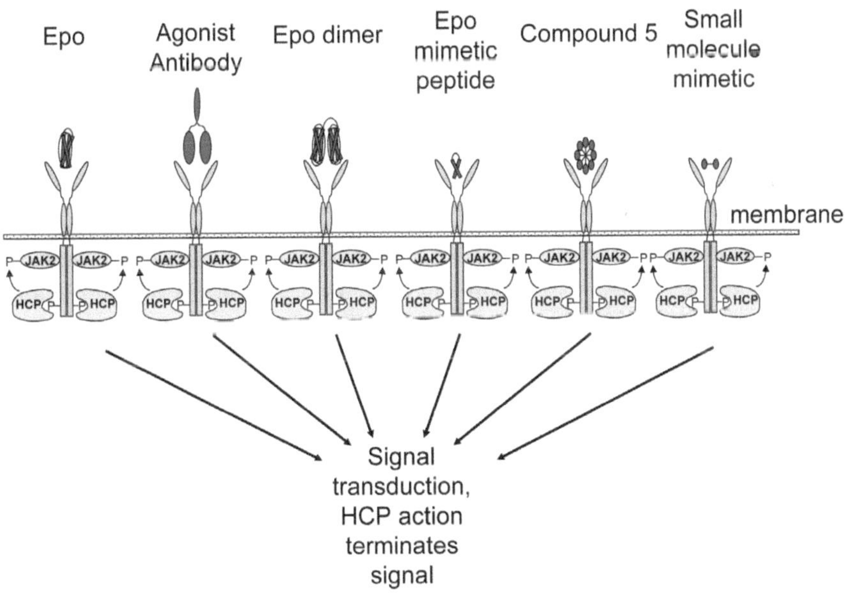

Figure 4. Mechanisms of EPOR Activation. EPOR are activated and when it is dimerized by binding of the two asymmetric receptor-binding sites by rHuEPO. EPO binding results in phosphorylation of EPOR, JAK2 recruitment, and phosphorylation of JAK2. The activation of JAK2 results in downstream signaling events. HCP can bind the activated (phosphorylated) receptor resulting in dephosphorylation of JAK2, thereby terminating signal transduction. The EPO mimetic compounds; agonist antibody [19], EPO dimer [67, 69], EPO mimetic peptide [72], compound 5 [81], and small molecule mimetics [79, 81] can all dimerize and activate the EPOR in a manner similar to that of rHuEPO.

to EPO. It self associated into dimers to form a bivalent molecule that could dimerize EPOR in a 2:2 mimetic:EPOR complex [73]. Additional sequence modification of this peptide resulted in EMP1, a 20-amino acid peptide with an approximate 50-fold increase in affinity over the starting peptide (Tab. 2). The affinity was increased more by covalent linkage of the 2 peptides [74]. The activity of this peptide was still significantly lower (500-fold) than that of rHuEPO when tested *in vitro*. In addition, the *in vivo* activity was very low (25,000-fold less than rHuEPO). However, this work demonstrated that a molecule smaller than rHuEPO could successfully dimerize and activate the receptor. An attempt to discover other EPO agonist peptides by another group was also successful [75]. However, the *in vitro* potency was not increased (Tab. 2).

One explanation for the low *in vivo* activity of mimetic peptides is likely their rapid clearance. One group addressed this problem by creating a fusion protein between EMP1 and a larger protein, plasminogen activator inhibitor (PAI1) [76], resulting in an increase in molecular weight from 4,800 Da to 66,000 Da (Tab. 2). The *in vivo* activity was significantly increased (2500-fold); its *in vivo* activity, however, was still significantly less than that of rHuEPO (100-fold) and the ability to be delivered orally was compromised by the size increase.

In another approach, additional screening of peptide mimetics of EPO produced a molecule with increased stability and affinity for EPOR compared with EMP-1. PEGylation of this molecule reduced its clearance and it is currently in clinical development (Hematide; Affymax Inc Palo Alto, CA) [77, 78].

The peptide mimetics described are significantly larger (approximately 4,200 Da) than the preferred size of an orally bioavailable compound (<600 Da). To overcome this issue, the peptides were used in an attempt to design lead compounds of smaller size. Some nonpeptide small molecule agonists have been isolated based on the EMP1 structure [79]; however, their *in vitro* activities were low (Tab. 2). An independent approach was to directly screen for small molecule EPO mimetics that could dimerize EPOR. Small molecule libraries containing compounds with two-fold symmetry were screened to find dimerizing compounds that agonize the receptor [80, 81]. This strategy did not result in discovery of agonist compounds, however. A small molecule (compound 1, approximately 500 Da) that bound but did not agonize EPOR was discovered. Compound 1 was made active *in vitro* by oligomerizing it with a multivalent crosslinker resulting in a molecule (compound 5) containing eight compound-1 molecules joined together. Compound 5 binding to EPOR was increased somewhat (10-fold) over that of compound 1; however, the size (6,400 Da) was greater than that required to be orally bioavailable. In addition, its *in vitro* activity was low relative to rHuEPO and the compound had toxicity. The feasibility of discovery of small molecules that could agonize EPOR was demonstrated, however to date there are no reports of success in identification of molecules suitable for clinical development.

Table 2. Activity of erythropoietin and mimetics

	MW KDa	*In vitro* activity EC50	Binding IC50	*In vivo* activity Units/mg	Rel difference in activity molar	Notes	Ref
rHuEPO	18,200	5–10 pM	100 pM	>100,000	1	peptide monomer	19, 20, 24, 36
EMP1	4200	400 nM	200 nM	2	25,000	peptide dimer	72,73
EMP dimer	4800	approx 20 nM	approx 2 nM	200	250	peptide dimer	73
EMP-PAI1	63,000	NA	NA	5000	100	peptide dimer	76
ERP	3085	100 nM	approx 1 nM	NA	NA	peptide dimer	75
ERB	2170	approx 3 nM	45 nM	NA	NA	peptide monomer	82
EPO Mab 71	150,000	200 pM	1 nM	NA	NA	peptide dimer	19
Compound 1	500	0	60 μM	NA	NA	nonpeptide monomer	80, 81
Compound 5	6400	1–5 nM	4 μM	NA	NA	nonpeptide dimer	81
A1B10C1	2100	1–10 nM	NA	NA	NA	nonpeptide dimer	79

NA = not available

Stimulation of erythropoiesis through other mechanisms

Difficulties with development of small molecules that activate EPOR by dimerization can be bypassed by targeting a different mechanism or a different step in the erythropoiesis pathway. One report describes a peptide that activates EPOR by binding to a domain on EPOR similar to MHC peptides (Fig. 4). This 23-amino acid peptide (ERP) is reported to have both *in vitro* and *in vivo* EPO activity [82]. The molecule appears to activate by binding EPOR at a region distal to its point of dimerization (transmembrane domain), suggesting that it is activated by binding to a different site than that caused by rHuEPO. The mechanism may be similar to that of the virus envelope protein, gp55 that also activates EPOR by an interaction with the EPOR transmembrane region [83, 84].

Another approach to mimetic discovery is to modulate steps downstream from receptor activation such as inhibition of hematopoietic cell phosphatase (HCP) [85, 86]. HCP is an enzyme that dephosphorylates JAK2, a kinase that is part of the EPOR signal transduction cascade [15]. JAK2 is normally activated (phosphorylated) as a consequence of EPOR activation. HCP binds to activated (phosphorylated) EPOR and then dephosphorylates JAK2, terminating signal transduction. Truncated EPOR lack the HCP binding site, and thus HCP, cannot dephosporylate JAK2, resulting in hypersensitivity of the receptor to EPO. Naturally occurring truncations have been described in humans whereby the affected individuals have increased hemoglobin concentrations but very low EPO concentrations due to a hypersensitive EPOR [87–89]. These observations suggest that small molecule antagonists of HCP may result in increased EPOR activity that increases erythropoiesis in the absence of added EPO.

One concern of HCP inhibitors relates to observations associated with HCP mutations in mice. These mice (motheaten) have a defective *HCP* gene [90] and have multiple hematopoietic abnormalities, including increases in macrophages, lymphocytes, and erythrocytes. HCP is a negative regulator for several different cytokine receptors besides EPOR [85, 86, 91, 92]. Although HCP inhibitors may be effective at increasing erythroid cell number, increases in cell number of other hematopoietic cells may limit the usefulness of these compounds.

Hypoxia mimetics

A more direct approach in stimulation erythropoiesis would be to identify molecules that stimulate production of endogenous EPO. Recently the mechanism by which hypoxia regulates *EPO* gene transcription has been partially elucidated and those pathways may yield novel drugs [93]. One of the several transcription factors that promotes *EPO* gene transcription is hypoxia inducible factors (HIF). The active form includes HIFα and HIFβ subunits of which several forms have been described. The HIFα genes are transcribed constitutively but their levels are dependent on the pO2 of oxygen which affects the rate of metabolism and thus the level of active HIF complex. In normoxia, HIFα is rapidly degraded due to oxygen dependent hydroxylation of the alfa subunit of HIFα by a prolyl hydroxylase (HIF-PH). Hydroxylated HIFα associates with Von Hippel Lindau protein which promotes ligation of the protein with ubiquitin followed by rapid degradation by the proteosome. In the absence of oxygen (or iron), HIF-PH activity is reduced and *EPO* gene transcription increases due to accumulation of HIF.

Genetic alterations in one of the three known HIF-PH encoding genes have been reported to result in polycythemia as does certain mutations in VHL [94–97]. Early observations that cobaltous ions can stimulate erythropoiesis [98, 99] are now explained according to their ability to inhibit HIF-PH. Indeed the original erythropoietic biologic unit (cobalt unit) was defined as the erythropoietic response associated with administration of 1 ml of 5 mM cobalt chloride to a starved rat [100]. Thus, alteration of this pathway can stimulate erythropoiesis and attempts to identify clinical candidates are underway.

Initial attempts to stimulate erythropoiesis by administering cobalt clinically were successful at stimulating erythropoiesis in humans [101–103]; however, toxicities associated with cobalt limited its usefulness [104]. Another molecule that inhibits HIF-PH is desferrioximine (DFO) [99, 105]. It is thought to inhibit by reducing the availability of iron, which is also critical for HIF-PH activity [105]. Its poor pharmacokinetic properties and poor efficacy has prevented clinical use [106].

Despite issues with cobalt and DFO, results with these molecules supported additional work to identify molecules with improved safety/efficacy profiles. Other small molecule antagonists of HIF-PH have been described and these can stimulate erythropoiesis in animals [93, 107]. One molecule FG2216 (Fibrogen, San Francisco) has advanced to Phase I/II clinical trials and dose-dependent increases in hemoglobin have been demonstrated in patients with chronic renal insufficiency. Clinical development was halted when a patient died because of drug-related liver failure, but clinical trials have resumed. Attempts to identify other molecules with better safety/efficacy profiles continue.

Gene therapy

Controlled delivery of *EPO* genes to humans is another promising approach for EPO therapy. Early work in this field depended on direct injection of plasmid DNA containing constitutively active *EPO* genes into the muscles of mice [108], resulting in a measurable increase in hematocrit. Several concerns became apparent from these studies, including inefficient and variable delivery of the *EPO* gene and subsequent variations in EPO concentration. In addition EPO expression decreased over time. Improvements in expression systems and gene delivery methods have been reported [109–112]. Current *EPO* gene therapy protocols require repeated administration of *EPO* genes. In addition, there is concern that the therapy may be irreversible or the vector may integrate into the genome at positions that alter gene expression, resulting in tumorigenicity. Constitutive high level expression of *EPO* genes could also result in life-threatening erythrocytosis [113]. Monkey *EPO* genes introduced into Cynomolgus macaque monkeys with their anterior tibialis muscle cells engineered to express monkey EPO initially showed excessive erythrocytosis [114]. Some animals developed antibody-mediated PRCA through an unknown mechanism, creating concerns about safe delivery of gene-delivered EPO.

The efficiency and irreversibility concerns of *EPO* gene therapy have been addressed by developing implantable capsules containing EPO-expressing cells [115]. The capsules can be removed, halting EPO delivery. A further improvement is to control EPO expression with a small molecule such as tetracycline, enabling increased EPO expression in response to oral administration of the gene activator [116] such as tetracycline [109, 110, 117], mifepristone [118], or rapamycin [119]. Additional advances are the development of vectors where EPO expression is controlled by oxygen tension [116], or to target the kidney for gene transfer [120]. Despite these advances, safe and controlled EPO delivery using gene therapy methods suitable for human use remains a distant but tantalizing opportunity.

Clearance of hyperglycosylated and PEGylated derivatives

Despite several decades of work, the mechanism by which ESAs are cleared is not completely understood. In addition, while clearance of ESAs is altered by hyperglycosylation and PEGylation, the pathways impacted by these modifications are also not completely understood. Clearance of ESAs was first thought to be mediated primarily through liver hepatocytes that express an asialoglycoprotein receptor (ASGR) [30, 121, 122] or through the kidney [10, 123–125]. Clearance through the liver appears to be limited to asialo EPO, not fully glycosylated EPO protein, and evidence of clearance through the liver of ESAs is not supported by other studies [122, 126, 127]. Initial data suggesting a major elimination pathway for rHuEPO through the kidneys were not supported by subsequent studies showing that rHuEPO clearance is independent of kidney function [126, 128, 129]. In healthy men, only a small amount of intact radiolabeled epoetin beta (<5% of the dose) was found to be excreted in the urine, suggesting that rHuEPO is degraded elsewhere in the body [130]. Consequently, the role of the liver and kidney in rHuEPO clearance is thought to be minimal.

Binding of EPO to EPOR can lead to cellular internalization and intracellular metabolism [25, 26, 131–133]. The bone marrow where EPOR expressing cells reside is hypothesized to play a role in clearance through EPOR-mediated binding, endocytosis and intracellular degradation. Only indirect evidence, mostly derived from chemotherapy experiments [134–137], support a role for clearance by this pathway *in vivo*.

To explore the possible role of an EPOR-mediated clearance pathway, an engineered rHuEPO analog (NM385) that was devoid of detectable receptor binding but retained similar structure and carbohydrate content to rHuEPO, was constructed and its clearance properties were examined [138]. After a single intravenous dose of 30 μg/kg in rats or mice, NM385 had a slightly longer terminal half-life but similar clearance compared with rHuEPO, suggesting that ESAs may be cleared to some degree through this pathway. The impact of complete elimination of receptor-binding activity on clearance of NM385 was substantially less than that observed with hyperglycosylated or PEGylated ESAs, suggesting that derivatives with PEG or additional carbohydrate have reduced clearance due to their impact on other pathway(s). In further support of this proposal, PEGylated rHuEPO and PEGylated NM385 both had similar clearance properties, suggesting that EPOR-mediated clearance is minimally impacted by hyperglycosylation or PEGylation. Taken together, these observations indicate that while EPOR-mediated pathways may play some role in clearance of ESAs, nonEPOR-mediated pathways account for the majority of elimination from the body.

Hyperglycosylated and PEGylated ESAs have other biophysical characteristics such as increased hydrodynamic size. Thus reduced clearance of hyperglycosylated or PEGylated derivatives might be explained by stearic factors. One pathway that may be impacted by size is degradation or metabolism in the interstitium, such as by cells involved in the reticuloendothelial scavenging pathway or lymphatic system [139]. ESAs with larger hydrodynamic size may have reduced transport from the blood to the interstitial fluid where degradation takes place. Consistent with this hypothesis, the lymphatic system is believed to play an important role in the reduced bioavailability after subcutaneous administration of proteins [140]. Only small peptides or free ^{125}I and not intact material are detected in tissues after intravenous administration of ^{125}I-darbepoetin alfa, suggesting that degradation may occur in tissue [138]. Additional work is required to test this hypothesis.

New formulations and devices

Endogenous EPO level is exquisitely controlled in the body by rapid changes in expression. In contrast, protein therapeutics may be stored in a nonphysiologic environment for extended periods of time, which in some instances may be years. Safe storage in any formulation requires that conditions and formulations be designed to minimize formation and accumulation of unnatural breakdown products or alterations in EPO structure. Inappropriate formulations that do not maintain the integrity of the product can risk exposing the patient to an abnormal form of the protein.

Formulations containing rHuEPO have been successfully developed and used safely and effectively for more than a decade. Despite the success of current formulations, change is sometimes required to keep up with regulatory or safety concerns or to allow for new technologies, such as new devices or delivery systems. Such manufacturing and formulation changes included removal of excipients, such as human serum albumin or bovine-derived products.

Prolonged stimulation of erythropoiesis is one desirable property that may be addressed by new delivery systems, including devices that allow controlled release of rHuEPO over long time periods. This approach necessitates that the molecule remain stable in the device for prolonged periods of time. Another approach is to introduce rHuEPO into a biodegradable matrix that degrades slowly over time (slow release) [141, 142]. This strategy requires development of methods to immobilize rHuEPO in a matrix, such as microparticles, that breakdown at predictable rates and release the product in a controlled manner.

The use of an appropriate slow-release process has not been successful for several reasons. The main reason is a requirement that the protein remain intact and unchanged during both the processing of the material and during the prolonged exposure in the body. Protein integrity is a particularly difficult requirement in biodegradable matrices because the protein is in a concentrated hydrated state at physiologic temperatures for extended periods of time. Small amounts of contaminating rHuEPO aggregates, misfolded rHuEPO, or breakdown products may compromise not only efficacy but also safety (e.g., may introduce immunogenicity). Another concern is that rapid breakdown of the matrix may result in excessive delivery resulting in overdose.

Conclusions

More than two decades have passed from the heady days when the *EPO* gene was cloned and rHuEPO was first administered to a patient. Recombinant HuEPO has proven to be a breakthrough as EPO replacement therapy, setting the bar high for any improvements that may follow. Nonetheless, there remains a desire for better erythropoietic molecules, new formulations, or more useful delivery systems. The discovery, development and regulatory approval of darbepoetin alfa (Aranesp) shows that it is possible to discover and develop newer molecules that can be clinically approved to provide different dosing options and improve patient convenience, e.g., by providing flexible options for dosing frequency. Further progress is anticipated as new devices that can simplify the administration of these drugs are developed. Additional changes in rHuEPO or in rHuEPO formulations are anticipated that may need to be developed to allow more effective use of these delivery devices. The future will be exciting as small-molecule orally active EPO mimetics are discovered; though matching the safety profile of rHuEPO presents a substantial hurdle to any small molecule program. Permanent correction of anemia may occur through gene therapy, thereby allowing additional treatment opportunities. These developments may require extensive research and testing; however, many believe that these developments are not a question of if, but when.

References

1. Winearls CG, Forman E, Wiffen P et al. Recombinant human erythropoietin treatment in patients on maintenance home haemodialysis. *Lancet* 1989;2:569.
2. Eschbach JW, Egrie JC, Downing MR et al. Correction of the anemia of end-stage renal disease with recombinant human erythropoietin. Results of a combined phase I and II clinical trial. *N Engl J Med* 1987;316:73–78.
3. Eschbach JW, Abdulhadi MH, Browne JK et al. Recombinant human erythropoietin in anemic patients with end-stage renal disease. Results of a phase III multicenter clinical trial. *Ann Intern Med* 1989;111:992–1000.
4. Eschbach JW, Kelly MR, Haley NR et al. Treatment of the anemia of progressive renal failure with recombinant human erythropoietin. *N Engl J Med* 1989;321:158–163.
5. Delano BG. Improvements in quality of life following treatment with r-HuEPO in anemic hemodialysis patients. *Am J Kidney Dis* 1989;14(2 Suppl 1):14–18.
6. Moreno F, Aracil FJ, Perez R et al. Controlled study on the improvement of quality of life in elderly hemodialysis patients after correcting end-stage renal disease-related anemia with erythropoietin. *Am J Kidney Dis* 1996;27:548–556.
7. Guthrie M, Cardenas D, Eschbach JW et al. Effects of erythropoietin on strength and functional status of patients on hemodialysis. *Clin Nephrol* 1993;39:97–102.
8. Combe C, Tredree RL, Schellekens H. Biosimilar epoetins: An analysis based on recently implemented European Medicines Evaluation Agency guidelines on comparability of biopharmaceutical proteins. *Pharmacotherapy* 2005;25:954–62.
9. Cleland JL, Powell MF, Shire SJ. The development of stable protein formulations: a close look at protein aggregation, deamidation, and oxidation. *Crit Rev Ther Drug Carrier Syst* 1993; 10:307–377.
10. Lasne F, Martin L, Crepin N et al. Detection of isoelectric profiles of erythropoietin in urine: Differentiation of natural and administered recombinant hormones. *Anal Biochem* 2002;311:119–126.
11. Storring PL, Tiplady RJ, Gaines Das RE et al. Epoetin alfa and beta differ in their erythropoietin isoform compositions and biological properties. *Br J Haematol* 1998;100:79–89.
12. Martin KJ. The first human cell line-derived erythropoietin, epoetin-delta (Dynepo), in the management of anemia in patients with chronic kidney disease. *Clin Nephrol* 2007;68:26–31.
13. Casadevall N, Nataf J, Viron B et al. Pure red-cell aplasia and antierythropoietin antibodies in patients treated with recombinant erythropoietin. *N Engl J Med* 2002;346:469–475.
14. Smalling R, Foote M, Molineux G et al. Drug-induced and antibody-mediated pure red cell aplasia: A review of literature and current knowledge. *Biotechnol Ann Rev* 2004;10:237–249.
15. Youssoufian H, Longmore G, Neumann D et al. Structure, function, and activation of the erythropoietin receptor. *Blood* 1993;81:2223–2236.
16. Constantinescu SN, Keren T, Socolovsky M et al. Ligand-independent oligomerization of cell-surface erythropoietin receptor is mediated by the transmembrane domain. *Proc Natl Acad Sci USA* 2001;98:4379–4384.
17. Seubert N, Royer Y, Staerk J et al. Active and inactive orientations of the transmembrane and cytosolic domains of the erythropoietin receptor dimer. *Molecular Cell* 2003;12:1239–1250.
18. Syed RS, Reid SW, Li C et al. Efficiency of signalling through cytokine receptors depends critically on receptor orientation. *Nature* 1998;395:511–516.
19. Elliott S, Lorenzini T, Yanagihara D et al. Activation of the erythropoietin (EPO) receptor by bivalent anti-EPO receptor antibodies. *J Biol Chem* 1996;271:24691–24697.
20. Elliott S, Lorenzini T, Chang D et al. Mapping of the active site of recombinant human erythropoietin. *Blood* 1997;89:493–502.
21. Watowich SS. Activation of erythropoietin signaling by receptor dimerization. Internl J. *Biochem Cell Biol* 1999;31:1075–1088.

22. Philo JS, Aoki KH, Arakawa T et al. Dimerization of the extracellular domain of the erythropoietin (EPO) receptor by EPO: one high-affinity and one low-affinity interaction. *Biochemistry* 1996;35:1681–1691.
23. Matthews DJ, Topping RS, Cass RT et al. A sequential dimerization mechanism for erythropoietin receptor activation. *Proc Natl Acad Sci USA* 1996;93:9471–9476.
24. Elliott S, Egrie J, Browne J et al. Control of rHuEPO biological activity: the role of carbohydrate. *Exp Hematol* 2004;32:1146–1155.
25. Sawada K, Krantz SB, Sawyer ST et al. Quantitation of specific binding of erythropoietin to human erythroid colony-forming cells. *J Cell Physiol* 1988;137:337–345.
26. Sawyer ST, Krantz SB, Goldwasser E. Binding and receptor-mediated endocytosis of erythropoietin in Friend virus-infected erythroid cells. *J Biol Chem* 1987;262:5554–5562.
27. Sawyer ST. Introduction: the erythropoietin receptor and signal transduction. *Ann NY Acad Sci* 1994;718:185–190.
28. Delorme E, Lorenzini T, Giffin J et al. Role of glycosylation on the secretion and biological activity of erythropoietin. *Biochemistry* 1992;31:9871–9876.
29. Takeuchi M, Takasaki S, Shimada M et al. Role of sugar chains in the *in vitro* biological activity of human erythropoietin produced in recombinant Chinese hamster ovary cells. *J Biol Chem* 1990;265:12127–12130.
30. Fukuda MN, Sasaki H, Lopez L et al. Survival of recombinant erythropoietin in the circulation: the role of carbohydrates. *Blood* 1989;73:84–89.
31. Egrie JC, Dwyer E, Browne JK et al. Darbepoetin alfa has a longer circulating half-life and greater *in vivo* potency than recombinant human erythropoietin. *Exp Hematol* 2003;31:290–299.
32. Jarsch M, Brandt M, Lanzendorfer M et al. Comparative erythropoietin receptor binding kinetics of C.E.R.A. and epoetin-beta determined by surface plasmon resonance and competition binding assay. *Pharmacology* 2008;81:63–69.
33. Schneider H, Chaovapong W, Matthews DJ et al. Homodimerization of erythropoietin receptor by a bivalent monoclonal antibody triggers cell proliferation and differentiation of erythroid precursors. *Blood* 1997;89:473–482.
34. Uchida E, Morimoto K, Kawasaki N et al. Effect of active oxygen radicals on protein and carbohydrate moieties of recombinant human erythropoietin. *Free Radical Res* 1997;27:311–323.
35. Elliott S, Lorenzini T, Asher S et al. Enhancement of therapeutic protein *in vivo* activities through glycoengineering. *Nat Biotechnol* 2003;21:414–421.
36. Lai PH, Everett R, Wang FF et al. Structural characterization of human erythropoietin. *J Biol Chem* 1986;261:3116–3121.
37. Sasaki H, Bothner B, Dell A et al. Carbohydrate structure of erythropoietin expressed in Chinese hamster ovary cells by a human erythropoietin cDNA. *J Biol Chem* 1987;262:12059–12076.
38. Rush RS, Derby PL, Smith DM et al. Microheterogeneity of erythropoietin carbohydrate structure. *Anal Chem* 1995;67:1442–1452.
39. Egrie JC, Browne JK. Development and characterization of novel erythropoiesis stimulating protein (NESP). *Nephrol Dialy Transplant* 2001;16:Suppl-13.
40. Roitsch T, Lehle L. Structural requirements for protein N-glycosylation. Influence of acceptor peptides on cotranslational glycosylation of yeast invertase and site-directed mutagenesis around a sequon sequence. *Eur J Biochem* 1989;181:525–529.
41. Elliott S, Chang D, Delorme E et al. Structural requirements for additional N-linked carbohydrate on recombinant human erythropoietin. *J Biol Chem* 2004;279:16854–16862.
42. Dennis JW, Granovsky M, Warren CE. Protein glycosylation in development and disease. *Bioessays* 1999;21:412–421.
43. Elliott S, Chang D, Delorme E et al. Isolation and characterization of conformation sensitive antierythropoietin monoclonal antibodies: effect of disulfide bonds and carbohydrate on recombinant human erythropoietin structure. *Blood* 1996;87:2714–2722.
44. Elliott S, Lorenzini T, Chang D et al. Fine-structure epitope mapping of antierythropoietin monoclonal antibodies reveals a model of recombinant human erythropoietin structure. *Blood* 1996;87:2702–2713.
45. Sasu BA, Hartley C, McElroy T et al. Has the unit of erythropoietic activity outlived its usefulness? *Blood* 2004;102:4386.
46. Macdougall IC, Gray SJ, Elston O et al. Pharmacokinetics of novel erythropoiesis stimulating protein compared with epoetin alfa in dialysis patients. *J Am Soc Nephrol* 1999;10:2392–2395.
47. Nissenson AR. Novel erythropoiesis stimulating protein for managing the anemia of chronic kid-

ney disease. *Am J Kid Dis* 2001;38:1390–1397.
48. Nissenson AR, Swan SK, Lindberg JS et al. Randomized, controlled trial of darbepoetin alfa for the treatment of anemia in hemodialysis patients. *Am J Kidney Dis* 2002;40:110–118.
49. Smith RE Jr, Tchekmedyian NS, Chan D et al. A dose- and schedule-finding study of darbepoetin alpha for the treatment of chronic anaemia of cancer. *Br J Cancer* 2003;88:1851–1858.
50. Ling B, Walczyk M, Agarwal A et al. Darbepoetin alfa administered once monthly maintains hemoglobin concentrations in patients with chronic kidney disease. *Clin Nephrol* 2005;63:327–334.
51. Canon JL, Vansteenkiste J, Bodoky G et al. Randomized, double-blind, active-controlled trial of every-3-week darbepoetin alfa for the treatment of chemotherapy-induced anemia. *J Natl Cancer Inst* 2006;98:273–284.
52. Narhi LO, Arakawa T, Aoki KH et al. The effect of carbohydrate on the structure and stability of erythropoietin. *J Biol Chem* 1991;266:23022–23026.
53. Endo Y, Nagai H, Watanabe Y et al. Heat-induced aggregation of recombinant erythropoietin in the intact and deglycosylated states as monitored by gel permeation chromatography combined with a low-angle laser light scattering technique. *J Biol Chem* 1992;112:700–706.
54. Lis H, Sharon N. Protein glycosylation. Structural and functional aspects. *Eur J Biochem* 1993;218:1–27.
55. Elliott S, Aoki K, Agoram B et al. Evaluation of hyperglycosylated erythropoiesis stimulating proteins developed using glycoengineering. *Proc Am Assoc Cancer Res* 2006;47:abstract 2176.
56. Osterborg A, De Boer R, Clemens M et al. A novel erythropoiesis-stimulating agent (AMG114) with 131-hour half-life effectively treats chemotherapy-induced anemia when administered as 200 mcg every 3 weeks. *J Clin Oncol* 2006;24:8626.
57. Delgado C, Francis GE, Fisher D. The uses and properties of PEG-linked proteins. *Crit Rev Therap Drug Carrier Syst* 1992;9:249–304.
58. Yamaoka T, Tabata Y, Ikada Y. Distribution and tissue uptake of poly(ethylene glycol) with different molecular weights after intravenous administration to mice. *J Pharm Sci* 1994;83:601–606.
59. Francis GE, Fisher D, Delgado C et al. PEGylation of cytokines and other therapeutic proteins and peptides: the importance of biological optimisation of coupling techniques. *Int J Hematol* 1998;68:1–18.
60. Goodson RJ, Katre NV. Site-directed pegylation of recombinant interleukin-2 at its glycosylation site. *Bio-Technology* 1990;8:343–346.
61. He XH, Shaw PC, Tam SC. Reducing the immunogenicity and improving the *in vivo* activity of trichosanthin by site-directed pegylation. *Life Sci* 1999;65:355–368.
62. Long DL, Doherty DH, Eisenberg SP et al. Design of homogeneous, monopegylated erythropoietin analogs with preserved *in vitro* bioactivity. *Exp Hematol* 2006;34:697–704.
63. Chen SY, Cressman S, Mao F et al. Synthetic erythropoietic proteins: tuning biological performance by site-specific polymer attachment. *Chem Biol* 2005;12:371–383.
64. Kochendoerfer G. Chemical and biological properties of polymer-modified proteins. *Expert Opin Biol Thera* 2003;3:1253–1261.
65. Weich NS, Tullai J, Guido E et al. Interleukin-3/erythropoietin fusion proteins: *in vitro* effects on hematopoietic cells. *Exp Hematol* 1993;21:647–655.
66. Coscarella A, Carloni C, Liddi R et al. Production of recombinant human GM-CSF-EPO hybrid proteins: *in vitro* biological characterization. *Eur J Haematol* 1997;59:238–246.
67. Sytkowski AJ, Lunn ED, Davis KL et al. Human erythropoietin dimers with markedly enhanced *in vivo* activity. *Proc Natl Acad Sci USA* 1998;95:1184–1188.
68. Sytkowski AJ, Lunn ED, Risinger MA et al. An erythropoietin fusion protein comprised of identical repeating domains exhibits enhanced biological properties. *J Biol Chem* 1999;274:24773–24778.
69. Dalle B, Henri A, Rouyer-Fessard P et al. Dimeric erythropoietin fusion protein with enhanced erythropoietic activity *in vitro* and *in vivo*. *Blood* 2001;97:3776–3782.
70. Coscarella A, Liddi R, Di Loreto M et al. The rhGM-CSF-EPO hybrid protein MEN 11300 induces anti-EPO antibodies and severe anaemia in rhesus monkeys. *Cytokine* 1998;10:964–969.
71. Yoshimura A, Longmore G, Lodish HF. Point mutation in the exoplasmic domain of the erythropoietin receptor resulting in hormone-independent activation and tumorigenicity. *Nature* 1990;348:647–649.
72. Wrighton NC, Farrell FX, Chang R et al. Small peptides as potent mimetics of the protein hormone erythropoietin. *Science* 1996;273:458–464.

73. Livnah O, Stura EA, Johnson DL et al. Functional mimicry of a protein hormone by a peptide agonist: the EPO receptor complex at 2.8 A. *Science* 1996;273:464–471.
74. Wrighton NC, Balasubramanian P, Barbone FP et al. Increased potency of an erythropoietin peptide mimetic through covalent dimerization. *Nat Biotechnol* 1997;15:1261–1265.
75. McConnell SJ, Dinh T, Le MH et al. Isolation of erythropoietin receptor agonist peptides using evolved phage libraries. *Bio Chem* 1998;379:1279–1286.
76. Kuai L, Wu C, Qiu Q et al. Plasminogen activator inhibitor-1 fused with erythropoietin (EPO) mimetic peptide (EMP) enhances the EPO activity of EMP. *J Peptide Res* 2000;56:59–62.
77. Fan Q, Leuther KK, Holmes CP et al. Preclinical evaluation of Hematide, a novel erythropoiesis stimulating agent, for the treatment of anemia. *Exp Hematol* 2006;34:1303–1311.
78. Stead RB, Lambert J, Wessels D et al. Evaluation of the safety and pharmacodynamics of Hematide, a novel erythropoietic agent, in a phase 1, double-blind, placebo-controlled, dose-escalation study in healthy volunteers. *Blood* 2006;108:1830–1834.
79. Goldberg J, Jin Q, Ambroise Y et al. Erythropoietin mimetics derived from solution phase combinatorial libraries. *J Am Chem Soc* 2002;124:544–555.
80. Biazzo DE, Motamedi H, Mark DF et al. A high-throughput assay to identify compounds that can induce dimerization of the erythropoietin receptor. *Anal Biochem* 2000;278:39–45.
81. Qureshi SA, Kim RM, Konteatis Z et al. Mimicry of erythropoietin by a nonpeptide molecule. *Proc Natl Acad Sci USA* 1999;96:12156–12161.
82. Naranda T, Kaufman RI, Li J et al. Activation of erythropoietin receptor through a novel extracellular binding site. *Endocrinology* 2002;143:2293–2302.
83. Constantinescu SN, Liu X, Beyer W et al. Activation of the erythropoietin receptor by the gp55-P viral envelope protein is determined by a single amino acid in its transmembrane domain. *EMBO J* 1999;18:3334–3347.
84. Li JP, Hu HO, Niu QT et al. Cell surface activation of the erythropoietin receptor by Friend spleen focus-forming virus gp55. *J Virol* 1995;69:1714–1719.
85. Sharlow ER, Pacifici R, Crouse J et al. Hematopoietic cell phosphatase negatively regulates erythropoietin-induced hemoglobinization in erythroleukemic SKT6 cells. *Blood* 1997;90:2175–2187.
86. Yi T, Zhang J, Miura O et al. Hematopoietic cell phosphatase associates with erythropoietin (Epo) receptor after Epo-induced receptor tyrosine phosphorylation: identification of potential binding sites. *Blood* 1995;85:87–95.
87. Watowich SS, Xie X, Klingmuller U et al. Erythropoietin receptor mutations associated with familial erythrocytosis cause hypersensitivity to erythropoietin in the heterozygous state. *Blood* 1999;94:2530–2532.
88. Prchal JT, Sokol L. 'Benign erythrocytosis' and other familial and congenital polycythemias. *Eur J Haematol* 1996;57:263–268.
89. de la Chapelle A, Traskelin AL, Juvonen E. Truncated erythropoietin receptor causes dominantly inherited benign human erythrocytosis. *Proc Natl Acad Sci USA* 1993;90:4495–4499.
90. Tsui FW, Tsui HW. Molecular basis of the motheaten phenotype. *Immunol Rev* 1994;138:185–206.
91. Lankester AC, van Schijndel GM, van Lier RA. Hematopoietic cell phosphatase is recruited to CD22 following B cell antigen receptor ligation. *J Biol Chem* 1995;270:20305–20308.
92. Tapley P, Shevde NK, Schweitzer PA et al. Increased G-CSF responsiveness of bone marrow cells from hematopoietic cell phosphatase deficient viable motheaten mice. *Exp Hematol* 1997;25:122–131.
93. Nangaku M, Kojima I, Tanaka T et al. Novel drugs and the response to hypoxia: HIF stabilizers and prolyl hydroxylase. Recent Patents *Cardiovas Drug Discovery* 2006;1:129–139.
94. Percy MJ, Furlow PW, Beer PA et al. A novel erythrocytosis-associated PHD2 mutation suggests the location of a HIF binding groove. *Blood* 2007;110:2193–2196.
95. Minamishima YA, Moslehi J, Bardeesy N et al. Somatic inactivation of the PHD2 prolyl hydroxylase causes polycythemia and congestive heart failure. *Blood* 2008;111:3236–3244.
96. Gordeuk VR, Sergueeva AI, Miasnikova GY et al. Congenital disorder of oxygen sensing: association of the homozygous Chuvash polycythemia VHL mutation with thrombosis and vascular abnormalities but not tumors. *Blood* 2004;103:3924–3932.
97. Al-Sheikh M, Moradkhani K, Lopez M et al. Disturbance in the HIF-1alpha pathway associated with erythrocytosis: further evidences brought by frameshift and nonsense mutations in the prolyl hydroxylase domain protein 2 (PHD2) gene. *Blood Cells Molec Dis* 2008;40:160–165.

98. Katsuoka Y, Beckman B, George WJ et al. Increased levels of erythropoietin in kidney extracts of rats treated with cobalt and hypoxia. *Am J Physiol* 1983;244:F129–F133.
99. Hirsila M, Koivunen P, Xu L et al. Effect of desferrioxamine and metals on the hydroxylases in the oxygen sensing pathway. *FASEB J* 2005;19:1308–1310.
100. Goldwasser E, White WF. Purification of sheep erythropoietin. *Fed Proc* 1959;18:236.
101. Curtis JR, Goode GC, Herrington J et al. Possible cobalt toxicity in maintenance hemodialysis patients after treatment with cobaltous chloride: a study of blood and tissue cobalt concentrations in normal subjects and patients with terminal and renal failure. *Clin Nephrol* 1976;5:61–65.
102. Wolf J, Levy IJ. Treatment of sickle cell anemia with cobalt chloride. *Arch Intern Med* 1954;93:387–396.
103. Shen SC, Homberger F. The anemia of cancer patients and its relation to metastases to the bone marrow. *J Lab Clin Med* 1951;37:182–198.
104. Barceloux DG. Cobalt. *J Toxicol* 1999;37:201–216.
105. Tielemans C, Collart F, Wens R et al. Improvement of anemia with deferoxamine in hemodialysis patients with aluminum-induced bone disease. *Clin Nephrol* 1985;24:237–241.
106. Davis BA, Porter JB. Results of long term iron chelation treatment with deferoxamine. Adv Exper. *Med Biol* 2002;509:91–125.
107. Hsieh MM, Linde NS, Wynter A et al. HIF prolyl hydroxylase inhibition results in endogenous erythropoietin induction, erythrocytosis, and modest fetal hemoglobin expression in rhesus macaques. *Blood* 2007;110:2140–2147.
108. Naffakh N, Henri A, Villeval JL et al. Sustained delivery of erythropoietin in mice by genetically modified skin fibroblasts. *Proc Natl Acad Sci USA* 1995;92:3194–3198.
109. Bohl D, Naffakh N, Heard JM. Long-term control of erythropoietin secretion by doxycycline in mice transplanted with engineered primary myoblasts. *Nat Med* 1997;3:299–305.
110. Naffakh N, Pinset C, Montarras D et al. Long-term secretion of therapeutic proteins from genetically modified skeletal muscles. *Hum Gene Ther* 1996;7:11–21.
111. Naffakh N, Danos O. Gene transfer for erythropoiesis enhancement. *Mol Med Today* 1996;2:343–348.
112. Meyer F, Finer M. Gene therapy: progress and challenges. *Cell Molec Biol* 2001;47(8):1277–1294.
113. Zhou S, Murphy JE, Escobedo JA et al. Adeno-associated virus-mediated delivery of erythropoietin leads to sustained elevation of hematocrit in nonhuman primates. *Gene Ther* 1998;5:665–670.
114. Chenuaud P, Larcher T, Rabinowitz JE et al. Autoimmune anemia in macaques following erythropoietin gene therapy. *Blood* 2004;103:3303–3304.
115. Dalle B, Payen E, Regulier E et al. Improvement of mouse beta-thalassemia upon erythropoietin delivery by encapsulated myoblasts. *Gene Ther* 1999;6:157–161.
116. Payen E, Bettan M, Henri A et al. Oxygen tension and a pharmacological switch in the regulation of transgene expression for gene therapy. *J Gene Med* 2001;3:498–504.
117. Bohl D, Heard JM. Delivering erythropoietin through genetically engineered cells. *J Am Soc Nephrol* 2000;11:Suppl-62.
118. Serguera C, Bohl D, Rolland E et al. Control of erythropoietin secretion by doxycycline or mifepristone in mice bearing polymer-encapsulated engineered cells. *Hum Gene Ther* 1999;10:375–383.
119. Ye X, Rivera VM, Zoltick P et al. Regulated delivery of therapeutic proteins after *in vivo* somatic cell gene transfer. *Science* 1999;283:88–91.
120. Maruyama H, Higuchi N, Nishikawa Y et al. Kidney-targeted naked DNA transfer by retrograde renal vein injection in rats. *Hum Gene Ther* 2002;13:455–468.
121. Spivak JL, Hogans BB. The *in vivo* metabolism of recombinant human erythropoietin in the rat. *Blood* 1989;73:90–99.
122. Dinkelaar RB, Engels EY, Hart AA et al. Metabolic studies on erythropoietin (EP): II. The role of liver and kidney in the metabolism of Ep. *Exp Hematol* 1981;9:796–803.
123. Fu J-S, Lertora JJL, Brookins J et al. Pharmacokinetics of erythropoietin in intact and anephric dogs. *J Lab Clin Med* 1988;111:669–676.
124. Breidbach A, Catlin DH, Green GA et al. Detection of recombinant human erythropoietin in urine by isoelectric focusing. *Clin Chem* 2003;49:901–907.
125. Catlin DH, Breidbach A, Elliott S et al. Comparison of the isoelectric focusing patterns of darbepoetin alfa, recombinant human erythropoietin, and endogenous erythropoietin from human

urine. *Clin Chem* 2002;48:2057–2059.

126. Widness JA, Veng-Pedersen P, Schmidt RL et al. *In vivo* ^{125}I-erythropoietin pharmacokinetics are unchanged after anesthesia, nephrectomy and hepatectomy in sheep. *J Pharmacol Exp Ther* 1996;279:1205–1210.
127. Jensen JD, Jensen LW, Madsen JK et al. The metabolism of erythropoietin in liver cirrhosis patients compared with healthy volunteers. *Eur J Haematol* 1995;54:111–116.
128. Kindler J, Eckardt KU, Ehmer B et al. Single-dose pharmacokinetics of recombinant human erythropoietin in patients with various degrees of renal failure. *Nephrol Dialy Transplant* 1989;4:345–349.
129. Macdougall IC, Roberts DE, Coles GA et al. Clinical pharmacokinetics of epoetin (recombinant human erythropoietin). *Clin Pharmacokinetics* 1991;20:99–113.
130. Yoon WH, Park SJ, Kim IC et al. Pharmacokinetics of recombinant human erythropoietin in rabbits and 3/4 nephrectomized rats. *Res Commun Mol Pathol Pharmacol* 1997;96:227–240.
131. Bao H, Jacobs-Helber SM, Lawson AE et al. Protein kinase B (c-Akt), phosphatidylinositol 3-kinase, and STAT5 are activated by erythropoietin (EPO) in HCD57 erythroid cells but are constitutively active in an EPO-independent, apoptosis-resistant subclone (HCD57-SREI cells). *Blood* 1999;93:3757–3773.
132. Fraser JK, Lin FK, Berridge MV. Expression of high affinity receptors for erythropoietin on human bone marrow cells and on the human erythroleukemic cell line, HEL. *Exp Hematol* 1988;16:836–842.
133. Gross AW, Lodish HF. Cellular trafficking and degradation of erythropoietin and Novel Erythropoiesis Stimulating Protein (NESP). *J Biol Chem* 2006;281:2024–2032.
134. Chapel S, Veng-Pedersen P, Hohl RJ et al. Changes in erythropoietin pharmacokinetics following busulfan-induced bone marrow ablation in sheep: evidence for bone marrow as a major erythropoietin elimination pathway. *J Pharmacol Exp Ther* 2001;298:820–824.
135. Veng-Pedersen P, Chapel S, Al-Huniti NH et al. Pharmacokinetic tracer kinetics analysis of changes in erythropoietin receptor population in phlebotomy-induced anemia and bone marrow ablation. *Biopharm Drug Disposition* 2004;25:149–156.
136. Hartley C, Elliott S, Begley CG et al. Kinetics of haematopoietic recovery after dose-intensive chemo/radiotherapy in mice: optimized erythroid support with darbepoetin alpha. *Br J Haematol* 2003;122:623–636.
137. Glaspy J, Henry D, Patel R et al. Effects of chemotherapy on endogenous erythropoietin levels and the pharmacokinetics and erythropoietic response of darbepoetin alfa: a randomised clinical trial of synchronous *versus* asynchronous dosing of darbepoetin alfa. *Eur J Cancer* 2005;41:1140–1149.
138. Agoram B, Molineux G, Jang G et al. Effects of altered receptor binding activity on the clearance of erythropoiesis-stimulating proteins: a minor role of erythropoietin receptor-mediated pathways. *Nephrol Dial Transplant* 2006;21:303–304.
139. Flaharty KK. Clinical pharmacology of recombinant human erythropoietin (r-HuEPO). *Pharmacotherapy* 1990;10:9S–14S.
140. Porter CJ, Charman SA. Lymphatic transport of proteins after subcutaneous administration. *J Pharm Sci* 2000;89:297–310.
141. Morlock M, Kissel T, Li YX et al. Erythropoietin loaded microspheres prepared from biodegradable LPLG-PEO-LPLG triblock copolymers: protein stabilization and *in vitro* release properties. *J Control Release* 1998;56:105–115.
142. Pistel KF, Bittner B, Koll H et al. Biodegradable recombinant human erythropoietin loaded microspheres prepared from linear and star-branched block copolymers: influence of encapsulation technique and polymer composition on particle characteristics. *J Control Release* 1999;59:309–325.

Erythropoietin receptor

Erythropoietins, Erythropoietic Factors and Erythropoiesis
Edited by S.G. Elliott, M.A. Foote, and G. Molineux

Structural basis for the signal transduction of erythropoietin

Timothy D. Osslund

Amgen, Thousand Oaks, California, USA

Introduction

The structures of recombinant erythropoietin (r-HuEPO), the extracellular ligand-binding portion of erythropoietin receptor (EPOR) and several novel peptides have been extensively studied using a variety of methods including X-ray crystallography and nuclear magnetic resonance (NMR) [1, 2]˙ The receptor for EPO has been classified as a member of the class 1 cytokine receptor super family [3, 4]. EPOR is a transmembrane receptor that is activated when EPO binds to the extracellular portion of the receptor causing dimerization [5]. The dimerization of the extracellular portion of the receptor induces tyrosine kinase proteins associated with the cytoplasmic side of the receptor to initiate the signal transduction events that eventually cause the committed erythroid–colony-forming units (CFU–E) progenitor cells to eventually mature into red blood cells [6] (Fig. 1).

Studies on molecules that cause 2 EPOR to form a dimer have demonstrated that EPOR can be bound and activated by structurally unrelated molecules. This oligomerization of EPOR can be caused by an engineered extracellular disulfide bond in EPOR that can stimulate constitutive signal transduction [7]. Antibodies with epitopes that react with the extracellular domain of EPOR also stimulate the EPOR [8] as well as several different peptides that demonstrate both agonist as well as antagonist activity [9–12]. In addition to r-HuEPO, a novel hyperglycosylated analog of r-HuEPO (Aranesp, darbepoeitin alfa) with improved pharmacologic properties, has demonstrated activation of EPOR. The common modality for all the molecules that cause signal transduction is that they all result in the formation of an r-HuEPO receptor homodimer. It was initially speculated that EPO functioned mechanistically as a cross-linker that brings the intracellular portion of EPOR into close proximity producing the signal transduction events that eventually allow the cell to mature into a red blood cell. However, recent data suggest that EPOR receptor may initially form a dimer on the cell surface [13], and that EPO provides the proper orientation of EPOR outside the cell, thereby allowing the intracellular kinases to be in the precise geometry for biologic activity.

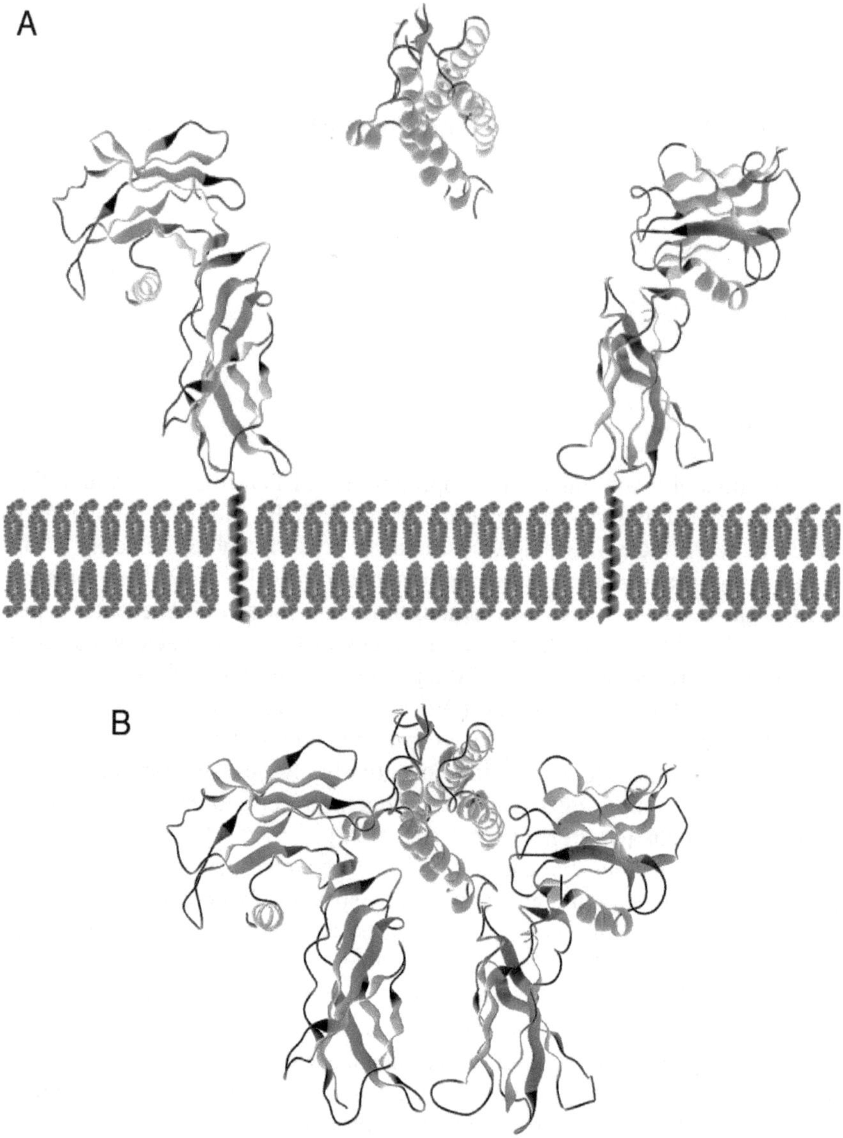

Figure 1. (a) Model depicting the structures of the two EPO receptors imbedded into the cell membrane. EPO, which is produced in the kidney and released into the bloodstream, functions by binding to two receptors, causing dimerization and receptor activation. The complex of EPO and two EPO receptors allows the precise geometry required by the intracellular portion of the EPOR (not shown in this diagram) for biological activity. The relative proximity of the two intracellular regions caused by the mechanical association of EPO with the EPO receptors allow the association and phosporylation of kinases which is the initiating event which eventually allows the cell to mature into a red blood cell. (b) The X-ray crystallographic structure of EPO: EPOR. The structure resolved by Syed et al. (1EER.pdb) shows the interaction between the 'sides' of the EPO alfa helical bundle and two EPO receptors.

With the introduction of pegylated epoetin in Europe, which is epoetin beta pegylated with a 30 kDa methoxy-polyethyleneglycol polymer connected to the amino terminus of rHuEPO and the σ-amino groups of Lys45 or Lys52 [14, 15], a third derivatized protein is being introduced into the clinic. The concept of pegylating a protein to improve its pharmacokinetic/pharmacodynamic properties of therapeutic proteins is well documented in the development of pegfilgrastrim (Neulasta) [16], which is a pegylated form of recombinant human granulocyte colony-stimulating factor (rHuG-CSF) and Pegasys (pegylated interferon alfa-2a) [17]. The binding of the pegylated epoetin (CERA, continuous erythropoietin receptor activator) is greatly reduced to approximately 50- to 100-fold lower affinity for EPOR binding sites. However, with respect to Lys45, it is anticipated that pegylation at this site would greatly reduce receptor binding affinity, as this residue has been mapped out to be within the high-affinity-binding site for binding to EPOR [2].

EPO shares with other cytokines the general property of binding to its receptor with high affinity (100–200 pM), relatively short half life (10 h), and low blood circulation levels (10 pg/mL). In addition, it takes only a small fraction of the possible EPOR (10%) to be occupied for the EPO signal transduction pathway to be stimulated. What becomes strikingly evident by the continued introduction of additional polymers, either synthetic as is the case of the pegylated epoetin beta or through glycosylation with respect to darbepoetin alfa, is that in the clinical setting the absolute binding affinity of the these molecules to EPOR is less important than the additional half life these molecules have with respect to the naturally occurring EPO.

Erythropoietin structure

The fundamental issue in resolving the X-ray crystallographic structure of the r-HuEPO ligand and receptor was obtaining a construct of r-HuEPO that was suitable for crystallization.

Proteins which are highly glycosylated such as r-HuEPO are often not good candidates for crystallization. When proteins are expressed in bacteria, the protein is not glycosylated and often forms aggregates presumably because the residues which were covered by the carbohydrate structures are now exposed [18]. Multiple bacterially expressed constructs were put through the crystal screening process until a suitable crystallization candidate could be determined. Among the variety of constructs that were produced included N24K, N38K, N83K, P121N, and P122S. The Pro mutations were suggested based on the possible cis-trans conformation heterogeneity that had been reported in other unrelated protein structures [19].

The first significant structural data came when the structure of the receptor and an EPO peptide mimetic was first revealed [10]. Within only a few years after this breakthrough structure, the structures of the r-HuEPO/EPOR2 complex [1] and the NMR structure of the unbound ligand [2] were published. The

NMR and crystal structures were very similar with the exception of the region around the Pro mutations at residue 121 and 122. An artifact of these mutations caused a small alpha helix to be formed in the crystal structure which is not found in the NMR structure. These data provided a further insight into the structure of the ligand and its structural relationship to the receptor.

r-HuEPO has a structural topology that is similar to other members of the cytokine hormone family. This family of proteins shares a four helical bundle motif in which the first two helices go in the opposite direction relative to the second set of helices. This 'up-up-down-down' motif requires two long loops between the first and second helix, a small loop between the second and third helix, and another long loop between the third and forth helix [20]. The class of proteins can be subdivided into proteins with long alpha helices and short crossing angles between the helices that are found in G-CSF and human growth hormone (HGH). Granulocyte-macrophage colony-stimulating factor (GM-CSF) is representative of a cytokine structure where the helices are significantly shorter and the crossing angles between helices are relative large. Often the cytokines with the shorter helices have a small beta strand connecting the long loops between helix A and B and C and D.

After the structure of EPO was elucidated, it was found that the EPO structure was a hybrid between the two classes of four helical bundles. The characteristic topology of r-HuEPO, and cytokines in general, is a core structure composed of a four helical bundle (Fig. 2). There is one pair of antiparallel long helices, A (residues 8 to 26) and D (residues 137 to 161), and another pair of shorter helices, B (residues 56 to 83) and C (residues 90 to 112). A long loop connects the first and second helix (AB) and a short loop connects the second and third helix (BC) and another long loop connects the third and fourth helix (CD) [21]. The core bundle has a left-handed twist and the crossing angles of the helices (–148° to 167°) [2] also provide a further classification of the cytokine family. Before the elucidation of the structure of r-HuEPO, most cytokines could be classified as either having long helices and short crossing angles (e.g., G-CSF) or short helices with long crossing angles (e.g., GM-CSF) [20]. Often the short helical cytokines have two antiparallel beta strands that structurally link the long loops AB and CD. The r-HuEPO structure illustrates a hybrid of the two classifications. r-HuEPO has longer helices and shorter crossing angles than the other shorter helical cytokines, but inconsistent with the longer helical classification, r-HuEPO has an antiparallel beta sheet: β1 (residues 39 to 41) and β2 (residues 133 to 135).

The first of two disulfide bonds structurally link together the amino terminus and the carboxy terminus at residues Cys7 to Cys161. The proximity of this disulfide bond relative to the amino terminus and carboxy terminus is presumed to hold the four helical bundle together at one end while a beta strand and second disulfide bond stabilize the quaternary structure of the protein at the opposing end. The second disulfide, Cys29 to Cys33, links the end of helix A with the AB loop. The hydrophobic core residues on helix D pack against

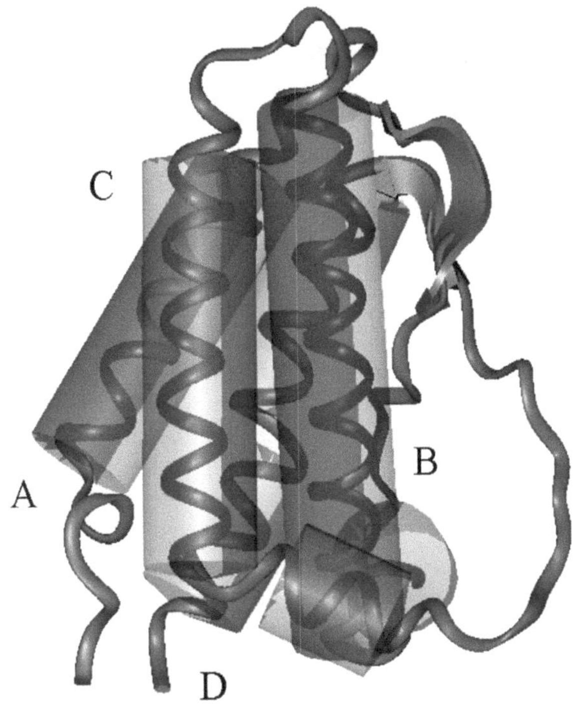

Figure 2. A secondary structure depiction of EPO. EPO has a structural topology which is similar to other members of the cytokine hormone family. The characteristic topology of EPO, and cytokines in general, is a core structure composed of a four helical bundle. There is one pair of antiparallel long helices, A (residues 8 to 26) and D (residues 137 to 161), and another pair of shorter helices, B (residues 56 to 83) and C (residues 90 to 112).

the hydrophobic residues of helices A, B and C that reflect the pattern of the approximate 3.5 residues per turn in a standard alpha helix.

In the X-ray structure of r-HuEPO bound to the EPOR, there are two additional short helices, the B' helix (residues 47 to 52) orthogonal to B and the minihelix C' (residues 114 to 121) following C with a 90° tilt beginning at Gly113. The second minihelix is not found in the NMR structure and may be an artifact of the mutations made around this area to develop a protein suitable for crystallization.

Structure of EPOR

The extracellular domain of EPORr has two β sandwich domains, designated D1 (*N*-terminal) and D2 (*C*-terminal), each of which shares an overall topology with the fibronectin type III fold. The D1/D2 domains are almost perpendicular to one another and form an inverted 'L' with the ligand-binding site located between the monomers. As is found in other members of the cytokine family of receptors, EPOR has a WSXWS, box that is a sequence of amino acids containing tryptophans and serines, which are essential for the biological activity of the receptor [22].

In addition to the predominantly β sheet structure, there is an α-helix found at the amino terminus which may have important structural implications (Fig. 3). Adopting the position inside the 'elbow' of the D1 and D2 domains, the α-helix makes extensive contacts with both D1 and D2 domains of the receptor, with a buried surface area of approximately 200 $Å^2$. The α-helix is amphipathic, having both hydrophobic and hydrophilic regions, and presents several residues towards a hydrophobic patch on the D1 domain of the receptor. Phe11 and Leu18 of the α-helix form the basis of hydrophobic interactions with Phe39, Leu27 and Phe29 in the D1 domain. *N*-terminal to the WSXWS box in domain D2, Phe208 also makes contacts with the hydrophobic pocket. Two electrostatic interactions occur between the α-helix and other regions of

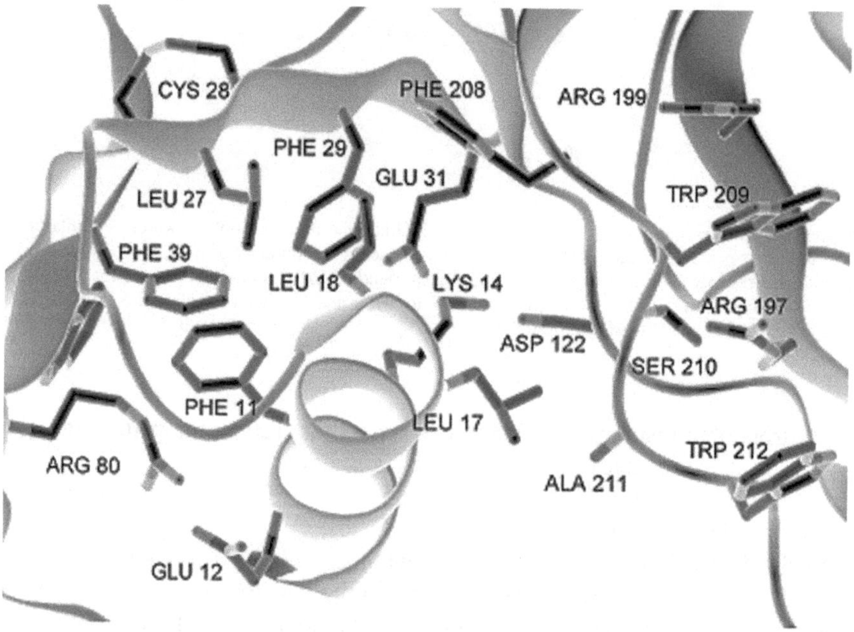

Figure 3. Amino Acid Contacts of the amino terminal alpha helix of the EPO receptor. Ribbon representation of the soluble EPOR receptor (EPOR). Selected residues at the helical interface are a depicted using a stick model. Residues 209–212 represent the putative cytokine family WSXWS box. The EPOR N-terminal α-helix is amphipathic and presents several hydrophobic residues towards a hydrophobic patch on the D1 domain of the receptor. PHE11 and LEU18 of the α-helix form the basis of hydrophobic interactions with PHE39, LEU27 and PHE29 in the D1 domain. N-terminal to the WSXWS box on domain D2, PHE208 also makes contacts with this hydrophobic pocket. Two electrostatic interactions occur between the α-helix and the rest of the protein. One salt bridge is formed between GLU12 and ARG80 and the other salt bridge links all three domains via the electrostatic interactions of LYS14 with GLU31 (on domain D1) and ASP122 (on domain D2).

EPOR. Several salt bridges function to link the α-helix with D1 and D2 including the electrostatic interactions between Lys14 with Glu31 (on domain D1) and Asp122 (on domain D2).

Alteration of the WSXWS sequence disrupts r-HuEPO binding and receptor signaling. The WSXWS motif has been shown to be critical for the folding and transport of the receptor to the cell surface, and an A211E mutation further improved the efficiency of the processes [22]. In the structure of an r-HuEPO/EPOR complex, Glu211 occupies the X position. The side-chain of Glu211 in the WSXWS motif is closest (<4.5 Å) to Leu17 of the *N*-terminal receptor helix. Additionally, the side chains of the tryptophans in the WSXWS motif, residues Trp209 and Trp212, sandwich the hydrophobic side chain of Arg197 in the receptor fold, while Ser210 and Ser213 are within hydrogen-bonding distance of Ala198 and Val196, respectively (not shown in figure). The observation that the WSXWS motif interacts with the *N*-terminal helix and the β-sheet residues Val196, Arg197, and Ala198 in the r-HuEPO/EPOR complex, suggests that the *N*-terminal helix may be important in stabilizing the folded EPOR through an interaction with WSXWS. The interaction between WSXWS and the α-helix may therefore be essential in holding the two domains, D1 and D2, in the precise geometry allowing dimerization and signal transduction by EPOR.

The r-HuEPO/EPOR complex is a 2:1 stoichiometry and the receptor molecules are held together through two distinct surface areas on r-HuEPO that are located on opposing faces of the molecule. These r-HuEPO/EPOR interfaces have been identified as high affinity (kDa approximately 1 nM) and low affinity (kDa approximately 1 μM) binding sites and are referred to as site 1 and site 2, respectively [23]. Site 1 is comprises r-HuEPO residues from helices A, B', D, and part of the AB loop, whereas site two comprises amino acids in the A and C helices exclusively. Presumably site 1 binds more tightly than site 2 because of the larger number of residues participating in the interaction with EPOR of the first site relative to the second. Site 1 is characterized by a hydrophobic-binding region in the middle of the binding site with flanking amino acids that allow electrostatic interactions with EPOR. The EPOR loops L1, L5 and L6 interact with the hydrophobic r-HuEPO residues with Phe93 of EPOR as the dominant hydrophobic contact. Phe93 is a pivotal residue firmly held in place by hydrogen bonding to r-Hu-EPO residues Thr44 and Asn147. Mutagenesis has shown that Phe93 on the receptor is a critical r-HuEPO-binding determinant, and mutation of this residue eliminates any detectable binding [24]. The site-1 binding site of r-HuEPO also has contributions from the A-B linker polypeptide residues (Thr44–Phe48) of r-HuEPO. The hydrophobic binding surface of site 1 itself is surrounded by hydrophilic interactions with Asn147 of r-HuEPO that functions as a major contributor to the electrostatic interactions. Asn147 binds to EPOR using three hydrogen bonds, one of which is between its Nδ atom and the carbonyl oxygen of Phe93. On either side of the Asn147 in r-HuEPO are arginine residues, Arg143 and Arg150, that form salt bridges with acids Glu60 and Glu117 of r-HuEPO, respectively.

r-HuEPO site 2 interactions with EPOR are less extensive than site 1, which explains its lower affinity. Most of the site-2 interactions are between residues of the C helix of r-HuEPO and the L3 loop of EPOR. The hydrophobic surface is created on EPOR by residues Phe93, Phe205, and Met150, although because of the positioning of the D1 domain relative to the site 2 interactions with r-HuEPO, there is a relatively flat EPOR surface that r-HuEPO residues can interact with. The closest nonpolar contact in site 2 includes r-HuEPO Leu108 and r-HuEPOR Phe93 Cγ3.9 Å). The rest of the C helix residues are positioned at a greater distance from EPOR hydrophobic surface than at the site 1 interface. Phe93 is again the major contributor to the central hydrophobic binding pocket (Leu5, Val11, Tyr15, Ser104, Thr107, and Leu108 of r-HuEPO) are within 4.5 Å of EPOR Phe93. Met150 of EPOR, however, is more buried than in site 1, making van der Waals contacts with Arg10, Val11, and Arg14 of r-HuEPO.

In general, the side-chain interactions at both interfaces are predominantly between positively charged lysines and arginines of r-HuEPO and negatively charged aspartate and glutamate side-chains of EPOR. Site 1 contains almost twice as many side-chain–side-chain interactions as site 2. However, the number of hydrophilic contacts involving main-chain atoms from either r-HuEPO or EPOR is equal for both sites. In addition, site 1 contains two hydrogen bonds that involve main-chain atoms from both r-HuEPO and EPOR. Hydrophilic interactions from three EPOR residues are common to both interfaces, two of which involve hydrogen bonding from a main-chain atom. Although most of residues that interact across both interfaces are polar, the expected contribution to binding affinity is greatest from the hydrophobic contacts.

The sequence identity of EPO between different mammals is reportedly about 60% [25]. Inspection of the two bindings sites with comparison to other species suggest why cross species reactivity is so high as 100% of the binding site residues on EPO are conserved within the mammalian species studied. The sequence (NFLRGK) of the high-affinity binding site is evolutionarily conserved also among vertebrates and is found in EPO sequence of the green puffer fish (*Tetraodon nigroviridis*).

Structure of an r-HuEPO peptide mimetic bound to EPOR

An elegant series of experiments resolved the structure of a soluble EPOR complexed with an agonistic peptide, which reportedly had no sequence identity to the naturally occurring ligand [10] (Fig. 4). Included in this body of work was the bacterial expression, purification, and refolding of a soluble form of EPOR (EBP) and the isolation of the self-dimerizing 20 amino acid peptide EMP1 (GGTYSCHFGPLTWVCKPQGG). The peptide has β-sheet structure and a disulfide bond constraining the peptide. The stoichiometry of the EBP/EMP1 complex is one peptide homodimer bound to two r-HuEPO receptor monomers. EMP1 was originally identified using peptide phage display technology. The 20 amino acid peptide mimicked the biologic activity of

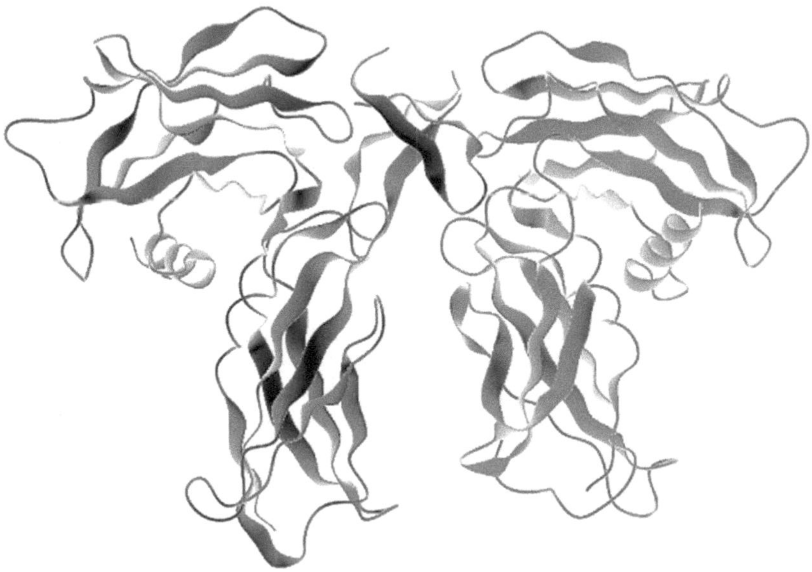

Figure 4. Ribbon depiction of the complex between the extracellular domain of Erythropoietin (r-Hu-EPO) and an agonist peptide. The X-ray crystallographic structure of the r-Hu-EPO receptor complexed with an agonist peptide was resolved by Livnah et al. The two-fold symmetry of the peptide allows the dimerization of 2 EPOR. The functional mimicry of EPO by a 20-residue peptide with a totally unrelated sequence was of considerable value in understanding the structure of the receptor and its signal transduction mechanisms.

EPO and functioned as an agonist. The binding affinity of the peptide compared with r-HuEPO binds about a thousand fold less at 200 nM, compared with 200 pM.

The interface between the peptide and EPOR is mainly hydrophobic in nature with very few invaginations and charged residues that would be characteristic of a naturally occurring binding pocket. The three residues on EPOR contributing the most binding energy to EMP1 are Phe93, Met150, and Phe205. These amino acids make contact with the Phe10 and Tyr4 of the peptide. The dimeric and symmetrical nature of the peptide is reflected in the nature of the binding interaction with EPOR. Both interaction sites on the EMP1 dimer are identical, therefore, the two receptors adopt orientations and relative affinities of the two binding sites that are different than that found in EPOR dimerized by r-HuEPO. If the two complexes (r-HuEPO/EPOR and EMP1/EPOR) are viewed along a plane that is perpendicular to the membrane, there is almost a 60° difference between the angles of the D1 domain of EPOR with respect to the r-HuEPO and peptide complexes. In addition there is a significant difference of about 45° in the angle of the hinge region between D1 and D2 with respect to the binding of the r-HuEPO and EMP1.

Conclusion

A number of different molecules can effect the dimerization and signal transduction of EPOR. Recent data suggest that the receptor may actually be self-associated on the cell surface in the absence of the natural ligand, suggesting that it may not merely be EPOR dimerization but an even more subtle conformational change that occurs only after the natural ligand, r-HuEPO binds.

The quality and scope of the structural biology research in understanding the interactions of EPOR with several different molecules has provided a detailed understanding of the role of individual amino acids within the receptor. These structures provide a rationale in the design of the next generation of erythropoiesis-stimulating molecules and may, in the future, provide the basis for a true small molecule mimetic that could effectively allow the activation of EPOR required for a signal transduction comparable in effect to the natural ligand.

References

1. Syed RS, REid SW, Li C et al. Efficiency of signalling through cytokine receptors depends critically on receptor orientation. *Nature* 1998;395:511–516.
2. Cheetham JC, Smith DM, Aoki KH et al. NMR structure of human erythropoietin and a comparison with its receptor bound conformation. *Nat Struct Mol Biol* 1998;5:861–866.
3. Bazan JF. Structural design and molecular evolution of a cytokine receptor superfamily. *Proc Natl Acad Sci USA* 1990;87:6934–6938.
4. D'Andrea AD, Lodish HF, Wong GG. Expression cloning of the murine erythrorpoietin receptor. *Cell* 1989;57:277–285.
5. Philo JS, Aoki KH, Arakawa T et al. Dimerization of the extracellular domain of the erythropoietin (EPO) receptor by EPO: One high-affinity and one low-affinity interaction. *Biochemistry* 1996;35:1681–1691.
6. Damen JEK, Krystal G. Early events in erythropoietin-induced signalling. *Exp Hematol* 1996;24:1455–1459.
7. Yoshimura A, Longmore G, Lodish HF. Point mutation in the exoplasmic domain of the erythropoietin receptor resulting in hormone-independent activation and tumorigenicity. *Nature* 1990;348:647–649.
8. Elliott, S, Lorenzini T, Yangihara D et al. Activation of the erythropoietin (EPO) receptor by bivalent anti-EPO receptor antibodies. *J Biol Chem* 1996;271:24691–24697.
9. Wrighton NC, Farrell FX, Johnson DL et al. Small peptides as potent mimetics of the protein hormone erythropoietin. *Science* 1996;273:458–463.
10. Livnah O, Stura EA, Johnson DL et al. Functional mimicry of a protein hormone by a peptide agonist: The EPO receptor complex at 2.8 A. *Science* 1996;273:464–471.
11. Wrighton NC, Balasubramanian P, Barbone FP et al. Increased potency of an erythropoietin peptide mimetic through covalent dimerization. *Nat Biotech* 1997;15:1261–1265.
12. Naranda T, Wong K, Kaufman RI et al. Activation of erythropoietin receptor through a novel extracellular binding site. *Endocrinology* 2002;143:2293–2302.
13. Constantinescu SN, Keren T, Socolovsky M et al. Ligand-independent oligomerization of cell-surface erythropoietin receptor is mediated by the transmembrane domain. *Proc Natl Acad Sci USA* 2001;98:4379–4384.
14. Macdougall IC, Eckardt K-U. Novel strategies for stimulating erythropoiesis and potential new treatments for anaemia. *The Lancet* 2006;368:947–953.
15. Jarsch, M, Brandt M, Lanzendorfer M et al. Comparative erythropoietin receptor binding kinetics of C.E.R.A. and epoetin-b determined by surface plasmon resonance and competition binding

assay. *Pharmacology* 2008;81:63–69.
16. Molineux G. The design and development of pegfilgrastim (PEG-rmetHuG-CSF, Neulasta). *Curr Pharmaceut Design* 2004;10:1235–1244.
17. Harris JM, Martin NE, Modi M. Pegylation: a novel process for modifying pharmacokinetics. *Clin Pharmacokinetics* 2001;40:539–551.
18. Narhi LO, Arakawa T, Aoki K et al. Asn to Lys mutations at three sites which are N-glycosylated in the mammalian protein decrease the aggregation of Escherichia coli-derived erythropoietin. *Protein Eng* 2001;14:135–140.
19. Ibrahimi OA, Yeh BK, Eliseenkova AV et al. Analysis of mutations in fibroblast growth factor (FGF) and a pathogenic mutation in FGF receptor (FGFR) provides direct evidence for the symmetric two-end model for FGFR dimerization. *Mol Cell Biol* 2005;25:671–684.
20. Hill CP, Osslund TD, Eisenberg D. The structure of granulocyte-colony-stimulating factor and its relationship to other growth factors. *Proc Natl Acad Sci USA* 1993;90:5167–5171.
21. Bazan JF. Haemopoietic receptors and helical cytokines. *Immunol Today* 1990;11:350–354.
22. Hilton DJ, Watowich SS, Katz L et al. Saturation mutagenesis of the WSXWS motif of the erythropoietin receptor. *J Biol Chem* 1996;271:4699–4708.
23. Philo JS, Wen J, Wypych J et al. Human stem cell factor dimer forms a complex with two molecules of the extracellular domain of its receptor, *kit*. *J Biol Chem* 1996;271:6895–6902.
24. Barbone FP, Middleton SA, Johnson DL et al. Mutagenesis studies of the human erythropoietin receptor. Establishment of structure-function relationships. *J Biol Chem* 1997;272:4985–4992.
25. Wen D, Bossiel JP, Tracy TE et al. Erythropoietin structure-function relationships: high degree of sequence homology among mammals. *Blood* 1993;82:1507–1516.

Erythropoietins, Erythropoietic Factors and Erythropoiesis
Edited by S.G. Elliott, M.A. Foote, and G. Molineux

Intracellular signaling by the erythropoietin receptor

Harvey F. Lodish[1], Saghi Ghaffari[2], Merav Socolovsky[3], Wei Tong[4] and Jing Zhang[5]

[1] *Whitehead Institute for Biomedical Research, Cambridge, MA and Department of Biology, Massachusetts Institute of Technology, Cambridge, MA, USA*
[2] *Department of Gene and Cell Medicine, Mount Sinai School of Medicine, New York, NY, USA*
[3] *Department of Pediatrics and Department of Cancer Biology, University of Massachusetts Medical School, Worcester, MA, USA*
[4] *Department of Pediatrics, University of Pennsylvania School of Medicine and Division of Hematology, Children's Hospital of Philadelphia, Philadelphia, PA, USA*
[5] *McArdle Laboratory for Cancer Research, University of Wisconsin, Madison, WI, USA*

Introduction

The erythropoietin receptor (EPOR) is crucial for promoting the survival, proliferation, and differentiation of mammalian erythroid progenitors [1]. Differentiation from colony-forming units–erythroid (CFU–E) to late basophilic erythroblasts is highly EPO dependent, whereas differentiation beyond this stage is no longer dependent on EPO, but requires adhesion of the cells to a fibronectin matrix, through α4β1 integrin [2].

EPOR and other cytokine receptors lack enzymatic activities and rely on the Janus kinase (JAK) family kinases for signal transduction that are bound through a conserved Box1 and Box2 domain in the receptor membrane-proximal portion. EPO binding triggers a conformational change in EPOR dimer, leading to trans-phosphorylation and activation of JAK. The cytoplasmic tail of EPOR contains multiple Tyr that, when phosphorylated by JAK, provide docking sites for SH2-containing cytosolic signaling molecules, including Stat5, phosphoinositide-3 kinase (PI3K), and the Shc and Grb2 proteins that activate the Ras/mitogen-activated kinase (MAPK) pathways. Importantly, phosphorylated receptors and JAK also recruit multiple signaling attenuators, including phosphatases (SHP) such as the protein tyrosine phosphatase SHP-1, several suppressor of cytokine signaling (SOCS) proteins, and members of the newly discovered Lnk adaptor family proteins.

The major function of EPOR is to prevent apoptosis of committed erythroid progenitors, to initiate a predetermined program of three to five-terminal cell divisions, and to activate a program of erythroid gene expression. EPOR activates the same signal transduction pathways that are activated by other cytokine receptors and many receptor tyrosine kinases. Thus, the response of a cell to a signal generated by EPOR is dependent on its properties and developmental history – presumably erythroid progenitors contain a unique constellation of transcription factors, such as GATA1 and FOG, and other proteins that allow them to interact with proteins, such as Stat5, that are activated in response to receptor signaling [3–6]. Such combinations of regulated and 'cell-specific' factors could regulate the highly specific pattern of erythroid gene expression [7].

We focus on the major signal transduction pathways activated by EPOR that are important for erythropoiesis (Fig. 1, see p. 159).

EPOR-activated STAT5 signaling

Signal transducer and activator of transcription (Stat) proteins were first discovered as mediators of interferon signaling [8] and are key mediators of signaling downstream of multiple cytokine receptors. Stat proteins are latent in the cytoplasm, but once phosphorylated through receptor activation, they relocalize to the nucleus where they act as transcriptional activators. Thus, they provide a rapid, single-step mechanism for cytokine-mediated gene induction.

Of the seven known mammalian Stat proteins, Stat3 and the two highly homologous Stat5a and Stat5b isoforms (here grouped simply as 'Stat5') have pleiotropic functions in multiple tissues. Within the hematopoietic system, Stat5 mediates survival and proliferative functions in hematopoietic stem cells as well as in lymphoid and myeloid progenitors. Stat5 is the principal Stat to be activated by EPOR. Although there have been reports of Stat1 and Stat3 activation by EPOR in some cell lines [9] their physiologic significance is uncertain.

STAT5 functional domains

The Stat proteins range in size between 750 and 900 amino acids and share conserved functional motifs, revealed by sequence comparisons and mutagenic and crystallographic studies. These motifs comprise the amino-terminal, the coiled coil, the DNA binding, the conserved linker, an SH2, a Tyr activation, and the *C*-terminal transactivation domains.

The SH2 domain of Stat5 binds phosphorylated Tyr on the activated EPOR that presumably increases the effective concentration of Stat5 in the vicinity of Jak2, facilitating its Tyr phosphorylation, presumably by Jak2. Of the eight EPOR cytoplasmic domain Tyr, Y343 and Y401 are sufficient for maximal Stat5 activation by EPOR, whereas Y429 and Y431 partially activate Stat5 [10]. The sequence surrounding these Tyr suggests that the Stat5 SH2 domain recognizes a YXXL motif on EPOR, a motif conserved in frog and fish EPOR [11].

EPOR-bound Stat5 becomes phosphorylated on Tyr 694, which then mediates Stat5 homodimerization through reciprocal phosphotyrosine-SH2 domain interactions. A mutant EPOR lacking all cytoplasmic phosphotyrosines can still mediate Stat5 tyrosine phosphorylation, albeit at reduced levels [10], consistent with the finding that an EPOR lacking all cytoplasmic Tyr can still generate survival signals in a Stat5-dependent manner [12]. Similarly, a knock-in mouse expressing a truncated EPOR lacking cytoplasmic tyrosines, EPOR-HM, is viable; we found that Stat5 becomes tyrosine-phosphorylated after EPO addition to primary erythroblasts of the EPOR-HM mouse, though with slower kinetics than wild-type erythroblasts (E. Porpiglia and M. Socolovsky, unpublished observation). One explanation of these observations is that Stat5 can bind directly to EPOR-activated Jak2, without mediation of EPOR cytoplasmic Tyr [13]. Potential mechanisms include an interaction between the *C*-terminal domain of Stat5 and the Jak2 pseudokinase (JH2) domain [14] or an interaction of the Stat5 SH2 domain with tyrosines in the Jak2 activation loop [15].

The Stat DNA-binding domain recognizes a family of palindromic 'GAS' enhancers of the general sequence TTTCCNGGAAA. The optimal sequence for both Stat5a and Stat5b is TTC(C/T)N(G/a)GAA [16]. However, Stat5 optimally binds to two tandem GAS enhancers, separated by six base pairs. These tandem sites bind optimally to Stat5 tetramers, while their binding of Stat5 dimers is weak or absent. Several such putative Stat5 tetramer-binding sites have been identified in known erythroid Stat5 target genes, including *bcl-*x_L and *pim1* [16], suggesting that Stat5 tetramers may induce a repertoire of genes that differs from that of Stat5 dimers.

The *N*-terminal domain is implicated in protein-protein interactions and is responsible for the formation of Stat5 tetramers. Tetramer formation depends on a conserved *N*-terminal Trp residue as well as on a Lys residue in this domain. Stat5 mutants that lack this Trp residue have reduced transcriptional activation properties, presumably due to their inability to form tetramers and cooperatively bind DNA GAS sites [13]. The *N*-terminal domain, together with the adjacent coiled coil domain, participate in a radical spatial reorientation of the Stat1 [18] and Stat5 dimers [17], that stabilizes the dephosphorylated, inactive form of these Stats in an antiparallel homodimer. Tyr phosphorylation results in a flipping of this structure into a parallel homodimer that is stabilized by reciprocal SH2-phosphotyrosine interactions [13, 18, 19].

The transactivation domain (TAD) at the *C*-terminus was originally identified in Stat1, where the splice variant Stat1β, lacking this domain, also lacks independent transcriptional activity. Similarly, a naturally occurring, protease-mediated *C*-terminal truncation of Stat5, Stat5Nsi, has a dominant negative effect on Stat5 transcription [20]. The Stat5 TAD has two Ser phosphorylation sites that are thought to enhance transcription. One of the Ser is within a PSP motif, recognized by MAP kinases [21] but the responsible kinase is not known. The transcriptional activation function of the TAD is at least in part due to its ability to complex with the coactivator histone acetyltransferases CBP/p300 [22].

Regulation of Stat5 tyrosine phosphorylation

Despite its apparent simplicity, gene induction by Stat proteins is subject to multiple regulators that determine the duration and termination of Stat signaling. These regulators affect JAK kinases that lead to Stat phosphorylation, phosphatases that dephosphorylate Stat proteins, and others that modify cytoplasmic/nuclear Stat transport.

Tyr phosphorylation of Stat proteins results in their redistribution from the cytoplasm to the nucleus, a process that is regulated differently for different Stat proteins. In the case of Stat5 and Stat3, but probably not Stat1, the inactive Stat molecules continuously shuttle in and out of the nucleus, although it is not known if they mediate a transcriptional function in this state. After cytokine stimulation, Stat5 nuclear import rate does not change, but the Tyr-phosphorylated dimer conformation may be resistant to nuclear export [23, 24].

Stat5-mediated function in erythropoiesis

Stat5 has a specific and essential role in EPO-mediated survival signaling in committed erythroid progenitors. We described a principal antiapoptotic pathway, whereby EPOR-activated Stat5 induces the immediate-early transcription of the antiapoptotic gene *bcl-x*$_L$ [25]. The bcl-x$_L$ protein is essential for erythropoiesis, and the *bcl-x*$_L$ knockout mouse dies at embryonic day 13 due to severe anemia [26]. Several transcription factors are thought to regulate EPO-independent expression of *bcl-x*$_L$ within erythroblasts, including GATA1. However, EPO signaling synergizes with GATA1 to substantially increase the amount of bcl-x$_L$ in erythroid precursors. The principal mediator of this effect is Stat5. Two Stat5 consensus binding sites are located within the first intron of the *bcl-x*$_L$ gene and that bind Stat5 directly after stimulation by EPO [26, 27]. Consistent with Stat5-mediated bcl-x$_L$ induction, cultured erythroid cells that have been deprived of EPO can be rescued from apoptosis by transduction with a constitutively active *Stat5* gene that maintains expression of *bcl-x*$_L$ [26].

To evaluate the role of Stat5 in erythropoiesis *in vivo*, we examined Stat5ab-deficient mice. The erythroid phenotype of these mice was initially overlooked [28–30]. However, Stat5ab-deficient mice do have an erythropoietic rate deficit manifesting as fetal and neonatal anemia. Anemia persists in approximately half the mice that survive into adulthood. The remainder attain a near-normal hematocrit, but are deficient in their response to acute phenylhydrazine-induced erythropoietic stress [25]. The anemia of Stat5ab-deficient mice is attributed to increased apoptosis of Stat5ab-deficient erythroblasts. Stat5ab-deficient fetal liver cells also undergo higher apoptosis and have a blunted response to EPO *in vitro* [25, 32]. Stat5ab-deficient mice were originally generated by deletion of the first exon each of the *Stat5a* and *Stat5b* genes [28], and are believed to express variable low levels of a transcription-

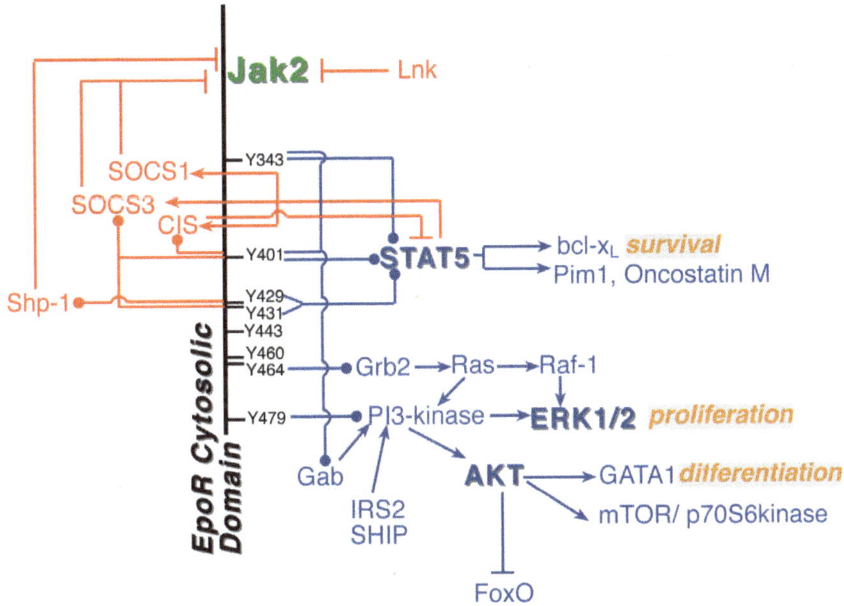

Figure 1. EPOR-activated signaling pathways. Summary of EPOR-activated pathways described in the text. The black vertical line represents the signaling domain of the EPOR. Jak2 (green) binds in the box1/box2 region (not marked) upstream of Tyr343, and its activation is the first step in the signaling cascade. Jak2 activation leads to phosphorylation of up to eight cytoplasmic tyrosines (numbered), which serve as docking sites for signaling molecules such as STAT5 and for negative regulators such as SOCS3 and CIS. Blue lines and arrows indicate positive signaling pathways that lead to survival, proliferation, and differentiation functions. Red lines and arrows indicate negative regulation of EPOR signaling. Lines terminating in a dot extending from specific tyrosines, mark docking sites for the indicated molecules. Arrows indicate activation by upstream kinases or transcriptional activation. Lines terminating in a horizontal line indicate inhibition.

ally active, *N*-terminally truncated Stat5. Stat5 null mice have been generated, where the entire *Stat5a* and *Stat5b* genes are deleted [31]. These mice have a qualitatively similar, but more severe erythroid phenotype, with fetal anemia leading to perinatal death of >99% of the mice.

To investigate the anemia in Stat5ab-deficient mice, we developed a flow-cytometric assay that identifies stage-specific erythroblasts directly in tissue [32]. Erythroblasts form the bulk of erythropoietic tissue and are classified by their morphology into a well-established developmental sequence. The Ter119 marker is specific to all erythroblasts. Ter119 positive cells were subdivided further into subsets of similar maturation by their level of expression of the transferrin receptor (CD71). Early erythroblasts are EPO dependent and express high levels of CD71, which subsequently declines with maturation (Fig. 2a).

Flow-cytometric analysis of Stat5ab-deficient hematopoietic tissue showed a dramatic increase in early ($Ter119^{high}CD71^{high}$) erythroblasts, which were

2a

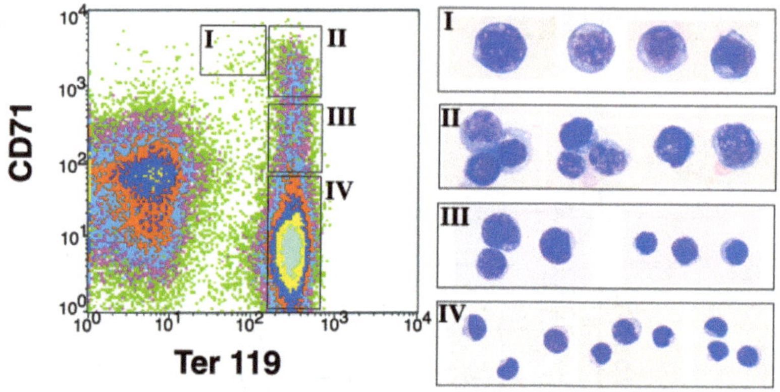

2b

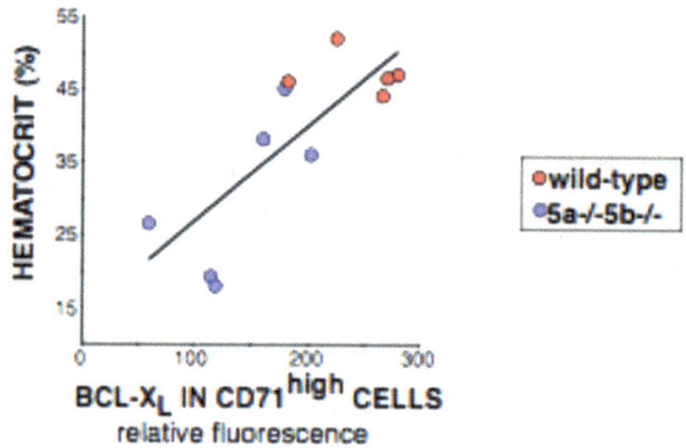

Figure 2. (a) CD71/Ter119 Flow-cytometric assay for stage-specific erythroblasts. Flow-cytometric profile of CD71 and Ter119-labeled spleen cells (right panel) and cytospins of cells sorted from regions I to IV (left panel). Originally published in [32]. (b) A correlation between bcl-x_L expression and hematocrit in $Stat5^{-/-}$ mice. Originally published in [32]. © the American Society of Hematology.

undergoing high rates of apoptosis and failing to differentiate efficiently into red cells. The amount of *bcl-*x_L in these early erythroblasts were reduced in a manner that correlated with the phenotype of the mice: the most severely anemic mice expressed the lowest amounts of *bcl-*x_L, whereas mice with near-normal hematocrit had near-normal *bcl-*x_L (Fig. 2b). These findings underlie the importance of the Stat5-bcl-x_L pathway for the maintenance of basal erythropoietic rate and normal hematocrit [32]. In its absence, increased apoptosis of early erythroblasts (ineffective erythropoiesis) results in anemia.

The role of STAT5 in the erythropoietic stress response

The erythropoietic system has a large reserve capacity, which can increase erythropoietic rate up to 10-fold in response to tissue hypoxia. EPO and EPOR signaling are principal mediators of this stress response. The Stat5ab-deficient mice mount a weak response to, and take longer to recover from, an acute phenylhydrazine induced erythropoietic stress [32]. Similarly, a knock-in mouse in which the wild-type EPOR is replaced with a truncated EPOR allele that lacks cytoplasmic domain Tyr (the 'HM' mouse), activates Stat5 weakly, and is similarly deficient in its response to acute phenylhydrazine-induced stress. These observations suggest that Stat5 plays a role in the erythropoietic response to stress. In addition to bcl-x_L, *in vitro* studies with HM mouse erythroblasts suggest that high amounts of EPO lead to Stat5-mediated induction of oncostatin-M and Pim-1 [33], but it is not known whether these molecules play a role in the erythropoietic stress response.

The abnormal response of the Stat5ab-deficient and the HM mouse models to stress does not, however, necessarily imply that Stat5 plays a role in the physiologic response to stress of the normal mouse. We found that wild-type mice injected with a single high dose of EPO respond with a transient increase in phosphorylated Stat5 in early erythroblasts in both bone marrow and spleen. We also found a transient increase in the bcl-x_L protein in the same cells, which peaked at 18 h and subsequently declined. By contrast, examination of mouse models of chronic erythropoietic stress does not show an increase in the amounts of either phosphorylated Stat5 or bcl-x_L above those found in erythroblasts in the basal state (E. Porpiglia, Y. Liv and M. Socolovsky, unpublished observation). These findings suggest that the Stat5-bcl-x_L pathway, in addition to its essential role in basal erythropoiesis, contributes to the acute response to erythropoietic stress. The response to chronic stress, however, does not use this pathway. We found that high concentrations of EPO result in suppression of Fas-mediated apoptosis in erythroblasts, both during fetal development and during the response to stress erythropoiesis in the adult mouse. Unlike the Stat5-bcl-xL pathway, EPO-mediated Fas and FasL suppression occurs in both acute and chronic erythropoietic stress [34, 35].

EPOR-stimulated PI3-kinase/AKT signaling

In response to EPO, the PI3K signaling pathway is rapidly activated in EPO-dependent cultured cells and in mouse and human erythroid progenitors. EPOR activation of PI3-kinase (PI3K) signaling is mediated by several distinct mechanisms. Phosphorylated EPOR Y479 binds directly to the p85 subunit of PI3K; this residue (Y479) is essential for EPOR binding to the P85 subunit of PI3K [36–38]. In addition, phosphorylation of Y479 in response to EPO is sufficient for activation of the PI3-K signaling pathway, as has been demonstrated by the ability of the F7Y479 EPOR mutant, in which all cytosolic tyrosines but Y479 are mutated to Phe, to activate PI3K and extracellular signal-regulated kinase (ERK) but not Stat5 signaling pathway in response to EPO in heterologous cells.

(Phospho) Y479 recruits PI3K to EPOR, resulting in RAS-independent activation of a MAPK cascade (ERK1/2) [36]. A role for protein kinase C (PKC) as an intermediate in this activation has also been suggested [39, 40]. It is not clear whether (phospho) Y479 also supports the activation of other MAPK pathways such as JNK or p38. Each of 2 Tyr, Y464 and Y479, alone can mediate a proliferative signal in EPOR-expressing BaF3 cells. However, only Y479 (but not Y464) is capable of inducing the differentiation of >85% of mature erythroid EPOR null CFU–E ectopically expressing mutant EPOR. (Phospho) Y479 is the only Tyr residue on EPOR that can support both proliferation and differentiation of erythroid cells to the same extent as the wild-type receptor [41], suggesting that pathways additional to ERK also may be activated by PI3K in response to EPO.

EPOR stimulates PI3K signaling pathway through several additional mechanisms. In the primitive hematopoietic UT7 cells and in human erythroid progenitor cells, EPOR is constitutively associated with Insulin Receptor Substrate (IRS)-2 [42]. After EPO stimulation, IRS-2 becomes rapidly Tyr phosphorylated and associates with the P85 subunit of PI3K and with phosphatidylinositol 3,4,5 triphosphate 5-phosphatase (SHIP) [43]. Thus, EPOR-associated IRS-2 provides an alternative mechanism for activation of PI3K in response to EPO. Furthermore, phosphorylated EPOR Tyr343 and Tyr401 recruit Gab proteins leading to their phosphorylation, association with, and activation of PI3K [42, 44]. Src tyrosine kinase may also mediate phosphorylation of EPOR resulting in its association with the P85 subunit of PI3K as has been observed in erythroleukemic K562 cells [45].

The existence of many isoforms and subunits of PI3K has precluded the assessment of the exact requirement for PI3K in the regulation of erythropoiesis. The activation of PI3K is essential for optimal erythropoiesis since the deletion of at least the p85α subunit of PI3K results in significant suppression of primitive and mature erythroid progenitor production [46].

Signaling by activated AKT (PKB) protein kinase

Three Akt (protein kinase B) forms are known in mammals (Akt-1, Akt-2, and Akt-3) and are generated from three genes. PI3K-generated phospholipids bind to the Akt pleckstrin homology (PH) domain. As a consequence, Akt translocates from the cytoplasm to the plasma membrane, where Akt is phosphorylated by regulatory kinases on two residues critical for its activation. These residues are a Thr (Thr308 in Akt1) in the activation loop within the catalytic domain and a Ser (Ser473 in Akt1) within the hydrophobic motif in the *C*-terminal region. PDK1 is the kinase phosphorylating Akt in the activation loop. The kinase phosphorylating the hydrophobic motif, named PDK2, has been identified as mammalian target of rapamycin complex 2 (mTORC2) [47]. Mammalian TOR (mTOR) is a large protein kinase that exists in two forms, associated with either rictor or raptor [48, 49]. In contrast to the raptor containing TOR (mTORC1), the rictor containing TOR (mTORC2) is not sensitive to rapamycin. The rictor-mTOR complex (mTORC2) directly phosphorylates Akt on Ser 473 and facilitates phosphorylation of Thr 308 by PDK1 [47].

Akt Ser Thr kinase promotes survival by several mechanisms, including phosphorylation and inhibition of several proapoptotic proteins such as caspase 9, Bad, and FoxO transcription factors or enhancement of ubiquitination of p53 through phosphorylation of Mdm2. Akt has many additional functions in the regulation of metabolism, gene expression, and cell fate specification. Akt also modulates cell growth and the protein synthesis machinery through regulation of mTOR/P70S6kinase signaling pathway [50]. The function of mTOR signaling in erythropoiesis is not known.

EPO stimulation of erythroid cells results in rapid phosphorylation and activation of Akt by PI3K in both primary human erythroid progenitors [51] and cultured erythroid cells [52, 53]. Several lines of evidence suggest that in erythroid cells, the PI3K/Akt signaling pathway mediates not only survival but also growth and differentiation. For example, inhibition of PI3K activation prevents differentiation of primary human erythroid progenitor cells in culture [51, 52]. In addition, Akt activity is critical for survival and maturation of primary mouse erythroid precursor cells [54, 55]. In response to EPO stimulation of cultured and primary erythroid cells, Akt phosphorylates the GATA-1 transcription factor at Ser310 and transactivates it [54, 56]. Mutation of Ser310 to Ala suppresses the transcriptional activity of GATA-1 by >30% in erythroid cells [54, 56]. Mice with the GATA-1 S310A mutation do not exhibit an erythroid phenotype, suggesting that molecular consequences associated with loss of phosphoS310 in GATA-1 are compensated *in vivo* [57]. Akt enhancement of GATA-1 activity is not limited to phosphorylating GATA-1S310 and is mediated by both phosphorylation-dependent and independent mechanisms [54]. These findings suggest that Akt regulation of erythropoiesis may be mediated by nonphosphorylation-dependent regulation of GATA-1 and regulation of erythropoiesis independently of GATA-1. One such mechanism may be through regulation of FoxO transcription factors.

Despite these results, mice lacking Akt1, Akt2, Akt3, or all Akt isoforms have not been reported to exhibit any erythroid phenotype [58–60] possibly due to functional redundancy of different Akt proteins. The erythroid phenotype in these mice has not been closely examined. Loss of Pim1/Pim2 leads to a mild hematopoietic phenotype. Mice lacking Akt1 in which genes for Pim1/Pim2 kinases are deleted have a severe erythroid phenotype, suggesting a role for Akt in the *in vivo* regulation of erythropoiesis [61].

Akt regulation of erythropoiesis may be mediated by its ability to phosphorylate and inhibit FoxO transcription factors [62]. FoxO transcription factors are mammalian homologs of DAF-16 in *C. elegans*. FoxO1, FoxO3, and

FoxO4 are ubiquitous, whereas FoxO6 is primarily expressed in the brain. EPO stimulation of cultured and primary erythroid cells leads to rapid PI3K/Akt-mediated phosphorylation of FoxO3 (the mouse homolog of human FOXO3a, FKHRL-1), a member of the FoxO family of Forkhead transcription factors, on three conserved residues (Thr32, Ser253, and Ser315 in FoxO3) [52, 63]. FoxO3 is the most abundant and active FoxO expressed in primary mouse erythroid precursor cells. While FoxO1 is expressed in fetal liver and bone marrow hematopoietic cells, including erythroid progenitors, FoxO1 expression is greatly downregulated during primary erythroid cell differentiation [64]. FoxO4 expression is barely detectable in primary erythroid cells. Phosphorylation of FoxO3 in erythroid cells leads to its cytosolic localization and suppression of its transcriptional activity [65]. FoxO3 regulates the expression of B-cell translocation gene 1 (BTG-1) in *p53* null erythroblasts, which may be important for erythroid cell maturation [66]. Mature CFU–E express the highest concentrations of EPOR; the expression of EPOR is downregulated on postprogenitor erythroid cells suggesting that EPOR signaling is gradually lost during erythroid maturation. Both FoxO3 expression and activity increase as primary erythroid precursors mature, suggesting that loss of EPOR signaling (and therefore Akt signaling) on erythroid precursor cells may be concomitant with enhanced FoxO3 activity [64]. The function of FoxO depends on the cell context. In erythroid cells, FoxO3 is essential for the regulation of lifespan, cell cycling, and maturation, and this regulation is mediated by controlling oxidative stress in these cells [64].

While EPO has been associated with regulation of cell mitosis, the exact downstream effectors are not known. Several cell cycle regulators and transcription factors are modulated by EPOR signaling in mouse and human erythroid progenitor cells including $p27^{KIP}$, Bcl6, $P21^{CIP}$, Cyclin D2, ataxia telangiectasia mutant (ATM), and Cyclin G2. Some of these may be regulated by the PI3K/Akt pathway in primary erythroid cells [64, 67–69]. In particular, $P27^{KIP}$ and Cyclin G2 may be direct transcriptional targets of FoxO3 in primary erythroid cells and presumably are regulated by the PI3K signaling pathway [64]. The function of ATM in erythropoiesis is not known. The activation of the PI3K/Akt pathway has been shown to override the DNA damage-induced cell cycle arrest of hematopoietic cells [70]. Whether this function is mediated by FoxO (and/or FoxO regulation of ATM) or is a direct consequence of Akt function is not known. Clearly further investigation is required to establish the exact role of EPO in the regulation of erythroid cell mitosis.

These studies establish that, by modulating the transcriptional activity of GATA-1 and FoxO transcription factors, Akt signaling participates in the regulation of erythroid cell differentiation and maturation. While it is clear that the PI3K/Akt signaling pathway plays important functions in the regulation of erythroid cell formation, further studies should identify downstream effectors and assess the exact mechanism of Akt function in response to EPO.

EPOR-stimulation of RAS/MAPK pathways

In mammalian cells, Ras activates groups of MAPK, including ERK, JNK, or SAPK (c-Jun *N*-terminal kinase) and p38 (reviewed in [71]). In various EPO-dependent cell lines, different groups of MAPK are activated in response to EPO [72–76]. Activation of the Ras/ERK pathway may occur through the canonical Grb2 binding site at (P)Y464 of EPOR [77] and cell proliferation promoted by this pathway in response to EPO requires Raf-1 [78]. Synergistic activation of ERK1/2 by EPO and stem cell factor (SCF) may be important for proliferation of human erythroid progenitors [79]. Alternatively, the ERK pathway may be activated independent of Ras through the PI3K pathway [36]. JNK, and p38 are implicated in inducing erythroid differentiation of EPO-dependent SKT6 cells [76] and the role of JNK and p38 in inducing apoptosis of erythroid cells remains controversial [76, 80, 81].

To date all the three *ras* genes (*H-*, *N-*, and *K-ras*) have been knocked-out in mice. Among them, only *K-ras* is essential for normal mouse development since *K-ras*$^{-/-}$ and *N-ras*$^{-/-}$; *K-ras*$^{+/-}$ mice die at mid-gestation stages with anemia [82]. The study of endogenous *K-ras* signaling in erythropoiesis has been greatly advanced by using a system based on highly purified erythroid progenitors from mouse fetal livers [83]. In this system, large amounts of erythroid progenitors and early erythroblasts can be purified through a single-step procedure from mouse fetal livers with approximately 75–85% purity. These cells express high amounts of receptors for EPO and stem cell factor (SCF); thus biochemical studies can be conducted using a pure primary cell population upon EPO or SCF stimulation. Purified erythroid progenitors can be cultured *in vitro* to go through their normal terminal proliferation and differentiation. During the two-day culture period, the cell number increases 15- to 20-fold, and erythroid differentiation can be monitored step-by-step and quantitatively using a flow cytometric analysis based on the expression of erythroid-specific TER119 and non-erythroid specific CD71 (transferrin receptor) surface proteins (Fig. 3).

In this system, overexpression of oncogenic Ras, mediated by retroviral vectors, leads to constitutive activation of its three major downstream signaling pathways, including the Raf/ERK, PI3K/Akt, and RalGEF/Ral pathways [84]. Consequently, it results in a severe block of terminal erythroid differentiation and cytokine-independent growth of erythroid progenitors [83].

In contrast to results obtained when oncogenic Ras was overexpressed from retroviral vectors, endogenous levels of oncogenic *K-ras* fail to constitutively activate the Stat5, Akt, and p44/42 MAPK pathways in primary erythroid progenitors. Rather, this hyperactivates these EPO-dependent signaling pathways [85], which explains previous observations that hematopoietic progenitors expressing oncogenic *K-ras* from its endogenous locus display hypersensitivity to cytokine stimulation in various colony assays [86, 87]. Consequently, expression of endogenous oncogenic *K-ras* leads only to a mild block of terminal erythroid differentiation and a mild hyperproliferation of primary ery-

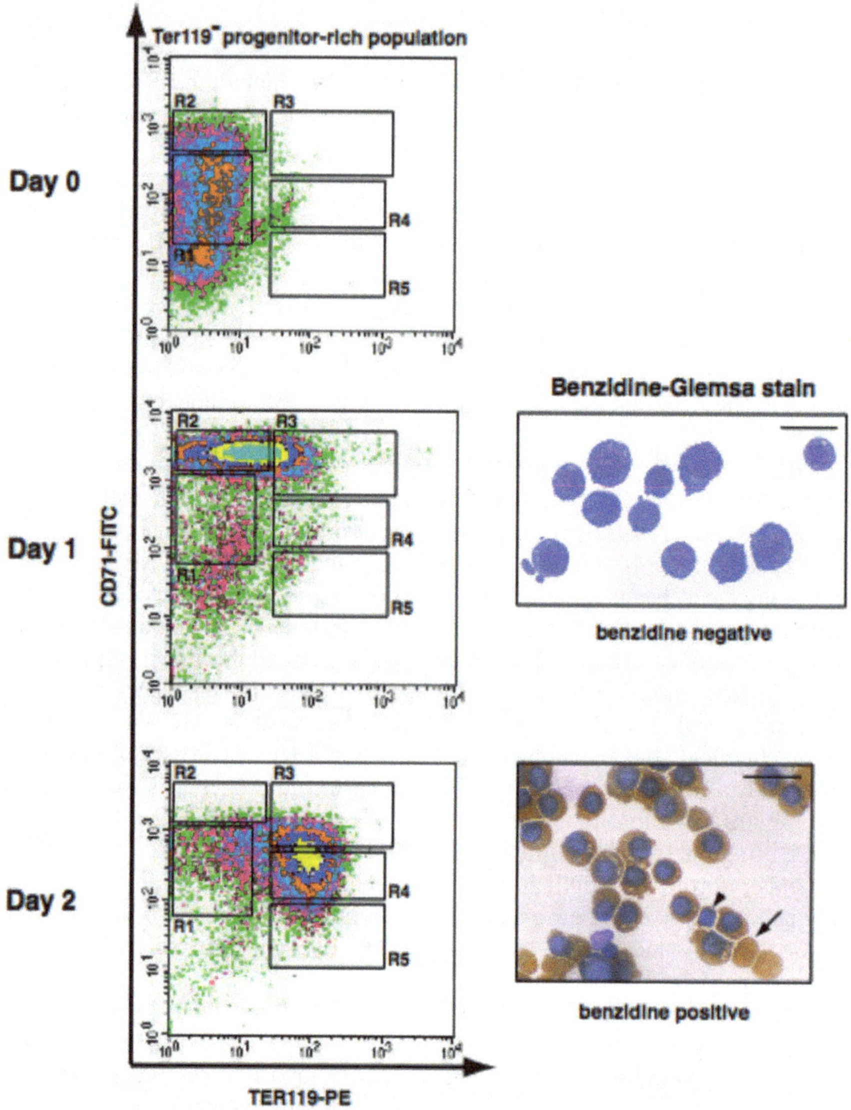

Figure 3. Flow cytometry analysis of fetal liver erythroblasts cultured *in vitro*. TER119-negative progenitor-rich cells were purified on day 0. Purified TER119-negative cells (approximately 70–80% R1 and R2 cells; essentially no R3–R5 cells) were cultured *in vitro* for 16–18 h on fibronectin-coated plates in medium containing serum and EPO. EPO was removed from culture later on. The differentiation profiles of cultured cells were examined by flow cytometry (left panels) and Benzidine-Giemsa stain (right panels). The arrowhead indicates an extruded nucleus and the arrow indicates an enucleated reticulocyte. Scale bar: 20 μm. Originally published in [83]. © the American Society of Hematology.

throid progenitors in the *in vitro* culture system [86]. Importantly, although endogenously expressed oncogenic *K-ras* does result in mild hyperproliferation of primary erythroid progenitors, it does not support cytokine-independent growth of these cells, in contrast to overexpression of oncogenic Ras [83].

Complimentary to the study of oncogenic *K-ras*, loss-of-*K-ras* function greatly reduces cytokine-dependent Akt activation [88]. Although EPO stimulation activates all the Ras isoforms, it primarily activates *N*- and *K-ras*. Moreover, it is *K-ras* but not *N-ras* that mainly regulates cytokine-dependent Akt activation [88]. The underlying mechanisms remain elusive, but may be due to different expression levels of Ras isoforms as well as their diversified *C*-termini, which greatly modulate their association with downstream effectors [89, 90]. Other studies showed that extensive reduction of Akt activation in *K-ras*-deficient erythroid progenitors leads to delayed erythroid differentiation, which only occurs during cytokine-regulated differentiation stages and is rather mild [88].

Taken together, both of gain and loss of *K-ras* signaling regulate erythroid differentiation through modulation of cytokine-dependent signaling (Fig. 4). Since the erythroid progenitors used in these studies are late progenitors mainly regulated by two cytokines (EPO and SCF), endogenous *K-ras* signaling only has mild effects on their differentiation.

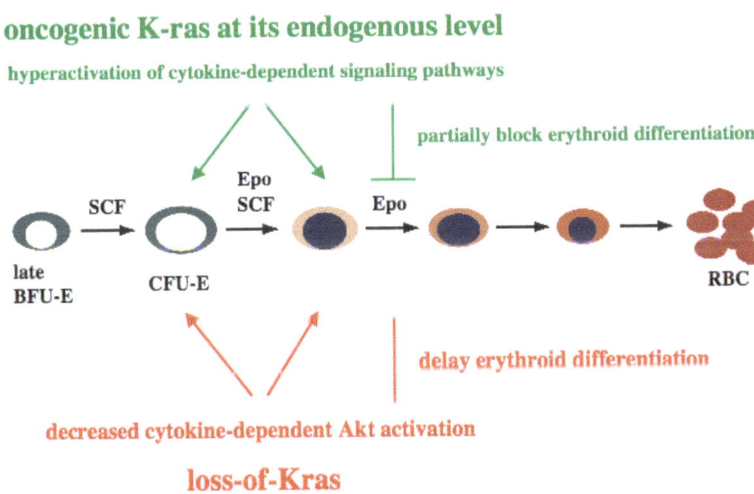

Figure 4. Endogenous *K-ras* signaling regulates erythroid differentiation through moderating cytokine-dependent signaling pathways. Reproduced from [110].

Negative regulators of EPO signaling

Erythroid progenitor cells possess considerable heterogeneity in their sensitivity to EPO, which is believed to provide flexibility to accelerate or decelerate red cell production in response to stress, such as anemia or polycythemia. GATA-1 promoter driven EPOR transgene expression rescues EPOR nullizygous mice from fetal anemia [91]. EPOR transgene expression levels correlate with their sensitivities to EPO in adult bone marrow CFU–E progenitors. Therefore, the magnitude of EPOR signaling pathways within erythroblasts results in different cellular responses to EPO. This intrinsic difference of erythroblasts in their sensitivity to EPO might be tightly regulated through several negative signaling molecules of EPO/EPOR signaling that control the threshold of EPO-dependent survival. Phosphorylated receptors and JAK recruit multiple signaling attenuators that provide checks and balances at multiple steps of cytokine-receptor signaling transduction to ensure tightly controlled cellular response to fit physiologic needs and prevent oncogenic transformation. In particular, the length of the phospho-Stat5 signal is determined by a number of negative regulators, and usually does not persist beyond a few hours.

One class of attenuators is the SH2 domain-containing protein tyrosine phosphatase Shp-1. Shp-1 is relatively limited in its distribution to hematopoietic cells, where it acts as a negative regulator [92]. In the basal state Shp-1 is inactive in the cytoplasm, due to inhibition of its catalytic loop by its *N*-terminal SH2 domain. Binding to phosphotyrosine 429 of EPOR cytosolic domain, through its SH2 domains, activates phosphatase activity. Shp-1 may negatively regulate EPOR signaling by binding the activated EPOR and dephosphorylating Jak2 [93, 94]. Shp-1 may also directly inhibit Stat5 by translocating to the nucleus where it dephosphorylates Stat5b [95]. Another cytoplasmic phosphatase, PTP1B, is a specific Stat5 phosphatase, and other nuclear phosphatases, including TC45, may also act on Stat5 [96, 97].

Another class of negative regulators is SOCS proteins. SOCS proteins are transcriptionally induced by JAK/Stat pathways upon cytokine stimulation, and act within a classical negative feedback loop both to inhibit activation of signaling pathways and to target signaling components for proteasomal degradation [98, 99]. SOCS1 and SOCS3 bind to the catalytic groove of JAK2 to inhibit its catalytic activity. Both SOCS proteins can target JAK2 for ubiquitination and degradation through their SOCS box ubiquitin E3 ligase interaction motif [100, 101]. Additionally, SOCS3 binds through its SH2 domain to Tyr residues of the EPOR: Y401, where STAT5 and SHP-2 also bind, and Y429/Y431, where SHP-1 also binds [102, 103]. This presumably reduces binding and activation of other SH2-containing signal transduction proteins. Importantly, SOCS1-deficient erythroid progenitors exhibit an enhanced sen-

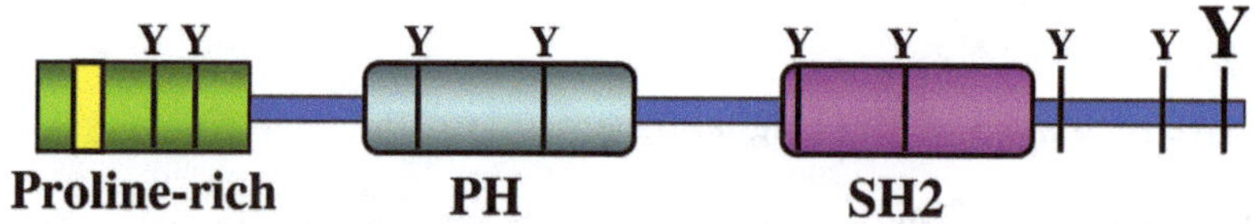

Figure 5. The protein structure of Lnk and its family members, APS and SH2-B. ‘Y’ indicates tyrosine residues, with the conserved tyrosine near the *C*-terminus shown in a large font. Positions of Y are drawn according to Lnk.

sitivity to EPO in forming erythroid colonies [104], consistent with its role as a principal negative regulator of the EPOR signaling pathway.

Recent work uncovered a new class of signal attenuators for rapidly down-regulating cytokine receptor signaling, the Lnk adaptor proteins [105, 106]. The Lnk family of proteins, which includes SH2-B and APS, are structurally distinct from the above-noted negative regulators in that they do not possess any known enzymatic activity, but contain other protein–protein interacting domains: a PH and a Src homology 2 domain (SH2) and a conserved Tyr near the *C*-terminus [107] (Fig. 5). Lnk-deficient mice have increased numbers of erythroid progenitors in the bone marrow and spleen. Bone marrow CFU–E progenitors are hypersensitive to EPO in generating red cell colonies. Despite having normal, steady-state red blood cell counts, $Lnk^{-/-}$ mice exhibit superior recovery after erythropoietic or myeloablative stress. In addition, Lnk deficiency resulted in enhanced EPO-induced signaling pathways, such as p42/44MAPK and Akt activation, in splenic erythroid progenitors.

Conversely, Lnk overexpression inhibits EPO-induced cell growth in 32D/EPOR cells. In primary culture of fetal liver cells, Lnk overexpression inhibits EPO-dependent erythroblast differentiation and induces apoptosis, without causing cell cycle arrest. Lnk blocks the three major signaling pathways – Stat5, Akt, and MAPK – induced by EPO in primary erythroblasts. In addition, the Lnk SH2 domain is essential for its inhibitory function, whereas the conserved tyrosine near the *C*-terminus and the PH domain of Lnk are not critical. Furthermore, wild type Lnk, but not the Lnk SH2 mutant, becomes Tyr-phosphorylated after EPO administration and inhibits EPOR phosphorylation and JAK2 activation. Hence, Lnk, through its SH2 domain, negatively modulates EPOR signaling by attenuating JAK2 activation, and negatively regulates EPO-mediated erythropoiesis [106].

Members of the protein inhibitor of activated STAT (PIAS) family have been implicated in the inhibition of STAT and other transcription factors by multiple mechanisms, including the recruitment of transcriptional repressors and sumoylation [108]. Little evidence to date shows that PIAS proteins regulate STAT5.

Unresolved issues

Whereas we know the identity of the major signal transduction pathways downstream of EPOR (Fig. 1), and have a good understanding of several transcription factors regulated by each pathway, many key issues remain unresolved. Many of the same signal transduction pathways are activated by other members of the type I cytokine receptor family, which includes the thrombopoietin receptor, granulocyte colony-stimulating factor, and prolactin receptor, as well as EPOR. They share homology in the membrane proximal region, which includes the Box1 and Box2 domains [109] and all signal by Jak2. Yet the same transcription factors, such as Stat5, regulate transcription of very different sets of genes in these different differentiation pathways. Thus it is important to identify the genes in erythroid progenitors whose promoter sites are occupied by erythroid-important transcriptional regulators such as STAT5 and FoxO3 (mouse homolog of human FOXO3a), at different stages of erythroid differentiation. One must determine other transcription factors, such as the erythroid-important factors GATA-1, and FOG with which these interact and trigger cell- specific gene expression. A related issue concerns the strength and duration of each of the signaling pathways activated downstream of EPOR. As examples, how do the amounts of STAT5 dimers and tetramers or phosphorylated and unphosphorylated Fox03A change with time after EPO stimulation, and as a consequence are different genes activated or repressed over time?

Acknowledgements
Research in HL's laboratory was supported by NIH grant PO1 HL 32262 and a grant from Amgen Corporation. MS is supported by NIH grant RO1 HL084168 and American Cancer Society grant RSG-06-051-01. WT is supported by Howard Temin Career Development award from NCI (K01-CA115679) and a grant from the MaCabe Foundation. This work was also supported by an American Cancer Society Research Scholarship (RSG LIB-110480), a Career Enhancement Award (K18 HL76510-01), a Black Family Exploratory Research Award, and Mount Sinai School of Medicine Research Funds to SG. JZ is supported by a Howard Temin Award from the National Cancer Institute.

References

1. Wu H, Liu X, Jaenisch R, Lodish HF. Generation of committed erythroid BFU–E and CFU–E progenitors does not require erythropoietin or the erythropoietin receptor. *Cell* 1995;83:59–67.
2. Eshghi S, Vogelezang M, Hynes R et al. a4β1 integrin and erythropoietin mediate temporally distinct steps in terminal erythroid differentiation and proliferation. *J Cell Biol* 2007;177:871–880.
3. Socolovsky M, Constantinescu SN, Bergelson S et al. Cytokines in hematopoiesis: specificity and redundancy in receptor function. *Adv Protein Chem* 1998;52:141–198.
4. Brisken C, Socolovsky M, Lodish HF et al. The signaling domain of the erythropoietin receptor rescues prolactin receptor-mutant mammary epithelium. *Proc Natl Acad Sci USA* 2002;99:14241–14245.
5. Socolovsky M, Fallon AEJ, Lodish HF. The prolactin receptor rescues EpoR –/– erythroid progenitors and replaces EpoR in a synergistic interaction with c-kit. *Blood* 1998;92:1491–1496.
6. Socolovsky M, Dusanter-Fourt I, Lodish HF. The prolactin receptor, as well as severely truncated erythropoietin receptors support differentiation of erythroid progenitors. *J Biol Chem* 1997;272:14009–14013.
7. Brivanlou AH, Darnell JE Jr, Signal transduction and the control of gene expression. *Science* 2002;295:813–818.
8. Levy DE, Darnell JE Jr, Stats: transcriptional control and biological impact. *Nat Rev Mol Cell Biol* 2002;3:651–662.
9. Kirito K, Nakajima K, Watanabe T et al. Identification of the human erythropoietin receptor region required for Stat1 and Stat3 activation. *Blood* 2002;99:102–110.
10. Klingmuller U, Bergelson S, Hsiao JG et al. Multiple tyrosine residues in the cytosolic domain of the erythropoietin receptor promote activation of STAT5. *Proc Natl Acad Sci USA* 1996;93:8324–8328.
11. Paffett-Lugassy N, Hsia N, Fraenkel PG et al. Functional conservation of erythropoietin signaling in zebrafish. *Blood* 2007;110:2718–2726.
12. Yoon D, Watowich SS. Hematopoietic cell survival signals are elicited through non-tyrosine-containing sequences in the membrane-proximal region of the erythropoietin receptor (EPOR) by a Stat5-dependent pathway. *Exp Hematol* 2003;31:1310–1316.

13. John S, Vinkemeier U, Soldaini E et al. The significance of tetramerization in promoter recruitment by Stat5. *Mol Cell Biol* 1999;19:1910–1918.
14. Fujitani Y, Hibi M, Fukada T et al. An alternative pathway for STAT activation that is mediated by the direct interaction between JAK and STAT. *Oncogene* 1997;14:751–761.
15. Barahmand-Pour F, Meinke A, Groner B et al. Jak2-Stat5 interactions analyzed in yeast. *J Biol Chem* 1998;273:12567–12575
16. Soldaini E, John S, Moro S et al. DNA binding site selection of dimeric and tetrameric Stat5 proteins reveals a large repertoire of divergent tetrameric Stat5a binding sites. *Mol Cell Biol* 2000,20.389–401.
17. Neculai D, Neculai AM, Verrier S et al. Structure of the unphosphorylated STAT5a dimer. *J Biol Chem* 2005;280:40782–40787.
18. Mao X, Ren Z, Parker GN et al. Structural bases of unphosphorylated STAT1 association and receptor binding. *Mol Cells* 2005;17:761–771.
19. Mertens C, Zhong M, Krishnaraj R et al. Dephosphorylation of phosphotyrosine on STAT1 dimers requires extensive spatial reorientation of the monomers facilitated by the N-terminal domain. *Genes Dev* 2006;20:3372–3381.
20. Azam M, Lee C, Strenhlow I et al. Functionally distinct isoforms of STAT5 are generated by protein processing. *Immunity* 1997;6:691–701.
21. Decker T, Kovarik P. Serine phosphorylation of STATs. *Oncogene* 2000;19:2628–2637.
22. Paulson M, Pisharody S, Pan L et al. Stat protein transactivation domains recruit p300/CBP through widely divergent sequences. *J Biol Chem* 1999;274:25343–25349.
23. Vinkemeier U. Getting the message across, STAT! Design principles of a molecular signaling circuit. *J Cell Biol* 2004;167:197–201.
24. Reich NC, Liu L. Tracking STAT nuclear traffic. *Nat Rev Immunol* 2006;6:602–612.
25. Socolovsky M, Fallon AEJ, Wang S et al. Fetal anemia and apoptosis of red cell progenitors in Stat5a$^{-/-}$5b$^{-/-}$ mice: a direct role for Stat5 in bcl-X_L induction. *Cell* 1999;98:181–191.
26. Motoyama N, Kimura T, Takahashi T et al. bcl-x prevents apoptotic cell death of both primitive and definitive erythrocytes at the end of maturation. *J Exp Med* 1999;189:1691–1698.
27. Moucadel V, Constantinescu SN. Differential STAT5 signaling by ligand-dependent and constitutively active cytokine receptors. *J Biol Chem* 2005;280:13364–13373.
28. Teglund S, McKay C, Schuetz E et al. Stat5a and Stat5b proteins have essential and nonessential, or redundant, roles in cytokine responses. *Cell* 1998;93:841–850.
29. Zang H, Sato K, Nakajima H et al. The distal region and receptor tyrosines of the Epo receptor are non-essential for *in vivo* erythropoiesis. *EMBO J* 2001;20:3156–3166.
30. Snow JW, Abraham N, Ma MC et al. STAT5 promotes multilineage hematolymphoid development *in vivo* through effects on early hematopoietic progenitor cells. *Blood* 2002;99:95–101.
31. Cui Y, Riedlinger G, Miyoshi K et al. Inactivation of Stat5 in mouse mammary epithelium during pregnancy reveals distinct functions in cell proliferation, survival, and differentiation. *Mol Cell Biol* 2004;24:8037–8047.
32. Socolovsky M, Nam H, Fleming MD et al. Ineffective erythropoiesis in Stat5a(–/–)5b(–/–) mice due to decreased survival of early erythroblasts. *Blood* 2001;98:3261–3273.
33. Menon MP, Karur V, Bogacheva O et al. Signals for stress erythropoiesis are integrated via an erythropoietin receptor-phosphotyrosine-343-Stat5 axis. *J Clin Invest* 2006;116:683–694.
34. Socolovsky M, Murrell M, Liu Y et al. Negative autoregulation by FAS mediates robust fetal erythropoiesis. *PLoS Biol* 2007;5:e252.
35. Liu Y, Pop R, Sadegh C et al. Suppression of Fas-FasL coexpression by erythropoietin mediates erythroblast expansion during the erythropoietic stress response *in vivo*. *Blood* 2006;108:123–133.
36. Klingmuller U, Wu H, Hsiao JG et al. Identification of a novel pathway important for proliferation and differentiation of primary erythroid progenitors. *Proc Natl Acad Sci USA* 1997;94:3016–3021.
37. Damen JE, Mui AL, Puil L et al. Phosphatidylinositol 3-kinase associates, via its Src homology 2 domains, with the activated erythropoietin receptor. *Blood* 1993;81:3204–3210.
38. Damen JE, Cutler RL, Jiao H et al. Phosphorylation of tyrosine 503 in the erythropoietin receptor (EpR) is essential for binding the P85 subunit of phosphatidylinositol (PI) 3-kinase and for EpR-associated PI 3-kinase activity. *J Biol Chem* 1995;270:23402–23408.
39. Barnache S, Mayeux P, Payrastre B et al. Alterations of the phosphoinositide 3-kinase and mitogen-activated protein kinase signaling pathways in the erythropoietin-independent Spi-1/PU.1

transgenic proerythroblasts. *Blood* 2001;98:2372–2381.
40. von Lindern M, Amelsvoort MP, van Dijk T et al. Protein kinase C alpha controls erythropoietin receptor signaling. *J Biol Chem* 2000;275:34719–34727.
41. Wu H, Klingmuller U, Acurio A et al. Functional interaction of erythropoietin and stem cell factor receptors is essential for erythroid colony formation. *Proc Natl Acad Sci USA* 1997;94:1806–1810.
42. Bouscary D, Pene F, Claessens YE et al. Critical role for PI 3-kinase in the control of erythropoietin-induced erythroid progenitor proliferation. *Blood* 2003;101:3436–3443.
43. Verdier F, Chretien S, Billat C et al. Erythropoietin induces the tyrosine phosphorylation of insulin receptor substrate-2. An alternate pathway for erythropoietin-induced phosphatidylinositol 3-kinase activation. *J Biol Chem* 1997;272:26173–26178.
44. Lecoq-Lafon C, Verdier F, Fichelson S et al. Erythropoietin induces the tyrosine phosphorylation of GAB1 and its association with SHC, SHP2, SHIP, and phosphatidylinositol 3-kinase. *Blood* 1999;93:2578–2585.
45. Kubota Y, Tanaka T, Kitanaka A et al. Src transduces erythropoietin-induced differentiation signals through phosphatidylinositol 3-kinase. *EMBO J* 2001;20:5666–5677.
46. Huddleston H, Tan B, Yang FC et al. Functional p85alpha gene is required for normal murine fetal erythropoiesis. *Blood* 2003;102:142–145.
47. Sarbassov DD, Guertin DA, Ali SM et al. Phosphorylation and regulation of Akt/PKB by the rictor-mTOR complex. *Science* 2005;307:1098–1101.
48. Kim DH, Sarbassov DD, Ali SM et al. mTOR interacts with raptor to form a nutrient-sensitive complex that signals to the cell growth machinery. *Cell* 2002;110:163–175.
49. Sarbassov DD, Ali SM, Kim DH et al. Rictor, a novel binding partner of mTOR, defines a rapamycin-insensitive and raptor-independent pathway that regulates the cytoskeleton. *Curr Biol* 2004;14:1296–1302.
50. Greer EL, Brunet A. FOXO transcription factors at the interface between longevity and tumor suppression. *Oncogene* 2005;24:7410–7425.
51. Haseyama Y, Sawada K, Oda A et al. Phosphatidylinositol 3-kinase is involved in the protection of primary cultured human erythroid precursor cells from apoptosis. *Blood* 1999;94:1568–1577.
52. Kashii Y, Uchida M, Kirito K et al. A member of Forkhead family transcription factor, FKHRL1, is one of the downstream molecules of phosphatidylinositol 3-kinase-Akt activation pathway in erythropoietin signal transduction. *Blood* 2000;96:941–949
53. Bao H, Jacobs-Helber SM, Lawson AE et al. Protein kinase B (c-Akt), phosphatidylinositol 3-kinase, and STAT5 are activated by erythropoietin (EPO) in HCD57 erythroid cells but are constitutively active in an EPO-independent, apoptosis-resistant subclone (HCD57-SREI cells). *Blood* 1999;93:3757–3773.
54. Zhao W, Kitidis C, Fleming MD et al. Erythropoietin stimulates phosphorylation and activation of GATA-1 via the PI3-kinase/AKT signaling pathway. *Blood* 2006;107:907–915.
55. Ghaffari S, Kitidis C, Zhao W et al. AKT induces erythroid-cell maturation of JAK2-deficient fetal liver progenitor cells and is required for Epo regulation of erythroid-cell differentiation. *Blood* 2006;107:1888–1891.
56. Kadri Z, Maouche-Chretien L, Rooke HM et al. Phosphatidylinositol 3-kinase/Akt induced by erythropoietin renders the erythroid differentiation factor GATA-1 competent for TIMP-1 gene transactivation. *Mol Cell Biol* 2005;25:7412–7422.
57. Rooke HM, Orkin SH. Phosphorylation of GATA1 at serine residues 72, 142, and 310 is not essential for hematopoiesis *in vivo*. *Blood* 2006;107:3527–3530.
58. Chen WS, Xu PZ, Gottlob K et al. Growth retardation and increased apoptosis in mice with homozygous disruption of the Akt1 gene. *Genes Dev* 2001;15:2203–2208.
59. Cho H, Thorvaldsen JL, Chu Q et al. Akt1/PKBalpha is required for normal growth but dispensable for maintenance of glucose homeostasis in mice. *J Biol Chem* 2001;276:38349–38352.
60. Cho H, Mu J, Kim JK et al. Insulin resistance and a diabetes mellitus-like syndrome in mice lacking the protein kinase Akt2 (PKB beta). *Science* 2001;292:1728–1731.
61. Hammerman PS, Fox CJ, Birnbaum MJ et al. Pim and Akt oncogenes are independent regulators of hematopoietic cell growth and survival. *Blood* 2005;105:4477–4483.
62. Brunet A, Bonni A, Zigmond MJ et al. Akt promotes cell survival by phosphorylating and inhibiting a Forkhead transcription factor. *Cell* 1999;96:857–868.
63. Mahmud DL, Deb DK, Platanias LC et al. Phosphorylation of forkhead transcription factors by erythropoietin and stem cell factor prevents acetylation and their interaction with coactivator p300

in erythroid progenitor cells. *Oncogene* 2002;21:1556–1562.
64. Marinkovic D, Zhang X, Yalcin S et al. FoxO3 is required for the regulation of oxidative stress in erythropoiesis. *J Clin Invest* 2007;117:2133–2144.
65. Ghaffari S, Jagani Z, Kitidis C et al. Cytokines and BCR-ABL mediate suppression of TRAIL-induced apoptosis through inhibition of forkhead FOXO3a transcription factor. *Proc Natl Acad Sci USA* 2003;100:6523–6528.
66. Bakker WJ, Blazquez-Domingo M, Kolbus A et al. FoxO3a regulates erythroid differentiation and induces BTG1, an activator of protein arginine methyl transferase 1. *J Cell Biol* 2004;164:175–184.
67. Sivertsen EA, Hystad ME, Gutzkow KB et al. PI3K/Akt-dependent Epo-induced signalling and target genes in human early erythroid progenitor cells. *Br J Haematol* 2006;135:117–128.
68. Fang J, Menon M, Kapelle W et al. EPO modulation of cell-cycle regulatory genes, and cell division, in primary bone marrow erythroblasts. *Blood* 2007;110:2361–2370.
69. Menon MP, Fang J, Wojchowski DM. Core erythropoietin receptor signals for late erythroblast development. *Blood* 2006;107:2662–2672.
70. Henry M, Lynch J, Eapen A et al. DNA damage-induced cell-cycle arrest of hematopoietic cells is overridden. *Blood* 2001;98:834–841.
71. Marshall CJ. Opportunities for pharmacological intervention in the ras pathway. *Ann Oncol* 1995;6:63–67.
72. Nagata Y, Nishida E, Todokoro K. Activation of JNK signaling pathway by erythropoietin, thrombopoietin, and interleukin-3. *Blood* 1997;89:2664–2669.
73. Nagata Y, Moriguchi T, Nishida E et al. Activation of p38 MAP kinase pathway by erythropoietin and interleukin-3. *Blood* 1997;90:929–934.
74. Gobert S, Duprez V, Lacombe C et al. The signal transduction pathway of erythropoietin involves three forms of mitogen-activated protein (MAP) kinase in UT7 erythroleukemia cells. *Eur J Biochem* 1995;234:75–83.
75. Nagata Y, Takahashi N, Davis RJ et al. Activation of p38 MAP kinase and JNK but not ERK is required for erythropoietin-induced erythroid differentiation. *Blood* 1998;92:1859–1869.
76. Nagata Y, Kiefer F, Watanabe T et al. Activation of hematopoietic progenitor kinase-1 by erythropoietin. *Blood* 1999;93:3347–3354.
77. Miura Y, Miura O, Ihle JN et al. Activation of the mitogen-activated protein kinase pathway by the erythropoietin receptor. *J Biol Chem* 1994;269:29962–29969.
78. Sakamoto H, Kitamura T, Yoshimura A. Mitogen-activated protein kinase plays an essential role in the erythropoietin-dependent proliferation of CTLL-2 cells. *J Biol Chem* 2000;275:35857–35862.
79. Sui X, Krantz SB, You M et al. Synergistic activation of MAP kinase (ERK1/2) by erythropoietin and stem cell factor is essential for expanded erythropoiesis. *Blood* 1998;92:1142–1149.
80. Jacobs-Helber SM, Ryan JJ, Sawyer ST. JNK and p38 are activated by erythropoietin (EPO) but are not induced in apoptosis following EPO withdrawal in EPO-dependent HCD57 cells. *Blood* 2000;96:933–940.
81. Shan R, Price JO, Gaarde WA et al. Distinct roles of JNKs/p38 MAP kinase and ERKs in apoptosis and survival of HCD-57 cells induced by withdrawal or addition of erythropoietin. *Blood* 1999;94:4067–4076.
82. Johnson L, Greenbaum D, Cichowski K et al. K-ras is an essential gene in the mouse with partial functional overlap with N-ras. *Genes Dev* 1997;11:2468–2481.
83. Zhang J, Socolovsky M, Gross AW et al. Role of Ras signaling in erythroid differentiation of mouse fetal liver cells: functional analysis by a flow cytometry-based novel culture system. *Blood* 2003;102:3938–3946.
84. Zhang J, Lodish HF. Constitutive activation of the MEK/ERK pathway mediates all effects of oncogenic H-ras expression in primary erythroid progenitors. *Blood* 2004;104:1679–1687.
85. Zhang J, Liu Y, Beard C et al. Expression of oncogenic K-ras from its endogenous promoter leads to a partial block of erythroid differentiation and hyperactivation of cytokine-dependent signaling pathways. *Blood* 2007;109:5238–5241.
86. Chan IT, Kutok JL, Williams IR et al. Conditional expression of oncogenic K-ras from its endogenous promoter induces a myeloproliferative disease. *J Clin Invest* 2004;113:528–538.
87. Braun BS, Tuveson DA, Kong N et al. Somatic activation of oncogenic Kras in hematopoietic cells initiates a rapidly fatal myeloproliferative disorder. *Proc Natl Acad Sci USA* 2004;101:597–602.

88. Zhang J, Lodish HF. Identification of K-ras as the major regulator for cytokine-dependent Akt activation in erythroid progenitors *in vivo*. *Proc Natl Acad Sci USA* 2005;102:14605–14610.
89. Jaumot M, Yan J, Clyde-Smith J et al. The linker domain of the Ha-Ras hypervariable region regulates interactions with exchange factors, Raf-1 and phosphoinositide 3-kinase. *J Biol Chem* 2002;277:272–278.
90. Booden MA, Sakaguchi DS, Buss JE. Mutation of Ha-Ras C terminus changes effector pathway utilization. *J Biol Chem* 2000;275:23559–23568.
91. Suzuki N, Ohneda O, Takahashi S et al. Erythroid-specific expression of the erythropoietin receptor rescued its null mutant mice from lethality. *Blood* 2002;100:2279–2288.
92. Neel BG, Gu H, Pao L. The 'Shp'ing news: SH2 domain-containing tyrosine phosphatases in cell signaling. *Trends Biochem Sci* 2003;28:284–293.
93. Klingmuller U, Lorenz U, Cantley LC et al. Specific recruitment of SH-PTP1 to the erythropoietin receptor causes inactivation of JAK2 and termination of proliferative signals. *Cell* 1995;80:729–738.
94. Yi T, Zhang J, Miura O et al. Hematopoietic cell phosphatase associates with erythropoietin (Epo) receptor after Epo-induced receptor tyrosine phosphorylation: identification of potential binding sites. *Blood* 1995;85:87–95.
95. Ram PA, Waxman DJ. Interaction of growth hormone-activated STATs with SH2-containing phosphotyrosine phosphatase SHP-1 and nuclear JAK2 tyrosine kinase. *J Biol Chem* 1997;272:17694–17702.
96. Aoki N, Matsuda T. A cytosolic protein-tyrosine phosphatase PTP1B specifically dephosphorylates and deactivates prolactin-activated STAT5a and STAT5b. *J Biol Chem* 2000;275:39718–39726.
97. Aoki N, Matsuda T. A nuclear protein tyrosine phosphatase TC-PTP is a potential negative regulator of the PRL-mediated signaling pathway: dephosphorylation and deactivation of signal transducer and activator of transcription 5a and 5b by TC-PTP in nucleus. *Mol Endocrinol* 2002;16:58–69.
98. Inagaki-Ohara K, Hanada T, Yoshimura A. Negative regulation of cytokine signaling and inflammatory diseases. *Curr Opin Pharmacol* 2003;3:435–442.
99. Krebs DL, Hilton DJ. SOCS: physiological suppressors of cytokine signaling. *J Cell Sci* 2000;113:2813–2819.
100. Frantsve J, Schwaller J, Sternberg DW et al. SOCS-1 inhibits TEL-JAK2-mediated transformation of hematopoietic cells through inhibition of JAK2 kinase activity and induction of proteasome-mediated degradation. *Mol Cell Biol* 2001;21:3547–3557.
101. Ungureanu D, Saharinen P, Junttila I et al. Regulation of Jak2 through the ubiquitin-proteasome pathway involves phosphorylation of Jak2 on Y1007 and interaction with SOCS-1. *Mol Cell Biol* 2002;22:3316–3326.
102. Hortner M, Nielsch U, Mayr LM et al. A new high affinity binding site for suppressor of cytokine signaling-3 on the erythropoietin receptor. *Eur J Biochem* 2002;269:2516–2526.
103. Sasaki A, Yasukawa H, Shouda T et al. CIS3/SOCS-3 suppresses erythropoietin (EPO) signaling by binding the EPO receptor and JAK2. *J Biol Chem* 2000;275:29338–29347.
104. Sarna MK, Ingley E, Busfield SJ et al. Differential regulation of SOCS genes in normal and transformed erythroid cells. *Oncogene* 2003;22:3221–3230.
105. Tong W, Lodish HF. Lnk inhibits Tpo-mpl signaling and Tpo-mediated megakaryocytopoiesis. *J Exp Med* 2004;200:569–580.
106. Tong W, Zhang J, Lodish HF. Lnk inhibits erythropoiesis and Epo-dependent JAK2 activation and downstream signaling pathways. *Blood* 2005;105:4604–4612.
107. Rudd CE. Lnk adaptor: novel negative regulator of B cell lymphopoiesis. *Sci STKE.* 2001;2001:PE1
108. Shuai K. Regulation of cytokine signaling pathways by PIAS proteins. *Cell Res* 2006;16:196–202.
109. Watowich SS, Wu H, Socolovsky M et al. Cytokine receptor signal transduction and the control of hematopoietic cell development. *Annu Rev Cell Dev Biol* 1996;12:91–128.
110. Zhang J, Lodish H. Endogenous K-ras signaling in erythroid differentiation. *Cell Cycle* 2007;6:1970–1973.

Erythropoietins, Erythropoietic Factors and Erythropoiesis
Edited by S.G. Elliott, M.A. Foote, and G. Molineux

Mechanism of erythropoietin receptor activation

Stefan N. Constantinescu

Ludwig Institute for Cancer Research and de Duve Institute, Université catholique de Louvain, Brussels B1200, Belgium

Introduction

Erythropoietin receptor (EPOR) is a founding member of the cytokine receptor superfamily [1]. A type I transmembrane protein that binds the ligand erythropoietin (EPO) with high affinity (kDa approximately 400 pM) on the surface of erythroid progenitors, EPOR is devoid of catalytic activity. Given that the extracellular domain of EPOR only contains two cytokine receptor homology modules (D1 and D2), EPOR is considered a simple (or short) receptor, unlike other superfamily members, like the receptors for interleukin (IL)-6 type cytokines or for granulocyte colony-stimulating factor (G-CSF), which possess other extracellular sequences, such as immunoglobulin-like domains [2].

Genetic evidence has established that EPOR is the sole receptor chain for EPO. Knock-out of *EPO* or of *EPOR* genes in mice results in a lethal anemia at day 12.5 of embryonic life [3]. The failure of definitive erythropoiesis in the fetal liver is due to apoptosis of proerythroblasts [3]. EPO and EPOR play essential nonredundant roles from the colony-forming unit–erythroid (CFU–E) stage to formation of red blood cells, but are not absolutely obligatory for generation of blast-forming unit–erythroid (BFU–E) and CFU–E cells [3]. The first protein in the signaling chain triggered by activation of EPOR by EPO is JAK2, a member of the Janus kinase family of tyrosine kinases [4]. JAK2 associates with EPOR and is tyrosine phosphorylated and activated after stimulation with EPO [4]. Knockout mouse embryos devoid of JAK2 also die at day 12.5, due to absence of definitive erythropoiesis [5, 6].

Functional challenges to EPOR from the unique features of the red blood cell compartment

As 'form follows function', before examining the mechanisms of activation of the receptor, it is of interest to put forward three essential ideas about EPOR function in red blood cell formation:

- The volume of blood occupied by red blood cells, defined as hematocrit, is approximately 40%. Increases in red blood cell numbers, which are regulated by EPO, cannot be of high amplitude since hematocrit increases above 15% to 20% lead to severe hemodynamic complications and are seen in disease states, such as myeloproliferative neoplasms;
- EPOR has apparently evolved sequences that make its cell surface localization inefficient, which is translated in low cell surface EPOR densities in erythroid progenitors and transfected cells; and
- Signaling by EPOR is very efficient, as occupancy of only a few receptors (5–10% receptor occupancy, detected at 10 pM EPO) suffices to induce a half maximal response, while half maximal binding is detected at much higher EPO concentrations, approximately 0.18 nM Epo [7]. Therefore, variations in EPO concentrations must therefore be translated into incremental and precisely controlled increases in red blood cell numbers.

EPO binding to EPOR

Epo binds to the receptor with a stoichiometry of 1:2 (EPO:EPOR) [7], a stoichiometry similar to that of the growth hormone binding to its receptor [8]. Binding of EPO to EPOR by two opposite sites: one site (called site 1) binds with high affinity (<1 nM) to the extracellular region of EPOR, while an opposite site (called site 2) binds with a much lower affinity (approximately 1 mM) to another extracellular receptor chain [9]. This binding is translated into an overall high affinity (kDa approximately 400 pM) of binding of EPO to intact receptors on living cells.

The numbers of cell surface EPOR molecules per cell are quite low. These range from 1,500 to 2,000 sites/cell surface in proerythroblasts and early basophilic erythroblasts, to 750 in late basophilic erythroblasts and 250 in chromatophilic and orthochromatophilic progenitors to <10 sites/cell surface in late orthochromatophilic erythroblasts and reticulocytes [10, 11]. In stably transfected hematopoietic cells, the maximal numbers are approximately 1,500 to 1,800 surface receptors/cell [12].

Structural features of EPOR extracellular domain

EPOR contains a number of features in the extracellular domain, conserved with other cytokine receptor superfamily members. Two pairs of cysteines in the extracellular domain forming intramolecular disulfide bonds, and a WSXWS motif (Fig. 1A) are absolutely required for folding of the receptor. Mutation of any of the Trp or Ser residues in the WSXWS motif abolishes traffic to the cell surface [13], but the nature of the X residue modulates folding and traffic. A substitution of the Ala residue normally present in EPOR by Glu (present in the growth hormone receptor) increases the number of cell surface

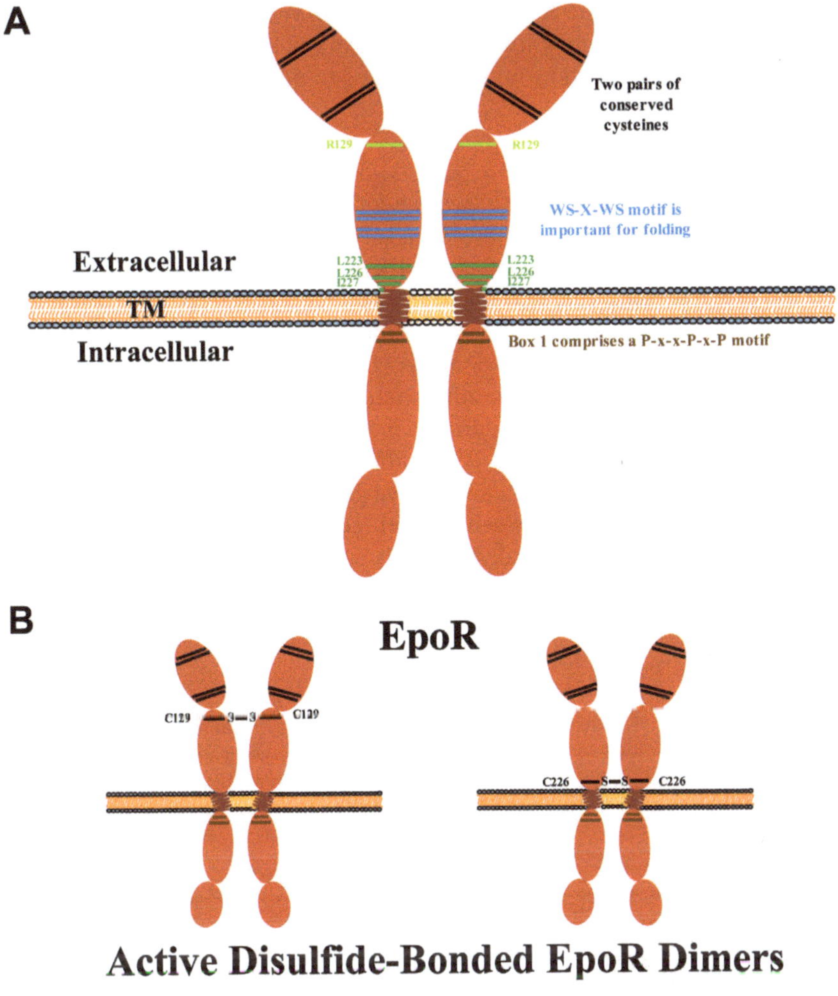

Figure 1. Structure of the erythropoietin receptor (EPOR). (A) EPOR extracellular domain contains two pairs of conserved Cys residues and a WSXWS motif, which are absolutely required for folding and forward traffic to the cell surface. The positions of extracellular residue R129 and juxtamembrane/transmembrane residues L223, L226, and I227 are shown, since mutation of any of these residues to Cys allows close apposition of monomers and formation of disulfide bonded active dimers. The intracellular domain of EPOR contains a Pro rich sequence (PxxPxP) denoted Box 1, which is required for binding the tyrosine kinase JAK2 (not shown, see Fig. 4). (B) Disulfide bond formation between EPOR monomers bearing mutations R129C or L226C results in dimerization in a productive conformation and activation in the absence of EPO. Similarly, when mutated to Cys, certain residues around R129, such as E132 and E133, as well as L223 and I227 at the extracellular-to-transmembrane domain junction yield constitutive active disulfide-bonded EPOR dimers.

receptors by approximately three-fold, most likely by enhancing folding rates in the lumen of the endoplasmic reticulum [12]. Alignment of EPOR sequences from several species suggested a selection against the presence of residues such as Glu at the X position of the WSXWS motif. Lower amounts of cell surface EPOR might allow finer tuning of red blood cell production by EPO, while higher amounts would be predicted to promote erythrocytosis.

Structure of ligand-bound EPOR extracellular domain

Several X-ray crystal structures have been solved for the extracellular domain of EPOR in isolation. The secondary and tertiary structures agree very well between all these studies, where the extracellular domain was crystallized in complex with a peptide mimetic, with EPO or without ligand [7, 14, 15]. The extracellular domain folds in two domains (D1 and D2) that form an L-shape and that are connected by a four residues helical linker. D1 and D2 are related in structure to fibronectin type III domains, but show differences between them with respect to fold subclass [14]. What differs between the structures of different ligand-EPOR complexes is the relative geometry, i.e., angle in the plane of the membrane surface, between the D1 domains of two monomers [7, 14–17]. An EPO mimetic peptide, which is a weak agonist, induces a 180° angle between D1 domains [14]. Strikingly, a dibromylated derivative of the mimetic peptide, which is an antagonist, induces a 165° angle between D1

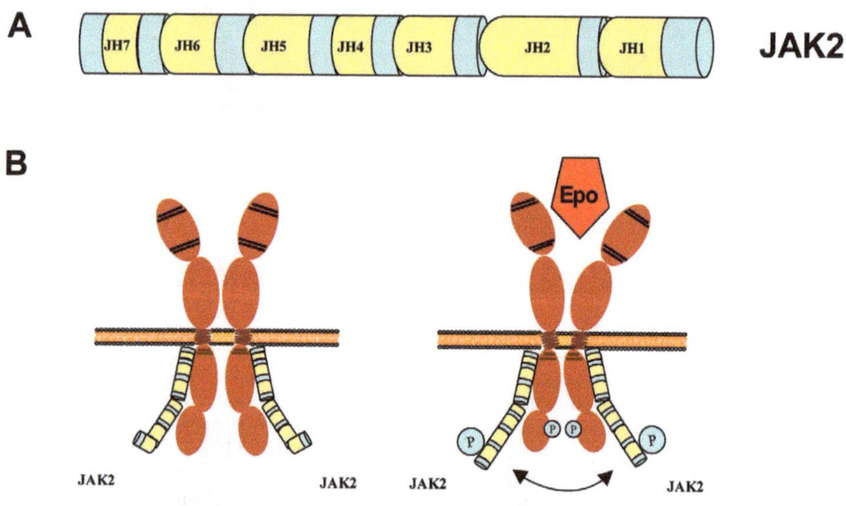

Figure 2. EPOR forms a complex with JAK2 before activation by EPO. (A) JAK2 shares seven JAK-homology (JH) regions with the other three members of the Janus kinase family. JH1, kinase domain, JH2, pseudokinase domain, JH3–JH5 contain an SH2-like sequence, while JH5–JH7 contain a FERM (Band-4.1, ezrin, radixin, moesin)-like domain that is required for attaching to the juxtamembrane cytosolic region of EPOR. (B) JAK2 is appended to the cytosolic domain of EPOR in the resting state, in the absence of EPO. In this conformation, the kinase domain (JH1) is inactive due to the inactive dimeric conformation of the receptor and to the inhibitory interactions between JH1 and JH2 domain. Upon ligand binding to EPOR extracellular domain, the dimer is re-oriented and that leads to transphosphorylation and activation of JAK2, and phosphorylation of cytosolic EPOR Tyr residues.

domains [16]. In contrast, the angle induced by the strong and physiologic agonist EPO is 120° [7]. These geometry differences are provocative, particularly when correlated with the drastically different biologic effects of EMP, of the antagonist derivative of EPO mimetic protein and of EPO [17]. However, caution must be exerted when literally interpreting these conformational differences, since the receptors are normally spanning the plasma membrane through transmembrane domains and are connected to cytosolic domains. Thus, the carboxyl ends of the extracellular domains are normally not free, as they are in these X-ray crystal structures, and the presence of the transmembrane-cytosolic domains may impose a different relative geometry to the two receptor monomers.

Structure of the unliganded EPOR extracellular domain

A surprising result was reported by the X-ray crystal structure of the unliganded EPOR [15]: the extracellular domain was crystallized as a dimer in a completely different conformation, that maintains the carboxyl ends of the extracellular domains (normally connected to the transmembrane domains) far away at a distance of >70 Å. At such a distance, it would be impossible for the JAK2 proteins appended to the intracellular domain to interact. However, whether such a dimeric conformation actually exists for the full-length EPOR is a matter of debate for several reasons: First, mutagenesis and functional studies [18–21], demonstrated that substituting several extracellular, or juxtamembrane-transmembrane residues with Cys leads to active disulfide-bonded dimers of unliganded EPOR (Fig. 1B), which indicates a much closer apposition of juxtamembrane and transmembrane residues for EPOR. Second, in buffers with physiologic salt concentration, the extracellular domain of EPOR is not a dimer, but a monomer, which can be dimerized by EPO addition [22, 23]. Third, the transmembrane domains of EPOR self-associate, and a viral envelope protein can specifically activate EPOR by specifically interacting with EPOR transmembrane domain. Nevertheless, this study [15] indicated that forcing the formation of EPOR extracellular domains dimers in the absence of EPO would lead to a conformation incompatible with signaling.

Epor cytosolic domain interacts with JAK2 in the absence of ligand

The intracellular domain of EPOR contains 239 amino acid residues and is devoid of catalytic activity. Recent evidence from several cytokine receptors indicates that JAK are appended to receptors before ligand binding [24–27] and that JAK are localized to membranes due to their recruitment by cytokine receptors [28]. Thus, the complete functional unit of a cytokine receptor includes the cognate JAK that is appended to that particular receptor (Fig. 2).

Janus kinases

The mammalian JAK family consists of four members, JAK1, JAK2, TYK2, and JAK3 [29]. JAK were initially named JAK for 'Just Another Kinase', as they were identified based on their homology to other tyrosine kinases [30, 31]. JAK2 was the third cloned member of the JAK family of protein tyrosine kinases [32]. Because JAKs proved to be crucial for signaling by all cytokines, which was first discovered for TYK2 and type I interferons [33], and because JAKs possess two kinase domains, one active and another inactive (pseudokinase domain), these enzymes were re-named Janus kinases after the Roman god custodian of the universe, the god of "beginnings and guardian of gates and doors", a god that has two heads and two faces. The metaphore holds up also at the functional level since receptor activation requires oligomerization of at least two JAK molecules, which face each other when receptor complexes are assembled.

JAKs have seven regions of homology, called JAK homology (JH) domains (JH1–7) [34]. From the carboxyl- to the amino-terminus, the JH domains are: the kinase (JH1); the pseudokinase (JH2) domains; JH3, and JH4 domains, which contain an SH2-like sequence; and then JH5–JH7, which contain a FERM (Band-4.1, ezrin, radixin and moesin)-like motif, which is crucial for appending JAK to cognate cytokine receptors (Fig. 2A). The JH1 domain contains all the necessary features of a catalytic tyrosine kinase, while the sequence of the JH2 pseudokinase domain lacks characteristic residues of active tyrosine kinases, making it catalytically inactive [31].

Sequences required for EPOR-JAK2 interaction

JAK2 is appended to the cytosolic domain of EPOR (Fig. 2B) [4]. Biochemical studies showed that certain cytosolic sequences are crucial for binding of JAK2 to the receptor. These included Box 1, a Pro-rich sequence (PxxPxP) located 11 amino acid residues downstream of the transmembrane domain, and Box 2, a charged and hydrophobic amino acid sequence located approximately 50 amino acid residues downstream of the transmembrane domain [35]. Both Box 1 and Box 2 are loosely conserved among type I cytokine receptors. A Trp residue (W282), located between Box 1 and Box 2, was shown early on to be required for physiologic JAK2 interaction and activation [35].

A higher resolution examination of the sequences required for JAK2 interaction with EPOR was possible after the observation that traffic of EPOR to the cell surface is dependent on JAK2 expression [24]. In the absence of JAK2, EPOR remains in an immature glycosylation (endoglycosidase H-sensitive) form in the endoplasmic reticulum. Binding of the FERM domain of JAK2 to EPOR is required for maturation of EPOR to the cell Golgi apparatus and cell surface localization [24]. Studies using chimeric JAK1–JAK2 proteins indicated that the FERM domain sequences and not the kinase or pseudokinase domains are crucial to impart specificity for receptor attachment [24]. This JAK2-dependency of EPOR cell surface localization provided a sensitive and specific assay for interrogating the sequences required for productive interaction.

Ala scanning mutagenesis established that not only Box 1, Box2 and W282, but also most other residues in the region between Box 1 and Box 2 mediate interaction of EPOR with JAK2 [24]. Interestingly, residues that precede Box 1, or D287 in the region between Box 1 and Box 2, are not required for promoting cell-surface localization of EPOR, but proved to be crucial for activating JAK2 [36].

The cell surface EPOR is a preformed dimer

Several lines of evidence established that the cell surface EPOR is a preformed inactive dimer in the absence of ligand. Such dimerization would be predicted to prevent EPOR from activation in the absence of ligand. In the same time, very low EPO concentrations would readily activate preformed receptor dimers, without the need to recruit a second subunit to a receptor monomer.

Fluorescence co-patching studies

Live cell surface imunofluorescence co-patching between receptors tagged at the amino-terminus demonstrated that EPOR exists on the cell surface as preformed dimers/oligomers in the absence of ligand [37]. In this technique [38], live cells at 4° C (to block internalization) are subjected to patching by a sandwich of antibodies (anti-tag and fluorescently-labeled secondary antibody), leading to formation of patches of cell surface transmembrane proteins, that are either fluorescent green or red, depending on the fluorochrom used for the secondary antibody. Labeling for two tags (e.g., HA and Myc or HA and FLAG) with different antibody sandwiches (primary antibodies must be of different species) will lead to the formation of separate patches, i.e., individual red and green patches, provided that proteins did not interact and did not form complexes before the patching procedure. Yellow patches, resulting form the superposition of red and green patches would be detected if the differently tagged transmembrane proteins were present in mutual complexes before the patching procedure was applied. When HA- and Myc-tagged EPOR proteins were co-expressed in 293-derived BOSC cells, it was noted that most HA-EpoR and Myc-EPOR were in mutual complexes on the cell surface (Fig. 3) [37]. Remarkably, these preassembled receptor complexes do not signal in the absence of ligand, indicating that preformed dimerization is most likely preventing activation.

HA-EPOR does not co-patch with FLAG-tagged thrombopoietin receptor or with FLAG-prolactin receptor (FLAG-PrlR), demonstrating the specificity

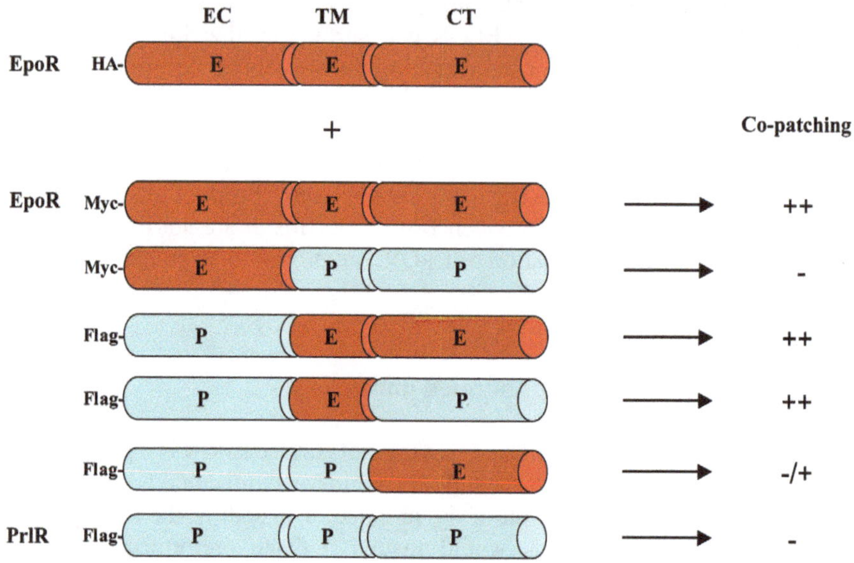

Figure 3. EPOR is a preformed dimer at the cell surface, as revealed by co-patching studies. See *Fluorescence co-patching studies* section in the text for further description.

of this assay. In contrast, chimeric EPOR-prolactin receptors, that contain the transmembrane domain of EPOR (PEP or PEE) do co-patch with EPOR, while those that only contain the extracellular or the cytosolic domain of EPOR do not (Fig. 3) [37]. Thus, the transmembrane sequences, and not the extracellular or the cytosolic sequences of EPOoR, mediate EPOR dimerization at the cell surface in the absence of ligand.

Fluorescence Resonance Energy Transfer (FRET) studies

FRET studies employed fusion proteins where yellow (YFP) and cyan (CFP) fluorescent protein sequences were fused at the amino-terminus of the EpoR. FRET measurements indicated that the majority of receptors are in close proximity (at a distance <100 Å) and oligomerized in the absence of ligand [39]. Like for the GH receptor, addition of ligand did not increase the proportion of dimerized receptors by FRET [39], suggesting that most EpoR chains already are part of dimer complexes. This result is consistent with that obtained by co-patching, where addition of Epo did not significantly enhance dimerization [37]. While EpoR and GH receptor were each shown to form dimers, they did not form heterodimers/heterooligomers, when co-expressed in the same cell, supporting the specificity of the FRET assay [39].

EPOR transmembrane domain strongly homodimerizes

Starting from the observation that heptad leucine motifs mediates self-recognition of artificial transmembrane segments, a data base search identified degenerate versions of these leucine motifs within transmembrane segments of several proteins, including EPOR [40]. In genetic bacterial TOXCAT assays, where activation of fusion proteins containing the ToxR transcription factor is induced by dimerization of test transmembrane sequences, EPOR transmembrane sequence self-oligomerizes to levels higher than those of the glycophorin A transmembrane sequence [40, 41]. These data provide a biochemical basis for the role of the transmembrane sequence in inducing oligomerization of the full-length EPOR. Furthermore, interrupting the α-helicity and dimerization of the EPOR transmembrane sequence, by introduction of a Gly-Pro sequence, impairs EPO-dependent signaling, suggesting a role for transmembrane domain dimerization in supporting ligand-dependent signaling [41].

Activation of EPOR by viral gp55 proteins through specific transmembrane domain interactions

The defective viral envelope protein gp55-P, coded by the polycythemia strain of spleen focus forming virus, binds to the murine EPOR co-expressed in the same cell and activates signaling [42]. Different degrees of EPOR activation can be induced, as a function of the precise transmembrane sequences present in gp55 proteins from different virus strains [43, 44]. For full EPOR activation, the transmembrane domain of the gp55 protein must be a dimer and must contain specific sequences that mediate interaction with the EPOR transmembrane domain. Dimerization of the domain of the strong activator gp55-P (polycythemia-inducing, SFFV strain 'P') was demonstrated in TOXCAT assays [40, 45]. In contrast, the transmembrane domain of the weak activator gp55-A (anemia inducing, SFFV strain 'A') was shown not to dimerize in TOXCAT assays [45]. Both gp55-P and gp55-A transmembrane domains mediate complex formation with EPOR, as shown by co-patching experiments [45]. Furthermore, sequence-specific interactions between the single transmembrane domains of EPOR and gp55-P were identified to be essential for receptor activation. Computational searches and mutagenesis data indicated that the face of the murine EPOR transmembrane domain containing S238 is predicted to interact specifically with gp55-P, and more specifically with Met390 of its transmembrane domain [44]. Since in the absence of EPO, the EPOR transmembrane domains stabilize an inactive dimeric receptor conformation, our model predicts that the transmembrane sequences of gp55 proteins specifically bind and disrupt the inactive EPOR transmembrane dimer, leading to different degrees of activation. For gp55-P, which is a dimer and contains specific transmembrane sequences like Met390, a conformational change will be induced in EPOR, leading to receptor activation and formation of red blood cells in the absence of EPO.

Protein fragment complementation assays on EPOR fusion proteins

The use of a protein fragment complementation assay in cultured cells provided evidence that transmembrane unliganded EPOR dimers likely exist in a conformation that prevents activation of JAK2 [46]. These assays were based on fusion proteins between the extracellular and transmembrane domains of EPOR to fragments of dihydrofolate reductase (DHFR). Only when the two fragments of DHFR are in close proximity (induced by dimerization) would fluorescently-labeled methotrexate be able to bind DHFR. For fusion proteins, where DHFR fragments were fused immediately downstream of the transmembrane domain, protein fragment complementation was detected only in the presence of the ligand EPO. This finding suggests that EPOR extracellular-transmembrane domains adopt a rigid structure that either impairs dimerization of intracellular DHFR fragments or locks them in a conformation where complementation cannot occur. When a 30-amino acid flexible linker was inserted between the EPOR transmembrane and the DHFR fragments, spontaneous complementation could be demonstrated in the absence of ligand. These functional data support the notion that dimerization of extracellular and transmembrane domains of EPOR prevents spontaneous activation. Interestingly, when DHFR fragments were fused after the carboxyl-terminus of the entire cytosolic domain, then spontaneous activation was also detected [46], suggesting that most of the intracellular domain might be highly flexible and possible that the full-length receptor is a dimer.

EPOR is activated as a dimer

Long before the X-ray crystal structure of the complex between EPO and the EPOR extracellular domain was available, several key functional studies have indicated that EPOR is activated as a dimer.

The R129C mutation

A single point mutation (R129C) in the extracellular domain (Fig. 1B) enables EPOR to induce EPO-independent cell growth and tumorigenesis [18]. Biochemical studies employing two-dimensional (nonreducing and, subsequently, reducing gel electrophoresis) demonstrated that the introduction of the Cys residue led to formation of disulfide bonds between two receptor monomers [19]. EPOR mutants containing R129S, R129E, and R129P mutations were functionally wild type, demonstrating that Cys, and not the loss of

Arg, was required for constitutive activation [19]. Both monomers and disulfide-linked EPOR R129C dimers were found at the cell surface, and were shown to exhibit normal EPO-binding characteristics. Thus, the EPOR R129C chains do not form disulfide bonds with 100% efficiency. Interestingly, mutations in the critical extracellular WSXWS motif abolished not only cell surface appearance of the receptor, but also the formation of the disulfide bond by the mutated EPOR R129C monomers [19]. These data indicate that disulfide bond formation at C129 require proper receptor folding and does not represent aggregation in the endoplasmic reticulum.

When the interface of the two EPOR monomers was modeled on the basis of the dimeric interface of the growth hormone receptor [8], it became evident that the R129C mutant will place the Cys residue in the interface of the dimer. Several other residues were predicted to be located in the interface. For some of those, E132 and E133, it was shown that their mutation to Cys also yields disulfide-bonded constitutive active receptors [47].

Truncated receptors where the entire cytosolic domain was deleted, were able to block, in a dominant manner, the activation of the wild type EPOR by EPO [47]. The dominant negative effect required excess expression levels of the truncated *versus* wild-type receptors, and was also induced, albeit less efficiently, by a less severely truncated receptor that conserved 60 cytosolic residues. These results established that activation of EPOR by EPO requires receptor dimerization.

Cysteine mutations at the juxtamembrane-transmembrane border of EPOR

Random and Cys scanning mutagenesis studies of the juxtamembrane and transmembrane domains of EPOR identified three EPOR point mutations, L223C, L226C, and I227C that promote constitutive activation (Fig. 1B) [20, 21]. Disulfide bonds are formed by Cys residues at positions 223, 226, and 227, that lock the receptor in an active dimeric conformation [20, 21].

When examining disulfide bond formation by single Cys mutants of EPOR in nonreducing gels or by chemical cross-linking, it became clear that mutation to Cys leads to formation of disulfide bonds for most residues in the interval 220 to 229. The extent of disulfide bond formation did not correlate with activity, in that the efficiency of disulfide bond formation was not higher for the three mutants, L223C, L226C, and I227C, that were constitutively active. The implication of these data is three-fold:

- Simply dimerizing the receptor and bringing in close apposition the transmembrane domains does not suffice for activation;
- Only some and not all dimeric conformations are active. A model where inactive receptors would be dimers that would have the transmembrane domains very far apart (>70 Å) [15] is inconsistent with these data since, in such a model, disulfide bond formation at positions 223, 226, and 227

would never form; and

- The lack of periodicity of disulfide bond formation for the entire region between residues 220–229 suggests that this region is not α-helical, and that some of it might be flexible.

Nuclear magnetic resonance studies suggest that residues 226–230 form a helix-cap structure, which prevents self-activation [20]. Consistent with this, when the residues predicted to form a helix-cap at the very start of the transmembrane domain are replaced by a stretch of Leu residues, which are known to form α-helices, spontaneous activation of the receptor was detected [20]. Therefore, the junction between the extracellular and transmembrane domains is key to preventing receptor self-activation and might emerge as a possible target for small molecule regulators of EPOR signaling.

Remarkably, this point is supported by the random mutagenesis study, where several other mutants were selected to induce activation, most of them being concentrated at the same residues at the junction between the extracellular and transmembrane domains, i.e., NKK, NRE, or NTH mutants of positions 226–228, single and double mutations at 226 and 227 [21]. Only one middle transmembrane domain double mutant LI→WK at 236–237 induced constitutive activation. For this WK mutation, individual mutants L236W and I237K were not active [21], emphasizing the power of randomly mutating groups of residues, which enabled the identification of unique combinations of mutations that lead to activation.

Dimer reorientation in EPO-induced EPOR activation

Rather than simply inducing dimerization of monomers, ligands for receptors like EPOR and growth hormone receptor are now presumed to induce reorientation of preformed inactive dimmers [37, 39]. The existence of these inactive and active dimeric orientations is supported by the observation that the transmembrane and juxtamembrane cytosolic regions of several cytokine receptors adopt a rigid α-helical orientation, and that intracellular signaling depends on the pitch (register) or these α-helical orientations [37, 39, 48].

Alanine scanning and alanine insertion mutagenesis

Ala scanning of the proximal 50 cytosolic juxtamembrane residues of EPOR was performed to identify key residues for JAK2 activation. Two assays were used as read-outs of this scan: the ability of EPOR Ala mutants to respond to EPO in cytokine-dependent Ba/F3 cells, and the ability of JAK2 to promote traffic of the mutant EPOR to the cell surface. Three classes of mutations were isolated:

- Mutations of Box 1, Box 2 and of residues between them, which lost both JAK2-dependent traffic and response to EPO (already discussed);
- mutations that did not affect either traffic or function; and
- mutations that did not abolish traffic, but blocked activation of JAK2 after EPO-dependent stimulation of EPOR. Residues belonging to this last group were denoted 'switch residues' as they are normally required to transmit the signal from the receptor to JAK2. They consist of hydrophobic residues preceding Box 1 (L253, I257, W258) and D287. The first three hydrophobic residues form a hydrophobic motif (hydrophobic-x-x-x- hydrophobic-aromatic/hydrophobic) that is conserved in several cytokine receptors [25, 26] and is required for signaling. The exact mechanism by which the 'switch residues' enable activation of JAK2 is not clear, but might involve interactions between JAK kinase/pseudokinase domains and receptors, while residues between Box 1 and Box 2 are required to bind the FERM domain residues of JAKs.

In conclusion, one key event that must occur upon activation is represented by a change in position of the 'switch' residues, which would in turn change the relative positions of JAK2 proteins, enabling them to transphosphorylate each other. This action would require close contact between the catalytic loop of one JAK molecule and the activation loop of the other.

To assess the influence of the helical orientation of the transmembrane region on EPOR signaling, extra Ala residues were inserted at position 252, after the transmembrane domain and the positive patch (^{248}SHRR^{251}) that is present at the junction between transmembrane and cytosolic domains. For each Ala insertion, a 109° rotation of the downstream helical residues is predicted to occur, so that insertion of 3 Ala would bring the dimeric interface close to the original position (327° clockwise), while insertion of 1 and 4 Ala would rotate the helix towards the opposite face. When signaling of the EPOR1A–4A mutants was examined, it was determined that EPOR3A signaled like the wild type receptor, while EPOR1A and EPOR 4A showed impaired signaling at physiologic EPO levels and at levels of receptor expression seen in erythroid progenitors. EPOR2A responded to EPO with an intermediate activity [36]. In such EPOR3A constructs the 'switch' residues required for activation of JAK2 would be 0.5 nm farther down from the membrane, but nevertheless signaling appears intact. On the background of one Ala insertion (EPOR1A), compensatory insertion of two extra Ala in the transmembrane domain upstream was able to restore full signaling [36]. The effect of compensation, revealing helical continuity, was lost when the 2 extra Ala were inserted before the transmembrane domain or within the first predicted transmembrane residues, suggesting that the first few residues of the predicted domain might not be helical or in register with the rest of the transmembrane domain. The interruption of helical continuity between the start of the domain (residue 226) and residues 231–234 is likely due to the fact that the first residues are not α-helical and may form a helix-cap [20].

An interesting difference has been noted between EPOR and the growth hormone receptor. Both seem to require dimer re-orientation for activation. In the case of the latter, insertion of Ala that would rotate the cytosolic domain also results in constitutive activation of the receptor [39]. For example, insertion of four extra Ala residues after the transmembrane domain (T4), or insertion of 1 Ala (J1) after the polar stretch that initiates the cytosolic domain (SKQQ, equivalent to ^{248}SHRR^{251} in EPOR) leads to constitutive receptor activation [39]. Those insertions are predicted to rotate the α-helical juxtamembrane domains towards an opposite interface. The 4Ala insertion in EPOR impaired activation by EPO (as the 1A also did), but did not induce constitutive activation [36]. This difference might be due to the fact that EPOR is activated by EPO in an asymmetric manner and that for activation, a 'rotation-only' conformational change might not suffice in the absence of a 'scissors'-like displacement.

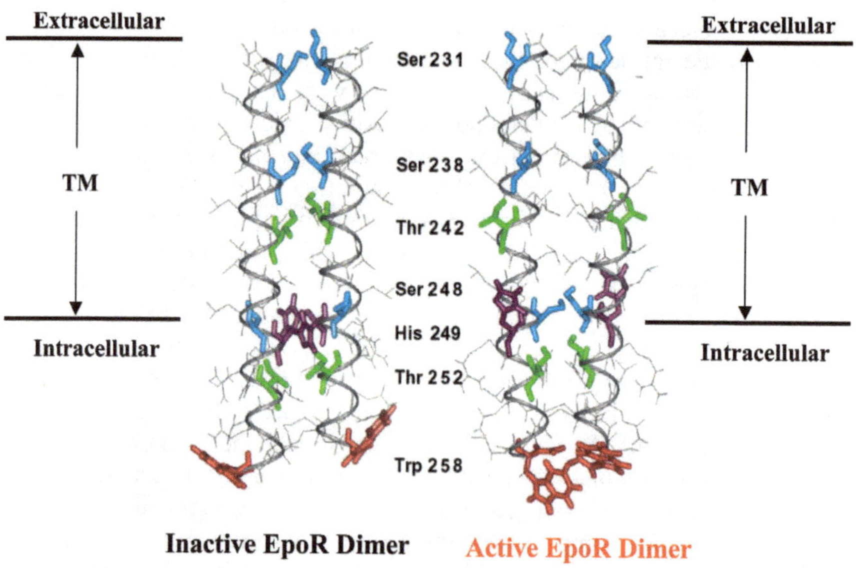

Figure 4. Model for the inactive and active transmembrane dimerization of EPOR. Independent computational searches identified two low energy structures for dimers containing the transmembrane sequence of EPOR. In one (left) dimerization is stabilized by hydrogen bonds between Ser, Thr, and His residues, and the cytosolic juxtamembrane W258 ('switch' residue preceding Box 1) is predicted to be outside the dimer interface. In the other (right), dimerization is stabilized by a Leu-zipper type of interaction, with several Leu residues in the interface. In this conformation, the cytosolic juxtamembrane switch residue W258 would be part of the dimeric interface. The latter structure was proposed to recapitulate the active EPOR conformation since coiled coil fusion proteins where this conformation was imposed were constitutively active, while the former was not.

Coiled coil-imposed orientation-specific dimerization

By replacing EPOR extracellular domain with a dimeric coiled coil, fusion proteins were engineered where the helical transmembrane domains were constrained into seven possible relative orientations [49]. One dimeric transmembrane conformation that imparted full activity to the cytosolic domain of the receptor was identified and shown to signal through JAK2, STAT5 and STAT3 proteins, and MAP kinase. In this dimeric conformation, transmembrane residues Leu241 and Leu 244 are predicted to be in the dimeric interface [20, 49], along with the hydrophobic motif 'switch' residue W258, which precedes Box 1 and is required for activation of JAK2 [24, 36] (Fig. 3). Another dimeric conformation appears to induce partial activation, with marginal JAK2 phosphorylation and preferential MAP kinase, but not STAT signaling. Independent computational searches identified two low energy left-handed dimers of the transmembrane domains, one where hydrogen bonds between Thr, Ser, and His residues stabilize the dimer (Fig. 4) and another where the dimeric interface is stabilized by Leu-zipper type of interactions and where W258 of the juxtamembrane region would be brought in the interface (Fig. 4). The latter corresponds to the active predicted structure of the coiled coil-EPOR fusion proteins cc-EPOR-III and cc-EPOR-VI, which induce activation of EPOR and where the dimeric interface contains 'switch' residue W258. The caveat of these results is represented by the symmetric nature of dimerization promoted by coiled coils, which is different from the asymmetric activation of EPOR by EPO. Nevertheless, the predicted position of particular residues in the individual coiled coil-EPOR fusion proteins was confirmed by Cys mutagenesis and dimer formation [20].

In conclusion, EPOR can be activated by the ligand EPO, by EPO-mimetic peptides [50], by disulfide bond formation between receptor monomers at extracellular (R129, E132, E133) [18, 47] or juxtamembrane residues [20, 21], by transmembrane domain mutations [21], by binding to the gp55 viral proteins [42], by fusions with coiled coils [49], or by rare agonistic monoclonal antibodies in certain conformations [51]. For the latter, it is relevant that only four of 96 monoclonal antibodies directed to the EPOR extracellular domain showed agonistic activity on cell proliferation, and of the four agonistic antibodies, only one induced robust erythroid differentiation [51]. Thus, all available evidence suggests that only some specific dimeric conformations are permissive for activation. The productive conformation would be required at several levels to fully activate JAK2; to access cytosolic receptor tyrosines for phosphorylation; and to exhibit the correct geometry of the phosphorylated tyrosine residues to promote recruitment of SH2- and PTB-domain containing signaling proteins [17]. Such conformational restrictions would be difficult to understand if the transmembrane and cytosolic domains of cytokine receptors would be completely flexible and unstructured [46]. Evidence presented for EPOR, growth hormone receptor, and gp130 [37, 39, 48] that the transmembrane and cytosolic juxtemembrane regions are in fact rigid and α-helically oriented provide a structural framework for understanding why certain conformations are fully active, others inactive and some show an intermediate level of activation.

EPOR dimer reorientation activates JAK2

The major target of activation for EPOR is JAK2. The key event that must occur for activation is represented by a change in position of the EPOR 'switch' residues, which would in turn change the relative positions of JAK2 proteins, enabling them to transphosphorylate each other (Fig. 2B). This would require close contact between the activation loop of 1 JAK molecule and the catalytic loop of the other. The first event leading to activation is phosphorylation of activation loop Y1007, which places the activation loop in the active conformation [52].

No crystal structure exists for a JAK protein or for a receptor-JAK complex. For the JH1 domain in isolation, X-ray crystal structures have been solved for JAK2 and JAK3 and, as expected, the classical kinase fold has been found for these domains [53, 54]. However, the key to understanding how JAK are activated by receptors will be based on the structures of the FERM-SH2 and pseudokinase domains and on exactly how a conformational change provoked

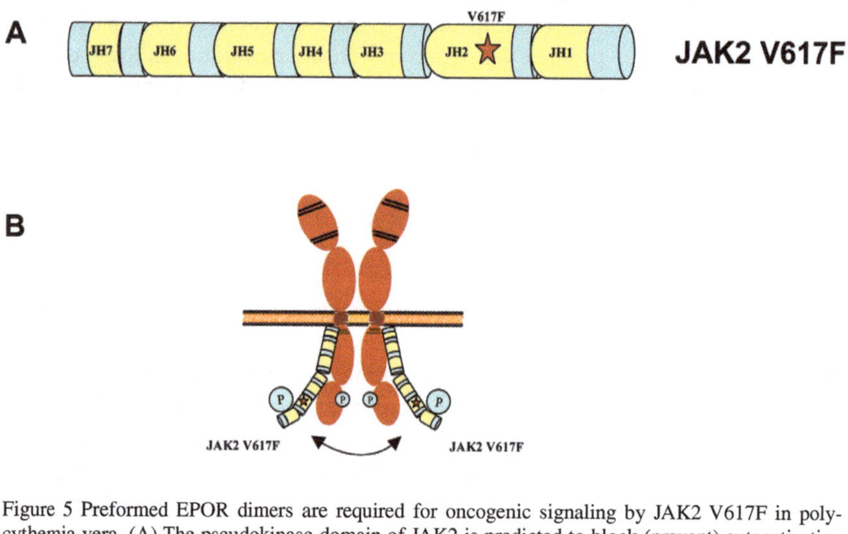

Figure 5 Preformed EPOR dimers are required for oncogenic signaling by JAK2 V617F in polycythemia vera. (A) The pseudokinase domain of JAK2 is predicted to block (prevent) autoactivation of the JH1 domain. A point mutation in the pseudokinase domain (V617F) renders the enzyme active, presumably by removing an intramolecular inhibitory interaction between the kinase and pseudokinase domains. This mutation is prevalent in human myeloproliferative neoplasms such as polycythemia vera, essential thrombocythemia, and primary myelofibrosis. (B) The mutated protein JAK2 V617F scaffolds to the cytosolic domain of EPOR, which is a preformed dimer. Receptor binding allows recruitment of JAK2 V617F to the membrane and further promotes activation by bringing to mutated proteins in close proximity. Unlike the wild type JAK2, which is not active when bound to an unliganded cytokine receptor, JAK2 V617F becomes fully active and initiates signaling.

by the receptor is transmitted to the JH1 kinase domain. Recall that the FERM domain is absolutely required for binding of JAK2 to the region encompassing Box 1 and Box 2.

Once activated, JAK2 is predicted to phosphorylate the cytosolic residues of EPOR, and of signaling proteins attracted to the receptor (such as STAT5, p85 subunit of phosphatydylinositol-3- kinase, and others), as well as proteins that would be attracted by the phosphorylated tyrosines and other motifs of JAK2 itself [55]. Notably, JAK2 contains a large number of tyrosine residues, some of which were shown to be phoshorylated after EPOR activation by EPO [56]. A FERM domain tyrosine, Y119 appears to exert, upon phosphorylation, a negative effect on the association of JAK2 with EPOR [57].

Preformed EPOR dimers are required for oncogenic signaling by the JAK2 V617F in polycythemia vera

Human myeloproliferative neoplasms, polycythemia perap, essential thrombocythemia, and primary (idiopathic) myelofibrosis are characterized by cytokine independent expansion of the myeloid compartment [58], especially the erythroid, megakaryocytic, and granulocytic lineages. A cardinal feature of these diseases, especially of polycythemia vera and essential thrombocytopenia is represented by *in vitro* cytokine-independent colony formation, for example EPO-independent erythroid colonies [59].

The major molecular determinant of myeloproliferative neoplasms is the presence of the *JAK2* V617F mutation, where the pseudokinase domain residue V617 is replaced by a phenylalanine (Fig. 5A); this mutation is present in >95% of patients with polycythemia vera and in >60% of patients with essential thrombocytopenia and primary myelofibrosis [60–63]. The V617F mutation appears to constitutively activate the kinase activity of JAK2 [64–66], most likely by removing an inhibitory effect of the pseudokinase domain on the kinase domain [67]. Indeed, previous studies suggested that the JH2 domain prevents self-activation of the JH1 domain [68]. In bone marrow reconstitution experiments, expression of JAK2 V617F in hematopoietic stem cells by retroviral transduction leads to polycythemia vera and primary myelofibrosis phenotypes [60, 69, 70]. For the essential thrombocytopenia phenotype, it appears that low levels of JAK2 V617F are required [69], as recently established by Pf4-Cre transgenic mouse models [71].

Saturation mutagenesis at position V617 indicated that large non-polar amino acids Trp, Leu, Ile, and Met also cause activation of JAK2 [72]. The strongest of these mutants was V617W, which was the only that stabilized Tyr phosphorylated suppressor of cytokine signaling 3 (SOCS3) [72], a mechanism by which JAK2 V617F overcomes inhibition by SOCS3 [73]. While JAK2 V617W would predictably be pathogenic in humans, the substitution of the Val codon, GTC, by TTG, the codon for Trp, would require three base pair changes, and for this reason, it might be unlikely to occur.

Expression of JAK2 V617F in hematopoietic cells expressing cytokine receptors leads to constitutive activation of STAT5, STAT3, MAP-kinase, and PI-3-kinase and hypersensitivity to ligands. To induce constitutive signaling, JAK2 V617F requires an intact FERM domain, indicating that it must be bound to a cytokine receptor. Introduction in the JAK2 FERM domain of a point mutation (Y114A), known to disrupt interaction with cytokine receptors [27] impairs constitutive signaling of JAK2 V617F in hematopoietic cells [74]. Furthermore, truncated JAK2 proteins that only contain the JH2–JH1 domains (pseudokinase and kinase) do not signal even when the V617F mutation is introduced in JH2 [75].

At high expression levels (six- to 10-fold higher than endogenous levels), JAK2 V617F leads to cytokine-independence of cell lines such as the IL-3-dependent Ba/F3 (murine proB) cells [60, 62, 66]. In such settings, it is presumed that JAK2 V617F binds to an endogenous cytokine receptor, such as IL-3 receptor β chain, which may be limiting, given the competition between JAK2 V617F and wild-type JAK2. However, at low levels of expression, comparable to or two-fold higher than endogenous levels, expression of JAK2 V617F does not suffice to induce autonomous growth of Ba/F3 or 32D cells [75, 76]. Expression of a dimeric type I cytokine receptor, such as EPOR is required for JAK2 V617F to induce autonomous growth under these settings. EPOR cytosolic tyrosine residues required for maximal STAT5 activation are required for promoting signaling by JAK2 V617F. In contrast, a mutated EPOR where residue W258 was mutated to Ala was able to support signaling by JAK2 V617F [75]. Recall that this 'switch' residue is crucial for activation of wild-type JAK2 by the EPO-activated EPOR. The model emerging from these studies suggests that scaffolding of the FERM domain of JAK2 V617F onto dimeric receptors such as EPOR would induce close apposition of two JAK2 V617F proteins. Unlike the wild-type JAK2, which remains inhibited when appended to dimeric receptors, JAK2 V617F becomes fully active and phoshorylates tyrosine residues on the receptor cytosolic tails, triggering ligand-independent signaling (Fig. 5B).

Conclusion

Activation of EPOR occurs when the native cell-surface dimeric conformation is reoriented towards a dimeric conformation, which is productive for signaling. The conformational change appears to involve both a relative rotation of monomers and a possible scissors-like displacement. Besides the physiologic ligand EPO, several other molecules can activate the receptor. A combination of structural and functional studies support the notion that only certain dimeric interfaces allow downstream signaling, that the unliganded receptor is an inactive dimer stabilized by transmembrane domain interactions and that the relative geometry induced by ligand-binding to the extracellular domain is important for the quality of signaling triggered by the cytosolic domains.

Several mutations at the junction between the extracellular and the transmembrane domain lead to constitutive receptor activation, indicating the critical role of this hinge region in regulating receptor activation. On the cytosolic juxtamembrane side, several residues, especially W258, were defined as critical for switching JAK2 catalytic activity on after ligand-activation of the receptor. Although the present body of knowledge about the mechanisms of activation of EPOR is extensive, the complete structure of the receptor-JAK2 complex in the presence and absence of EPO would be necessary for definitively ascribing dimer geometry to signaling output. Nevertheless, sequences around the start of the transmembrane domain may offer interesting targets for screening for small molecule regulators of EPOR signaling.

Acknowledgements
Generous support is acknowledged from the FNRS, Belgium; the Salus Sanguinis Foundation; the Action de Recherche Concertée MEXP31C1 of the University catholique de Louvain, Brussels; the Fondation contre le Cancer, Brussels; the PAI Program BCHM61B5, Belgium; and the Atlantic Philanthropies New York. The author thanks Steven O Smith and Carmen C Diaconu for generous contribution to figures.

References

1. D'Andrea AD, Lodish HF, Wong GG. Expression cloning of the murine erythropoietin receptor. *Cell* 1989;57:277–285.
2. Heinrich PC, Behrmann I, Haan S et al. Principles of interleukin (IL)-6-type cytokine signalling and its regulation. *Biochem J* 2003;374:1–20.
3. Wu H, Liu X, Jaenisch R et al. Generation of committed erythroid BFU–E and CFU–E progenitors does not require erythropoietin or the erythropoietin receptor. *Cell* 1995;83:59–67.
4. Witthuhn BA, Quelle FW, Silvennoinen O et al. JAK2 associates with the erythropoietin receptor and is tyrosine phosphorylated and activated following stimulation with erythropoietin. *Cell* 1993;74:227–236.
5. Neubauer H, Cumano A, Muller M et al. Jak2 deficiency defines an essential developmental checkpoint in definitive hematopoiesis. *Cell* 1998;93:397–409.
6. Parganas E, Wang D, Stravopodis D et al. Jak2 is essential for signaling through a variety of cytokine receptors. *Cell* 1998;93:385–395.
7. Syed RS, Reid SW, Li C et al. Efficiency of signalling through cytokine receptors depends critically on receptor orientation. *Nature* 1998;395:511–516.
8. de Vos AM, Ultsch M, Kossiakoff AA. Human growth hormone and extracellular domain of its receptor: crystal structure of the complex. *Science* 1992;255:306–312.
9. Philo JS, Aoki KH, Arakawa T et al. Dimerization of the extracellular domain of the erythropoietin (EPO) receptor by EPO: one high-affinity and one low-affinity interaction. *Biochemistry* 1996;35:1681–1691.
10. Broudy VC, Lin N, Brice M et al. Erythropoietin receptor characteristics on primary human erythroid cells. *Blood* 1991;77:2583–2590.
11. Zhang J, Socolovsky M, Gross AW et al. Role of Ras signaling in erythroid differentiation of mouse fetal liver cells: functional analysis by a flow cytometry-based novel culture system. *Blood* 2003;102:3938–3946.
12. Hilton DJ, Watowich SS, Murray PJ et al. Increased cell surface expression and enhanced folding in the endoplasmic reticulum of a mutant erythropoietin receptor. *Proc Natl Acad Sci USA* 1995;92:190–194.
13. Hilton DJ, Watowich SS, Katz L et al. Saturation mutagenesis of the WSXWS motif of the erythropoietin receptor. *J Biol Chem* 1996;271:4699–4708.
14. Livnah O, Stura EA, Johnson DL et al. Functional mimicry of a protein hormone by a peptide agonist: the EPO receptor complex at 2.8 A. *Science* 1996;273:464–471.

15. Livnah O, Stura EA, Middleton SA et al. Crystallographic evidence for preformed dimers of erythropoietin receptor before ligand activation. *Science* 1999;283:987–990.
16. Livnah O, Johnson DL, Stura EA et al. An antagonist peptide-EPO receptor complex suggests that receptor dimerization is not sufficient for activation. *Nature Struct Biol* 1998;5:993–1004.
17. Ballinger MD, Wells JA. Will any dimer do? *Nature Struct Biol* 1998;5:938–940.
18. Yoshimura A, Longmore G, Lodish HF. Point mutation in the exoplasmic domain of the erythropoietin receptor resulting in hormone-independent activation and tumorigenicity. *Nature* 1990;348:647–649.
19. Watowich SS, Yoshimura A, Longmore GD et al. Homodimerization and constitutive activation of the erythropoietin receptor. *Proc Natl Acad Sci USA* 1992;89:2140–2144.
20. Kubatzky KF, Liu W, Goldgraben K et al. Structural requirements of the extracellular to transmembrane domain junction for erythropoietin receptor function. *J Biol Chem* 2005;280:14844–14854.
21. Lu X, Gross AW, Lodish HF. Active conformation of the erythropoietin receptor: random and cysteine-scanning mutagenesis of the extracellular juxtamembrane and transmembrane domains. *J Biol Chem* 2006;281:7002–7011.
22. Matthews DJ, Topping RS, Cass RT et al. A sequential dimerization mechanism for erythropoietin receptor activation. *Proc Natl Acad Sci USA* 1996;93:9471–9476.
23. Qureshi SA, Kim RM, Konteatis Z et al. Mimicry of erythropoietin by a nonpeptide molecule. *Proc Natl Acad Sci USA* 1999;96:12156–12161.
24. Huang LJ, Constantinescu SN, Lodish HF. The N-terminal domain of Janus kinase 2 is required for Golgi processing and cell surface expression of erythropoietin receptor. *Mol Cells* 2001;8:1327–1338.
25. Radtke S, Hermanns HM, Haan C et al. Novel role for Janus kinase 1 in the regulation of oncostatin M receptor surface expression. *J Biol Chem* 2002;10:10.
26. Ragimbeau J, Dondi E, Alcover A et al. The tyrosine kinase Tyk2 controls IFNAR1 cell surface expression. *EMBO J* 2003;22:537–547.
27. Royer Y, Staerk J, Costuleanu M et al. Janus kinases affect thrombopoietin receptor cell surface localization and stability. *J Biol Chem* 2005;280:27251–27261.
28. Behrmann I, Smyczek T, Heinrich PC et al. Janus kinase (Jak) subcellular localization revisited: the exclusive membrane localization of endogenous Janus kinase 1 by cytokine receptor interaction uncovers the Jak.receptor complex to be equivalent to a receptor tyrosine kinase. *J Biol Chem* 2004;279:35486–35493.
29. O'Shea JJ, Gadina M, Schreiber RD. Cytokine signaling in 2002: new surprises in the Jak/Stat pathway. *Cell* 2002;109:S121–131.
30. Krolewski JJ, Lee R, Eddy R et al. Identification and chromosomal mapping of new human tyrosine kinase genes. *Oncogene* 1990;5:277–282.
31. Wilks AF, Harpur AG, Kurban RR et al. Two novel protein-tyrosine kinases, each with a second phosphotransferase-related catalytic domain, define a new class of protein kinase. *Mol Cell Biol* 1991;11:2057–2065.
32. Harpur AG, Andres AC, Ziemiecki A et al. JAK2, a third member of the JAK family of protein tyrosine kinases. *Oncogene* 1992;7:1347–1353.
33. Velazquez L, Fellous M, Stark GR et al. A protein tyrosine kinase in the interferon alpha/beta signaling pathway. *Cell* 1992;70:313–322.
34. Ziemiecki A, Harpur AG, Wilks AF. JAK protein tyrosine kinases: their role in cytokine signalling. *Trends Cell Biol* 1994;4:207–212.
35. Miura O, Cleveland JL, Ihle JN. Inactivation of erythropoietin receptor function by point mutations in a region having homology with other cytokine receptors. *Mol Cell Biol* 1993;13:1788–1795.
36. Constantinescu SN, Huang LJ, Nam H et al. The erythropoietin receptor cytosolic juxtamembrane domain contains an essential, precisely oriented, hydrophobic motif. *Mol Cells* 2001;7:377–385.
37. Constantinescu SN, Keren T, Socolovsky M et al. Ligand-independent oligomerization of cell-surface erythropoietin receptor is mediated by the transmembrane domain. *Proc Natl Acad Sci USA* 2001;98:4379–4384.
38. Henis YI, Moustakas A, Lin HY et al. The types II and III transforming growth factor-beta receptors form homo-oligomers. *J Cell Biol* 1994;126:139–154.
39. Brown RJ, Adams JJ, Pelekanos RA et al. Model for growth hormone receptor activation based on subunit rotation within a receptor dimer. *Nat Struct Mol Biol* 2005;12:814–821.

40. Gurezka R, Laage R, Brosig B et al. A heptad motif of leucine residues found in membrane proteins can drive self-assembly of artificial transmembrane segments. *J Biol Chem* 1999;274:9265–9270.
41. Kubatzky KF, Ruan W, Gurezka R et al. Self assembly of the transmembrane domain promotes signal transduction through the erythropoietin receptor. *Curr Biol* 2001;11:110–115.
42. Li JP, D'Andrea AD, Lodish HF et al. Activation of cell growth by binding of Friend spleen focus-forming virus gp55 glycoprotein to the erythropoietin receptor. *Nature* 1990;343:762–764.
43. Chung SW, Wolff L, Ruscetti SK. Transmembrane domain of the envelope gene of a polycythemia-inducing retrovirus determines erythropoietin-independent growth. *Proc Natl Acad Sci USA* 1989;86:7957–7960.
44. Constantinescu SN, Liu X, Beyer W et al. Activation of the erythropoietin receptor by the gp55-P viral envelope protein is determined by a single amino acid in its transmembrane domain. *EMBO J* 1999;18:3334–3347.
45. Constantinescu SN, Keren T, Russ WP et al. The erythropoietin receptor transmembrane domain mediates complex formation with viral anemic and polycythemic gp55 proteins. *J Biol Chem* 2003;278:43755–43763.
46. Remy I, Wilson IA, Michnick SW. Erythropoietin receptor activation by a ligand-induced conformation change. *Science* 1999;283:990–993.
47. Watowich SS, Hilton DJ, Lodish HF. Activation and inhibition of erythropoietin receptor function: role of receptor dimerization. *Mol Cell Biol* 1994;14:3535–3549.
48. Greiser JS, Stross C, Heinrich PC et al. Orientational constraints of the gp130 intracellular juxtamembrane domain for signaling. *J Biol Chem* 2002;277:26959–26965.
49. Seubert N, Royer Y, Staerk J et al. Active and inactive orientations of the transmembrane and cytosolic domains of the erythropoietin receptor dimer. *Mol Cells* 2003;12:1239–1250.
50. Wrighton NC, Farrell FX, Chang R et al. Small peptides as potent mimetics of the protein hormone erythropoietin. *Science* 1996;273:458–464.
51. Elliott S, Lorenzini T, Yanagihara D et al. Activation of the erythropoietin (EPO) receptor by bivalent anti-EPO receptor antibodies. *J Biol Chem* 1996;271:24691–24697.
52. Feng J, Witthuhn BA, Matsuda T et al. Activation of Jak2 catalytic activity requires phosphorylation of Y1007 in the kinase activation loop. *Mol Cell Biol* 1997;17:2497–2501.
53. Lucet IS, Fantino E, Styles M et al. The structural basis of Janus kinase 2 inhibition by a potent and specific pan-Janus kinase inhibitor. *Blood* 2006;107:176–183.
54. Boggon TJ, Li Y, Manley PW et al. Crystal structure of the Jak3 kinase domain in complex with a staurosporine analog. *Blood* 2005;106:996–1002.
55. Constantinescu SN, Ghaffari S, Lodish HF. The erythropoietin receptor: structure, activation and intracellular signal transduction. *TEM* 1999;10:18–23.
56. Matsuda T, Feng J, Witthuhn BA et al. Determination of the transphosphorylation sites of Jak2 kinase. *Biochem Biophys Res Commun* 2004;325:586–594.
57. Funakoshi-Tago M, Pelletier S, Matsuda T et al. Receptor specific downregulation of cytokine signaling by autophosphorylation in the FERM domain of Jak2. *EMBO J* 2006;25:4763–4772.
58. Spivak JL. Polycythemia vera: myths, mechanisms, and management. *Blood* 2002; 100:4272–4290.
59. Prchal JF, Axelrad AA. Bone-marrow responses in polycythemia vera. *N Engl J Med* 1974; 290:1382.
60. James C, Ugo V, Le Couedic JP et al. A unique clonal JAK2 mutation leading to constitutive signalling causes polycythaemia vera. *Nature* 2005;434:1144–1148.
61. Levine RL, Wadleigh M, Cools J et al. Activating mutation in the tyrosine kinase JAK2 in polycythemia vera, essential thrombocythemia, and myeloid metaplasia with myelofibrosis. *Cancer Cell* 2005;7:387–397.
62. Kralovics R, Passamonti F, Buser AS et al. A gain-of-function mutation of JAK2 in myeloproliferative disorders. *N Engl J Med* 2005;352:1779–1790.
63. Baxter EJ, Scott LM, Campbell PJ et al. Acquired mutation of the tyrosine kinase JAK2 in human myeloproliferative disorders. *Lancet* 2005;365:1054–1061.
64. Vainchenker W, Constantinescu SN. A unique activating mutation in JAK2 is at the origin of polycythemia vera and allows a new classification of myeloproliferative diseases. *Hematology* 2005;195–200.
65. Zhao R, Xing S, Li Z et al. Identification of an acquired JAK2 mutation in polycythemia vera. *J Biol Chem* 2005;280:22788–22792.

66. Staerk J, Kallin A, Demoulin JB et al. JAK1 and Tyk2 activation by the homologous polycythemia vera JAK2 V617F mutation: cross-talk with IGF1 receptor. *J Biol Chem* 2005;280:41893–41899.
67. Lindauer K, Loerting T, Liedl KR et al. Prediction of the structure of human Janus kinase 2 (JAK2) comprising the two carboxy-terminal domains reveals a mechanism for autoregulation. *Protein Eng* 2001;14:27–37.
68. Saharinen P, Silvennoinen O. The pseudokinase domain is required for suppression of basal activity of Jak2 and Jak3 tyrosine kinases and for cytokine-inducible activation of signal transduction. *J Biol Chem* 2002;277:47954–47963.
69. Lacout C, Pisani DF, Tulliez M et al. JAK2V617F expression in murine hematopoietic cells leads to MPD mimicking human PV with secondary myelofibrosis. *Blood* 2006;108:1652–1660.
70. Wernig G, Mercher T, Okabe R et al. Expression of Jak2V617F causes a polycythemia vera-like disease with associated myelofibrosis in a murine bone marrow transplant model. *Blood* 2006;107:4274–4281.
71. Tiedt R, Schomber T, Hao-Shen H et al. Pf4-Cre transgenic mice allow the generation of lineage-restricted gene knockouts for studying megakaryocyte and platelet function *in vivo*. *Blood* 2007;109:1503–1506.
72. Dusa A, Staerk J, Elliott J et al. Substitution of pseudokinase domain residue V617 by large non-polar amino acids causes activation of JAK2. *J Biol Chem* 2008
73. Hookham MB, Elliott J, Suessmuth Y et al. The myeloproliferative disorder-associated JAK2 V617F mutant escapes negative regulation by suppressor of cytokine signaling 3. *Blood* 2007;109:4924–4929.
74. Wernig G, Gonneville JR, Crowley BJ et al. The Jak2V617F oncogene associated with myeloproliferative diseases requires a functional FERM domain for transformation and for expression of the Myc and Pim proto-oncogenes. *Blood* 2008;111:3751–3759.
75. Lu X, Huang LJ, Lodish HF. Dimerization by a cytokine receptor is necessary for constitutive activation of JAK2V617F. *J Biol Chem* 2008;283:5258–5266.
76. Lu X, Levine R, Tong W et al. Expression of a homodimeric type I cytokine receptor is required for JAK2V617F-mediated transformation. *Proc Natl Acad Sci USA* 2005;102:18962–18967.

Clinical utility

Erythropoietins, Erythropoietic Factors and Erythropoiesis
Edited by S.G. Elliott, M.A. Foote, and G. Molineux

Pharmacokinetics of erythropoiesis-stimulating agents

Sameer Doshi[1], Juan Jose Perez-Ruixo[2], Graham R. Jang[1], Andrew T. Chow[1]

[1] *Amgen, One Amgen Center Drive, Thousand Oaks, CA 91320, USA,* [2] *Amgen Inc, Spain*

Introduction

Erythropoietin (EPO) is a 30.4 kDa glycoprotein hormone secreted by the kidneys in response to tissue hypoxia, which stimulates red blood cell production. To regulate erythrocyte production, recombinant human EPO (rHuEPO) activates the EPO receptor (EPOR) and stimulates the proliferation and differentiation of erythrocytic progenitors in the bone marrow, leading to reticulocytosis and increased erythrocyte numbers and hemoglobin concentration in the blood. Epoetin alfa, the first commercial form of rHuEPO marketed in the United States of America (USA) and European Union (EU), and epoetin beta marketed outside the USA are both expressed in Chinese hamster ovary (CHO) cells. Both epoetins have a 165-amino acid sequence identical to human urinary EPO, contain three sialic acid-containing *N*-linked and one *O*-linked carbohydrate chains [1] leading to the same biologic effects as endogenous EPO [2–4]. No important differences in clinical efficacy are apparent between epoetin alfa and epoetin beta, and they are generally used interchangeably [5]. Darbepoetin alfa is a hyperglycosylated rHuEPO analog with five amino acid changes and two additional *N*-linked carbohydrate chains, which has the same mechanism of action as rHuEPO. Darbepoetin alfa, however, has a three-fold increased serum half-life [6–8], and increased *in vivo* potency [9], allowing for more convenient modes of administration, including extended dosing intervals [10, 11]. A large methoxy-polyethylenglicol polymer chain was integrated into the epoetin beta molecule through amide bonds between the *N*-terminal amino group of ALA and the σ-amino groups off Lys (Lys45 or Lys52) by means of a succinimudyl butanoic acid linker [12], resulting in a PEGylated epoetin beta molecule that is marketed in EU. As with other PEGylated therapeutic proteins, the PEGylation of epoetin beta reduced EPOR binding affinity. This biologic disadvantage is counter balance with a half-life of approximately 130 h in humans and allows for extended dosing intervals, similarly to darbepoetin alfa. In this chapter, we review the clinical pharmacokinetics of the marketed protein-based erythropoiesis stimulating agents (ESAs).

Mechanism of action

The mechanism of EPO synthesis is regulated through the blood partial pressure of oxygen in the kidney and other EPO-producing organs [13]. Under conditions of hypoxia, *EPO* gene transcription is activated when hypoxia-inducible factor (HIF) binds to a hypoxia-responsive enhancer (HRE) located in 3' flanking sequence of the *EPO* gene. After EPO is produced, it acts on erythroid precursor cells through EPOR, which is a transmembrane protein that is a member of the type I cytokine receptor superfamily. Two regions of EPO contain two separate EPOR-binding sites [14, 15], and EPOR is activated when one rHuEPO molecule homodimerizes two EPOR molecules on the surface of the erythroid blast-forming unit (BFU–E) and colony-forming unit (CFU–E). Each CFU–E cell contained approximately 1,000 EPOR per cell [14, 16], and approximately 5–10% of these must be continuously occupied with ligand for the cells to survive and differentiate fully into erythrocytes [17, 18]. A high-affinity (1 nM) and low-affinity (4 μM) EPOR binding sites have been characterized, however, their clinical significance is not well understood [19, 20]. ESA binding to EPOR can lead either to internalization or subsequent dissociation of the molecule from the receptor. Once internalized along with the receptor complex, the ligand may either be degraded, or it may be recycled intact to the cell surface and released [21–23].

EPOR homodimerization activates a number of signal transduction pathways, including Janus kinase/signal transducers and activators of transcription (JAK-STAT), phosphatidylinositol 3-kinase (PI3K), and RAS-mitogen-activated protein kinase (MAPK) [24, 25]. Current research suggests that EPO exerts its control over erythropoiesis by preventing cell apoptosis (programmed cell death) thus allowing more red cell precursors to survive, proliferate, and/or induce erythroid-specific proteins [26]. Some of the apoptotic tendency of erythroid cells is attributed to proapoptotic molecules produced by hematopoietic cells, macrophages, inflammatory cytokines, and stromal cells. EPO prevents apoptosis by upregulation of antiapoptotic *Bcl* proteins (Bcl2, BclXL) and downregulation/inactivation of apoptotic proteins, such as BAD, a prodeath Bcl2 family member, tumor necrosis factor-related apoptosis-inducing ligand (TRAIL) forkhead and caspases [27–32].

Darbepoetin alfa and PEGylated epoetin beta stimulate erythropoiesis by the same mechanisms as endogenous EPO and rHuEPO. *In vitro*, the affinity of darbepoetin alfa and PEGylated epoetin beta for EPOR is three- to five-fold and 50- to 100-fold less than that of rHuEPO, respectively; however, the increase in mean residence time of darbepoetin alfa and PEGylated epoetin beta results in a prolonged period of time above an erythropoietic threshold that exceedingly compensates for the reduced receptor affinity, yielding an increased *in vivo* activity [7, 9, 33]. However, in dialysis patients with stable hemoglobin concentrations, no difference in efficacy has been observed between darbepoetin alfa and PEGylated epoetin beta [34].

Regulation of erythropoietin

The primary site of EPO synthesis in adults is the peritubular cells of the kidney [5, 24, 25]. The liver is a secondary site of EPO production, with synthesis occurring in both hepatocytes and fibroblastoid interstitial cells [35]. Astrocytes and neurons within the central nervous system have been reported to express EPO [36]. No preformed stores of EPO exist, and plasma EPO concentrations are maintained at a constant concentration by the homeostatic turnover, which consists in the basal production and elimination of the hormone [37]. Within a healthy individual, the serum EPO concentration tends to be controlled tightly; however, large interindividual variability is evident from the normal range, 5 to 35 IU/L [37]. Maintenance of normal serum concentrations of endogenous EPO requires the synthesis of about 2–3 IU/kg/day, or approximately 1,000–1,500 IU/week for a 70 kg man [38]. Sex differences [38] and regular-to-moderate athletic training [39] did not appear to affect endogenous EPO serum concentrations.

The blood flow in the kidney has a circadian rhythm in normal individuals [40], therefore, the endogenous production of EPO was expected to have diurnal variations with the highest levels of endogenous EPO in the evening and at night [41, 42]. The circadian rhythm was first quantified by Hayashi et al. [43], who reported a 10% amplitude of the baseline EPO production rate. Similar results were obtained by Olsson et al. [44], who also evidenced an asymmetric time course of endogenous EPO, with a broad peak in the evening and night, and relatively narrow trough in the morning.

The overexpression of EPO occurs in a number of adaptive and pathologic conditions, which needs to be considered when the ESA pharmacokinetics is investigated. In response to acute hypoxic stress, such as severe blood loss or severe anemia, EPO production rate can increase 100- to 1,000-fold. Numerous studies have shown an exponential increase in serum EPO, with increasing degrees of anemia [5, 45], although the maximal bone marrow response to such stimulation is only a four- to six-fold increase in red cell production rate [46]. Overproduction of EPO with accompanying erythrocytosis may be an adaptive response to conditions that produce chronic tissue hypoxia, such as living at high altitude, chronic respiratory diseases, cyanotic heart disease, sleep apnea, smoking, localized renal hypoxia, or hemoglobinopathies with increased oxygen affinity [24]. Paraneoplastic production of EPO from tumors and cysts, including renal carcinomas, benign renal tumors, Wilms' tumors, hepatomas, liver carcinomas, cerebellar hemangioblastomas, adrenal gland tumors, and leiomyomas, can also result in high plasma concentrations EPO.

In chronic kidney disease, up to 60% of patients have hemoglobin concentrations <11 g/dL before beginning dialysis [47]. Multiple mechanisms contribute to the low hemoglobin concentrations [25], but the most important is the inability of the diseased kidneys to produce an appropriate EPO response for the given degree of anemia or an inability to meet the increased red blood cell demands of uremic patients [48, 49]. In addition, the uremic state itself appears to blunt the bone marrow response to EPO, perhaps through polyamines, inflammatory cytokines, and/or parathyroid hormone mediators [25, 50]. It is of interest that serum EPO concentrations in chronically anemic dialysis patients increase to some extent in response to acute hypoxic stress (from either acute bleeding or systemic hypoxemia), suggesting that kidney failure does not result in a complete inability to produce EPO [51, 52].

The anemia of cancer is also of multifactorial etiology [37]. As with other anemias of chronic disease, including those associated with chronic infection and inflammatory disorders, there is decreased production of endogenous EPO [53], cytokine-induced suppression of bone marrow function, disordered iron absorption and metabolism [54], and decreased erythrocytes survival. Furthermore, the amount of endogenous EPO transiently increases up to six-fold within the 48 h after the administration of chemotherapy and returns to baseline within a week [55–57]. In the case of myeloablative chemotherapy, severe thrombocytopenia might contribute to a significant loss of red blood cells. It was demonstrated that ^{51}Cr-labeled control red blood cells administered to rats that had received 60 mg/kg of carboplatin were substantially lost during development of thrombocytopenia, probably due to internal hemorrhage, but not related to an increase in cell lysis susceptibility or reduction in red cell survival time [58]. Finally, the anemias associated with infant prematurity, pregnancy, allogeneic bone marrow transplantation, and HIV infection are often characterized by inappropriately low EPO concentrations [35].

Pharmacokinetics

Absorption

Due to gastrointestinal enzymatic degradation of proteins, protein-based ESAs are not administered by the oral route; either intravenous or subcutaneous routes are used. The route of administration is determined according to indication, patient, and healthcare provider preference, as well as regional considerations that affect medical practice. After a single dose of subcutaneous rHuEPO, absorption is slow, leading to peak serum concentrations at 5–30 h and a longer terminal half-life (24–79 h) than that obtained after intravenous administration. These results indicate the presence of flip-flop pharmacokinetics [59] where the rate of absorption is slower than the rate of elimination. Thus, absorption is rate limiting, and the observed terminal half-life after subcutaneous dosing reflects the absorption rather than elimination rate [60–66].

After subcutaneous administration, protein therapeutics typically enter into the systemic circulation through the blood capillaries or the lymphatic system. The lymphatic system is considered to be the primary route of absorption from the subcutaneous injection site for protein therapeutics greater than about 16 kDa due to restricted vascular access afforded by the continuous endothelial layer of blood capillaries [67–69]. The intercellular junctions within the lymphatic vessel wall create cleft-like openings that provide an alternative pathway from the interstitium into the lymph and indirectly into the systemic circulation by the thoracic duct [70]. The relative roles of the blood and lymphatic absorption pathways after subcutaneous administration of rHuEPO and darbepoetin alfa have been directly assessed using lymph cannulated animal models. In sheep, the cumulative amount of rHuEPO and darbepoetin alfa recovered in peripheral lymph was 84–90% of the administered dose, respectively, indicating almost complete absorption from the subcutaneous injection site (interdigital space) and minimal clearance during transit through the lymphatic system [71, 72]. A rat model with continuous lymph collection from thoracic lymph duct evidenced that <3% of the rHuEPO dose was recovered in the lymph. Therefore, the rHuEPO subcutaneous absorption in rat occurred mainly through direct uptake into the blood capillaries without substantial involvement of the lymphatic pathway [73]. These results are supported by studies using an immunocytochemical approach where it was shown that insulin, and even molecules as large as albumin, can penetrate blood capillary endothelium by the transcellular route [74]. It has been argued the discrepancies between the results obtained from rat and sheep model might be attributed to the utilization of different anatomical injection sites and doses administered [73]. However, it has been shown that the systemic availability of darbepoetin alfa from the interdigital (106%), shoulder (92%), and abdomen (85%) subcutaneous injection sites was almost complete relative to the intravenous group. However, the rate of absorption was 0.15, 0.08, and 0.02 h^{-1} for the interdigital, shoulder, and the abdomen injection sites, respectively, which showed significant variation on the absorption rate across injection site [75].

In humans, the absorption process after subcutaneous administration of rHuEPO is complex, with at least two apparent pathways of absorption. A dual absorption model describes the pharmacokinetics of rHuEPO after subcutaneous administration of 20–160 kIU to healthy subjects [44]. In this model, the faster process corresponded to a direct absorption from the injection site to the extravascular compartment, characterized by a zero-order process that lasted for 0.7 h, followed by a slow first-order process of entry into systemic circulation with an absorption rate of 0.034 h^{-1}. The delayed and long lasting (1.5 days) zero-order absorption corresponds to the absorption *via* the lymphatic system, which requires a lag-time of almost 3 h to account for the transit time from the injection site to the point of lymph collection [44]. A similar model was used before to describe the pharmacokinetics of rHuEPO after subcutaneous administration in rats [76], monkeys [77], and healthy volunteers [78]. In these studies, approximately a 67%, 65%, and 79% of the absorbed dose in rat, monkeys, and healthy volunteers, respectively, was attributed to entry into the circulation by the lymphatics. These values are slightly lower than the values reported in sheep [71]. In addition, it has been suggested that the fraction of dose that is absorbed through the lymphatic system in humans is increased at doses higher than 300 IU/kg [44].

The subcutanteous absorption of darbepoetin alfa in humans is slow, with peak concentrations reached at 34–58 h post-dose, followed by a generally monophasic decline. Similarly to rHuEPO, darbepoetin alfa also displays flip-flop pharmacokinetics, with a longer half-life after subcutaneous dosing than after intravenous dosing. The mean terminal half-life of darbepoetin alfa, 70 h, was associated with large between patient variability, consistent with variability observed for other ESAs [79]. Nevertheless, the typical profile of the dual absorption does not become evident from the darbepoetin alfa subcutaneous pharmacokinetic profiles, which has been characterized using a simple first-order absorption model [80]. One potential explanation for this finding is that the rate of absorption through the two different pathways becomes similar and, consequently, the dual absorption profile becomes just a first-order absorption process. A 20–30% reduction in the darbepoetin alfa absorption rate per decade of age was reported. The magnitude of this finding is consistent with the estimated effect of age on the rHuEPO absorption rate and reflects the longer terminal half-life and the larger exposure of both drugs in older patients. It has been hypothesized that the age-dependent reduction in lymphatic flow rate could be the physiologic reason behind this relationship [80]. The data available also suggest that the pharmacokinetic profile of rHuEPO and darbepoetin alfa after subcutaneous administration is similar in adults and children; however, subcutaneous absorption in children may be more rapid than in adults for both drugs [61].

It has been shown that PEGylation may increase the absorption half-life of subcutaneously administered agents [81]. The absorption process of PEGylated epoetin beta after subcutaneos injection was indeed slow, with peak concentrations reached at 42–120 h post-dose, followed by a generally monophasic decline. As observed for other ESA, PEGylated epoetin beta also displays flip-flop pharmacokinetics after subcutaneous administration. Actually, the terminal half-life after a single subcutaneous injection to healthy volunteers ranged from 102–216 h, which is slightly higher than the terminal half-life observed after intravenous administrations (70–122 h). Similarly, in patients with chronic kidney disease not receiving dialysis, the mean terminal half-life after subcutaneous administration was 142 h, which is about 84% higher than the terminal half-live observed after intravenous administration, 77 h. These results are consistent with the multiple ascending dose study in healthy volunteers and in patients with chronic kidney disease on dialysis, although the difference in terminal half-live between intravenous and subcutaneous administration were narrower [82]. As for darbepoetin alfa, the typical profile of the dual absorption does not become evident from the pharmacokinetic profiles obtained after subcutaneous administration of PEGylated epoetin beta [83]. No effect of age on PEGylated epoetin beta absorption has been reported.

After subcutaneous administration, absorption rates of rHuEPO vary according to the administration site, most likely reflecting regional differences in blood and lymph flow [84]. Macdougall et al. administered 125 IU/kg subcutaneously of radiolabeled rHuEPO to healthy volunteers in sites in the arm, abdomen, and thigh, demonstrating more rapid and extensive absorption when rHuEPO was injected into the thigh compared with the abdomen or arm [85]. The AUCs(0–∞) measured after injections into the arm and abdomen were 89% and 77%, respectively, of the AUC(0–∞) measured after injection into the thigh. Jensen et al. [84], however, concluded that the only notable difference in pharmacokinetics between epoetin beta when injected at 100 IU/kg into the thigh compared with the abdomen was a longer mean residence time (32.7 *versus* 26.2 h, respectively), which they attributed to slightly slower absorption. Similar results were obtained for PEGylated epotin beta. Actually, the maximum concentration (C_{max}) after subcutaneous injections of 3 µg/kg into the arm and abdomen were 95% and 86%, respectively, of the C_{max} measured after injection into the thigh. Similarly, the AUCs(0–last) after subcutaneous injections of 3 µg/kg into the arm and abdomen were 96% and 94%, respectively, of the AUCs(0–last) measured after the injection into the thigh [83]. Overall, the absorption rate of ESA varies according to the site of administration, however, the magnitude of the differences between the sites of administration was relatively small and not considered to be clinically relevant as the pharmacodynamic profile did not evidence any difference across the site of administration.

Bioavailability

Initial bioavailability estimates for rHuEPO after subcutaneous administration ranged from 15–40% [86, 87], and were similar for epoetin alfa and beta [88]. However, most assessments of rHuEPO subcutaneous bioavailability did not take into account the nonlinear disposition kinetics of rHuEPO and/or the time dependent clearance [89], and have estimated bioavailability relative to the same single intravenous dose administration with a comparison of AUC. In this situation, there is an underestimation of the bioavailability, especially at lower doses of rHuEPO, at which the nonlinearity in clearance is most evident. To accurately determine the bioavailability of a drug cleared in a nonlinear manner, it is necessary to determine the underlying intravenous pharmacokinetics and then estimate the subcutaneous bioavailability by modeling techniques [90].

In addition, when the pharmacokinetics of subcutaneous rHuEPO and darbepoetin alfa were studied over a wider dose range in healthy volunteers, exposure was found to increase more than proportional with dose, even when accounting for the nonlinear clearance of rHuEPO [44, 80, 91, 92]. Similar results were observed in dogs where the rHuEPO subcutaneous bioavailability for the 400 IU/kg dose, 26.8%, increased to 73% at doses of 1,000 IU/kg, and to 100% at doses >2,400 IU/kg [77]. More recently, rHuEPO subcutaneous bioavailability (F) in healthy subjects was described as a function of dose using a linear, (F, % = 38.8 + 0.0250 · Dose, IU/kg) [93], or hyperbolic (F, % = 29.1 + 64.9 · Dose, kIU/[63.2 + Dose, kIU]) [44] function. The linear function was also deemed appropriate to describe the increase of darbepoetin alfa subcutaneous bioavailability with dose (F, % = 44.8 + 0.0586 · Dose, µg) [80]. Assuming a conversion factor of 200 IU of rHuEPO per 1 µg of darbepoetin alfa, the slopes of the linear function for the subcutaneous bioavailability were very similar for rHuEPO and darbepoetin alfa. Furthermore, according to those models the subcutaneous bioavailability of darbepoetin alfa would increase from 57% to 69% when the 200 µg dose is increased up to 400 µg, while the subcutaneous bioavailability of rHuEPO would increase from 54% to 65% when the 40 kIU dose is increased up to 80 kIU.

The bioavailability of PEGylated epoetin beta in healthy subjects ranged from 40–60% in the dose range of 0.8–3.2 µg/kg, which is slightly lower than the bioavailability of epoetin and darbepoetin alfa at equivalent doses. These results are consistent with the findings observed in comparative studies conducted in rat and dog. In addition, the dose-dependent absolute bioavailability of PEGylated epoetin beta has not been reported. However, statistical analysis of a Phase I study where subcutaneous PEGylated epoetin beta was administered in ascending doses from 0.1–3.2 ug/kg evidenced a more than dose proportional increase in the AUC and C_{max}, which could be an indirect evidence of the dose-dependent absolute bioavailability in absence of saturable elimination process. Similar to other ESA, the absolute bioavailability of PEGylated epoetin evidenced a high degree of intersubject variability (53–92%) [82].

The estimates of subcutaneous bioavailability of ESA indicated incomplete drug absorption and/or degradation during the absorption process. Based on the negative correlation previously observed between the molecular size and bioavailability of heparins [94], it was suggested that the large molecular size of ESA might impede drug absorption from the subcutaneous injection site [95], and could also explain the lower absolute bioavailability of PEGylated epoetin beta with respect to epoetin beta in different studies. In addition, the apparent increase in subcutaneous bioavailability with dose for ESA might indicate saturable presystemic processes. In fact, saturable injection site loss has been reported for several proteins, and/or saturable degradation by proteolytic enzymes in the lymph may also contribute to this phenomenon [68]. Nevertheless, despite the apparent low bioavailability, subcutaneous administration of ESA produces equivalent or better efficacy to intravenous administration, and this is assumed to be due to the prolonged absorption leading to reduced receptor saturation and increased drug efficiency [96, 97].

Distribution

Serum rHuEPO concentrations rose rapidly during the intravenous infusion and then declined in a bi-exponential manner [44]. In early clinical studies, however, a single phase of rHuEPO disposition was reported after intravenous administration [87, 93], most likely because the early rapid disposition phase was not captured with the sampling schedule used [90]. After the intravenous administration of rHuEPO, peak serum rHuEPO concentrations correlated linearly with dose, and a dose of 50 IU/kg produced concentrations of about 1,000 mIU/mL 15 min after the end of the infusion [38]. As expected from its large molecular weight, rHuEPO's volume of distribution was similar to the plasma volume (40–60 mL/kg), suggesting confinement of EPO within the plasma circulation [44, 66, 98]. Nevertheless, many reports have shown that blood capillaries are permeable for a variety of macromolecules. Plasma and lymph concentrations of rHuEPO in rat were 6.6 and 5.2 IU/mL, respectively, after 1.5 h after intravenous administration of 400 IU/kg, respectively. The lymph concentrations were much higher after intravenous administration than those detected after subcutaneous injection, suggesting that rHuEPO is capable of redistributing from systemic circulation to lymph [73]. In addition, PEGylation seems to limit tissue penetration as the distribution volume of PEGyated epoetin beta is only half of that of epoetin beta in rat and dog [82]. Although limited extravascular distribution to peripheral compartments has been reported for ESA [44], and it is unclear what it represents, it could be nonspecific distribution into bone marrow, blood cells, or other tissue depending on availability for transport across the capillary endothelia or through the interstitial space [99, 100].

As expected from the similarity in the molecular structure between epoetins alfa and beta, similar results in the volume of distribution were found in the pharmacokinetic analysis of rHuEPO beta in healthy volunteers [101, 102]. Steady state volume of distribution (V_{ss}) of epoetins alfa or beta across different animal species, including human, can be expressed as a function of body weight (BW) as follows:

$$V_{ss}\,(\text{mL}) = 90 \cdot \text{BW(kg)}^{0.853}\ (r^2 = 0.922,\ p < 0.01)\ [103].$$

A population pharmacokinetic analysis of darbepoetin alfa in healthy volunteers indicated that darbepoetin alfa has a similar volume of distribution as rHuEPO [80]. Similar results were found in patients who were receiving continuous ambulatory peritoneal dialysis, hemodialysis, and chemotherapy [6, 104–106]. In addition, when the volume of distribution of ESA in healthy volunteers and patients receiving dialysis are compared across studies, they appear similar and close to the plasma volume [82, 87, 107–110]. The data available also suggest that the pharmacokinetic profile of rHuEPO and darbepoetin alfa after intravenous administration is similar in adults and children. However, no pharmacokinetic studies have been conducted in pediatric patients receiving PEGylated epoetin beta [82].

In addition, dose-proportional increases in cerebrospinal fluid (CSF) concentrations of EPO were observed as a consequence of the penetration of systemically administered rHuEPO into the central nervous system of patients with cancer and brain metastases [111]. The concentration of rHuEPO in the CSF increased after a period of slow equilibration (1–3 h), peaked after approximately 9–24 h, and declined bi-exponentially thereafter with similar terminal slopes in serum and CSF. The penetration of rHuEPO into the CSF ranged from 0.02–0.31% after single doses, which is similar to results observed in rats and macaques and consistent with the data obtained from PEGylated epoetin beta in rats [82, 112, 113]. Although the mechanism involved in the rHuEPO transport across the blood-brain barrier is not well understood, several hypotheses have been suggested and included receptor-mediated transport; carrier-mediated transport; fluid phase endocytosis; non-specific or receptor-mediated adsorptive endocytosis; and transmembrane diffusion [114]. The possibility that rHuEPO plays a role in neuroprotection is the subject of considerable debate. EPO has been shown to exert cytoprotective effects on erythroid progenitor cells as well as various nonerythroid cells. These protective effects have been largely attributed to antiapoptotic signalings of EPO. In a trial of patients who had had a stroke, EPO concentrations of approximately 15 mIU/mL in the CSF correlated with improved functional outcomes [115], consistent with concentrations achieved after a single dose of 1,500 IU/kg, where the peak rHuEPO concentrations in the CSF ranged from 11 mIU/mL to 40 mIU/mL. However, Katavetin et al. have argued that injured cells undergoing apoptosis are generally too severely damaged to function properly. Therefore, simply corrupting apoptotic pathway is unlikely to be an effective strategy, because the remaining damaged cells may not function appropriately, or they may eventually undergo necrotic cell death. Evidence suggests that EPO also provides cytoprotection by ameliorating oxidative stress, the principal cellular insult. EPO may exert its antioxidative effects directly by exploiting intracellular antioxidative mechanisms such as hemeoxygenase-1 and glutathione peroxidase. In addition, EPO may act indirectly by inducing iron depletion and thereby inhibiting iron-dependent oxidative injury. Increasing red blood cells by EPO may also indirectly reduce cellular oxidative stress, as red blood cells are loaded with a substantial amount of antioxidative enzymes [116]. Further investigation regarding the mechanisms of cellular antioxidative responses to EPO would provide a better insight to cytoprotective action of EPO, and would support the development of better cytoprotective drugs in the near future.

Elimination

Despite the long clinical experience with ESAs, the mechanism(s) of their clearance have not been fully elucidated and there is a paucity of information regarding which organ(s) and tissue(s) are important in the metabolism and elimination of these drugs. Pharmacokinetic studies of rHuEPO have demonstrated dose-dependent disposition in human adults [44, 99, 117, 122], premature infants [118], monkeys [77], sheep [119], and rats [76, 120] after intravenous administration. In these studies, nonlinear kinetics could be related to the limited number of EPOR located on the finite, but expandable, number of bone marrow erythroid progenitors that may play a role in clearance [127]. In addition, possibly nonerythroid EPOR pathways located in other tissues could be also involved in the receptor-mediated endocytosis and subsequent intracellular lysosomal degradation [5].

EPO concentrations remain abnormally high in dogs that have been sublethally irradiated, even after the cessation of hypoxia [121], suggesting that EPO concentrations were significantly "influenced by the functional state of the erythroid tissue of the marrow" [121]. Subsequently, it has been further documented that after bone marrow ablation [122–124], in patients with aplastic anemia [125], or in anemic patients with hypoplastic marrows [122, 126, 127], plasma EPO levels were disproportionately increased relative to slightly decreased hemoglobin concentrations. Conversely, individuals with hyperactive marrow owing to hemolytic anemia had disproportionately low plasma EPO concentrations [127–129] and rapid EPO plasma disappearance [130, 131].

Additional nonclinical evidence implicating the primary role of the bone marrow in EPO metabolism has been provided when after intravenous administration of ^{125}I-rHuEPO to rats, the highest organ-specific radioactivity uptake was observed in bone marrow and spleen [132, 133]. These tissues have also been shown to rapidly metabolize EPO with tissue uptake and clearance directly correlated with the number of CFU–E [134]. ^{125}I-rHuEPO uptake by these organs was transient and observed only at low doses. These data also suggest a saturable process for EPO elimination, which is consistent with nonclinical and clinical publications reporting nonlinear pharmacokinetics for EPO.

In sheep, an 80% reduction of rHuEPO clearance was observed after busulfan-induced bone marrow ablation and was associated with a markedly decreased bone marrow cellularity [135]. These findings suggest that some elimination may occur in the bone marrow. This suggestion is supported by *in vitro* studies demonstrating that EPO is rapidly internalized and degraded after binding to EPOR on erythroid progenitor cells [21, 22, 136].

Gross and Lodish investigated the trafficking and degradation of rHuEPO and darbepoetin alfa by EPOR-expressing cells in cell culture and found that rHuEPO and darbepoetin alfa were subjected to EPOR-mediated endocytosis followed by degradation in lysosomes [22]. Both drugs were degraded only by cultured cells (BsF3) that expressed EPOR, not the untransformed counterpart, and receptor binding, dissociation, and trafficking properties affected their relative rates of cellular uptake and intracellular degradation. rHuEPO had a faster EPOR on-rate than darbepoetin alfa ($k_{on} = 5.0 \cdot 10^8\ M^{-1} \cdot min^{-1}$ *versus* $1.1 \cdot 10^8\ M^{-1} \cdot min^{-1}$) and a slower off-rate ($k_{off} = 0.029\ min^{-1}$ *versus* $0.042\ min^{-1}$) resulting in an approximately five-fold higher affinity. Surface-bound rHuEPO and darbepoetin alfa were internalized at the same rate ($k_{in} = 0.06\ min^{-1}$), and after internalization 60% of each ligand was resecreted intact and 40% degraded. Therefore, rHuEPO was more likely to bind EPOR and internalize than darbepoetin alfa. As a consequence, rHuEPO was degraded faster than darbepoetin alfa through receptor-mediated endocytosis [22]. While this observation suggested that darbepoetin alfa may have reduced clearance *in vivo* because of reduced EPOR-mediated endocytosis and degradation, darbepoetin alfa has other biophysical characteristics, such as increased molecular size and decreased isoelectric point, opening the possibility that the reduced clearance could be explained by other mechanisms.

Additional clinical evidence that rHuEPO elimination occurs in the bone marrow comes from studies in patients with cancer who received myeloablative chemotherapy and hematopoietic stem cell transplantation [137]. A 34% reduction in clearance and a 95% increase in terminal half-life were demonstrated for rHuEPO after administration of high-dose chemotherapy. Chemotherapy-induced bone-marrow ablation results in a decline in EPOR carrying progenitor and loss of the nonlinear part of the clearance process, suggesting that the nonlinear clearance of rHuEPO may be due to elimination by EPOR binding and endocytosis in the bone marrow. Similar changes in the pharmacokinetics of darbepoetin alfa in patients after chemotherapy were reported, with a reduction of clearance up to 40% in patients with chemotherapy-induced anemia [105, 138]. A role for an EPOR-mediated pathway in clearance is also consistent with the significant increase in rHuEPO clearance observed during the early post-bone marrow transplant period [137].

The results described above clearly differ from those reported in rats where the EPO clearance and half-life were not significantly altered by hypo or hyperplastic marrow states [139]. Piroso et al. used cyclophosphamide to induce marrow hypoplasia and phenylhydrazine or bleeding to induce marrow hyperplasia. It has been suggested that the extremely high EPO concentrations observed among the phenylhydrazine and phlebotomized groups of study animals had saturated EPOR-mediated clearance and, as a consequence, the pre- and post-ablation differences in rHuEPO pharmacokinetics were not apparent [137]. Alternatively, the role of the EPOR-mediated pathway may not be as important a mechanism of clearance of rHuEPO as initially thought [140]. Nevertheless, the EPOR-mediated pathway is also consistent with the significant increase in rHuEPO clearance observed during the early post-transplant period [137]. The significant engraftment of erythroid progenitors approximately 1 week after the transplantation increases the amount of cell surface EPOR and, therefore, the rHuEPO clearance. A clinical implication of this finding is that the administration of rHuEPO relatively early after marrow ablation to patients undergoing peripheral stem cell transplantation may stimulate erythropoiesis before significant peripheral reticulocytosis is detectable [137].

In spite of the *in vitro* results suggesting the role of EPOR on ESA clearance, evidence of an *in vivo* role for this pathway is indirect and mostly arises from chemotherapy experiments. Chemotherapy-based approaches may result in nonspecific effects to EPOR-independent clearance mechanisms. If chemotherapy impacts one or more of the EPOR-independent elimination pathways, such as destruction of macrophages or neutrophils, then their diminished number may explain or at least contribute to the decrease in ESA clearance that is observed after chemotherapy treatment. For example, chemotherapy is known to cause pancytopenia, and a recent report showed that rHuEPO was cleared *in vitro* from cell culture medium containing unfractionated bone marrow-derived cells [141]. In addition, an rHuEPO analog with no detectable receptor binding activity was depleted from the culture medium at the same rate as rHuEPO, suggesting that EPOR-mediated endocytosis was not the primary route of rHuEPO degradation in this model system.

An alternative method to explore the role of the EPOR-mediated clearance in the elimination of ESAs is to investigate the pharmacokinetics of EPO analogs with different EPOR-binding activity. Using site directed mutagenesis [7, 142], three rHuEPO analogs were engineered: NM385, NM340, and NM294 [140]. NM385 was devoid of detectable receptor binding activity ($IC_{50} > 10$ µM), but retained similar structure and carbohydrate content as rHuEPO and, therefore, this molecule should not be cleared by the EPOR-mediated clearance pathway. Furthermore, if the EPOR-mediated pathway was dominant, NM385 would show a substantially reduced clearance and increased half-life compared with rHuEPO. Pharmacokinetic data obtained after a single intravenous dose of NM385 3 and 30 µg/kg in mice showed that, compared with rHuEPO, a 119% and 69% increase in half-life with a 99% and 34% reduction in clearance, respectively. The observation that the effect on clearance was more pronounced at lower nonpharmacologic doses is consistent with saturation of EPOR-mediated clearance at higher doses [140].

Compared with rHuEPO, the peptide backbone of the NM385 analog carries three additional negative charges. To control for the change in charge associated with the substitutions on NM385, the rHuEPO analog NM340 was examined. NM340 retained receptor binding activity ($IC_{50} = 135$ pM) and carried four additional negative charges, while retaining the same carbohydrate content as rHuEPO and NM385. Pharmacokinetic data obtained after a single intravenous dose of 30 µg/kg in rats showed that NM340 had only a 18% and 14% increase in terminal half-life and clearance, respectively, compared with rHuEPO. Hence, extra negative charges on the peptide backbone of NM385 *per se* are unlikely to explain the observed effects on its pharmacokinetic parameters. Taken together, these results suggest that EPOR-mediated pathways play some role in clearance of ESAs. Furthermore, the findings described above also show that EPOR-independent pathways play a substantial role in the clearance of ERA as none of the techniques used to decrease the number of receptor with chemotherapy or to block the EPOR pathway with analogs without binding activity were able to shut down completely the elimination of EPO.

Several pathways involved in EPOR-independent clearance have been proposed, including liver and renal clearance. *In vivo* studies demonstrated that the kidney and liver exert negligible effect on EPO *in vivo* elimination [85, 120, 143]. Urinary excretion contributed only in a minor way to rHuEPO disposition with only approximately 5% of the administered dose excreted unchanged into the urine [96, 144]. Recently, the pharmacokinetics of subcutaneous darbepoetin alfa was compared in adults with normal renal function, with chronic renal insufficiency, or requiring dialysis, and in pediatric patients with chronic kidney disease requiring hemodialysis or peritoneal dialysis [145], and renal function did not markedly affect exposure based on the area under the serum darbepoetin alfa-time curve, nor did it markedly affect the terminal half-life of darbepoetin alfa. These findings are consistent with data from rHuEPO and, therefore, suggest that renal excretion plays a minor role in elimination of ESAs and that altered renal function does not influence the rate of absorption.

Several studies have tested the hypothesis that EPO is cleared by an asialoglycoprotein receptor (ASGR) that is expressed in the liver. While the desialylated rHuEPO is cleared by hepatocytes by ASGR-mediated endocytosis in a matter of minutes [133, 146, 147], fully intact rHuEPO is cleared only after several hours. Clearance of rHuEPO by ASGR would require a two-step process, desialylation and subsequent clearance by ASGR in liver, but no evidence exists that desialylation of rHuEPO occurs [133, 147, 148]. In addition, hepatectomy experiments in sheep did not reveal reduced clearance of rHuEPO [148], and patients with liver disease showed no difference in rHuEPO clearance rates [149]. The rHuEPO redistributed from systemic circulation to lymph does not undergo significant clearance within the lymphatics [73].

It has been shown that carbohydrate side chains of EPO are necessary for persistence and *in vivo* biologic activity of the molecule, but not for *in vitro* receptor binding or stimulation of proliferation as carbohydrate can reduce the affinity for EPOR. Darbepoetin alfa is a hyperglycosylated analog of rHuEPO, with three- to five-fold lower affinity for the EPOR compared with rHuEPO [8, 9, 22, 150] but has three- to four-fold longer serum half-life and greater *in vivo* activity than rHuEPO. *In vitro* experiments have suggested that the lower affinity for EPOR of darbepoetin alfa relative to rHuEPO may contribute to its lower clearance [22]. To understand which clearance pathway was impacted by hyperglycosylation, the pharmacokinetics of EPO analogs with different EPOR binding activity and levels of glycosylation was examined. The analog NM294 has similar receptor binding activity to rHuEPO, but with the addition of one carbohydrate chain and, consequently, four additional negative charges. Similarly, the hyperglycosylated EPO analog, darbepoetin alfa, has eight extra negative charges due to the addition of two carbohydrate chains. If EPOR affinity and not glycosylation level was the key determinant of elimination, NM294 would be expected to have similar clearance to rHuEPO, while darbepoetin alfa should have lower clearance. However, after a single intravenous dose of 30 μg/kg in rats, NM294 and darbepoetin alfa had 51% and 62% lower clearance, respectively, than rHuEPO [140]. These findings suggest that hyperglycosylation impacts an EPOR-independent clearance pathway, and support the hypothesis that EPOR-mediated clearance may not play a dominant role on ESA elimination.

The attachment of polyethylene glycol chains to therapeutic proteins can improve the pharmacokinetic, pharmacodynamics, and immunologic properties of protein pharmaceuticals [151]. The reduced clearance of PEGylated molecules can be produced by altering the different clearance pathways, including reduced renal clearance, reduced clearance by the immune system, reduced proteolysis and/or steric hindrance to receptor-mediated clearance alterations. It has been argued that steric hindrance resulting from the PEGylation may influence the binding affinity of the therapeutic protein to cellular receptors, resulting in changes in the bioactivity of the agent [152]. In addition, these pharmacokinetic and pharmacodynamic changes may translate to greater exposure of a drug, and therefore may allow either a decrease in the dose level or a decrease in the frequency of dosing, or both [153]. PEGylation of epoetin alfa (PEG-EPO) with succinimidyl propionate esters of 5 kDa PEG resulted in a lower *in vitro* receptor binding activity and a longer half-life *in vivo* [154, 155]. In other studies, PEGylation of rHuEPO and darbepoetin alfa resulted in a 58- and eight-fold lower receptor binding activity, respectively [22]. Jarsch et al. [156] confirmed that equilibrium dissociation constants for PEGylated epoetin beta and epoetin beta were 140 and 2.9 nM, respectively. In addition, the concentrations that inhibit 50% of the EPOR-receptor binding values were 200 and 1.5 nM for PEGylated epoetin beta and epoetin beta. Compared with epoetin beta, PEGylated epoetin beta activity at the receptor level is characterized by a slower association with and faster dissociation from EPOR, which evidence a 50- to 100-fold lower affinity for EPOR-binding sites and reduced specific activity *in vitro*. No data are available to quantify the role of the steric hindrance on the internalization rate of PEGylated epoetin beta after binding to the receptor.

Pharmacokinetic data obtained after a single intravenous dose of 30 µg/kg in rats showed that the PEGylation of rHuEPO, NM385, and darbepoetin alfa resulted in 91%, 93%, and 46% reduction in clearance [140]. The reduction in clearance and receptor-binding activity correlated with the molecular weight and total mass of the PEG that was attached. While rHuEPO and NM385 carried a 30 kDa PEG, darbepoetin alfa carried a 5–10 kDa PEG. The pharmacokinetic results are also consistent with those published by Mehvar et al. who showed that after intravenous administration in mice, the half-life of 50 kDa PEG was 987 min, substantially longer than that observed for 6 kDa PEG (18 min). These findings suggest that PEGylation affects EPOR-independent clearance pathway(s), and support the hypothesis that EPOR-mediated clearance may not play a dominant role on ESA elimination.

Taken together it appears that there are at least two ESA clearance pathways: a capacity-limited clearance pathway using EPOR-mediated endocytosis by erythroid progenitor cells or other cells expressing EPOR; and a linear clearance that is EPOR-independent, reflecting other mechanism(s) involved, although still not well understood. However, most of the previous analyses where rHuEPO saturable clearance was reported were based on limited datasets, often covering a narrow dose-range, which may have prohibited a clear understanding of the nonlinear nature of rHuEPO pharmacokinetics. In fact, Veng-Pedersen et al. [117] proposed that rHuEPO is cleared solely by nonlinear, saturable mechanism(s), while Kato et al. [120] reported both saturable and nonsaturable clearance components in rats, which was also observed in sheep [157], monkeys [77], infants [158], and healthy subjects [93]. A population pharmacokinetic meta-analysis of rHuEPO in 533 healthy subjects using data from 16 clinical studies in which a wide range of intravenous and subcutaneous rHuEPO doses were administered have helped in defining the two separate elimination pathways and understanding the influence of demographic characteristics and other covariates on the pharmacokinetic parameters of rHuEPO [44]. At low concentrations, including the endogenous concentrations observed at baseline or in ESA-untreated states, the nonlinear clearance operates at full capacity, giving a total clearance of about 0.9 L/h. As concentrations increase, the nonlinearity of pharmacokinetics becomes more important and, at the highest doses, the nonlinear clearance of rHuEPO was nearly fully saturated, so that total clearance decreased to almost one third, being mainly represented by the linear component estimated to be 0.3 L/h. At these doses or concentrations, rHuEPO pharmacokinetics is approximately dose linear. The ratio of nonlinear (saturable) and linear clearance mechanisms is similar to what has been published in humans [137], but appears to vary between species [120], making comparisons between studies in different species difficult.

The dose-dependent clearance appears to be independent of the type of rHuEPO (epoetin alfa *versus* epoetin beta) or population (healthy subjects or patients with chronic renal failure). Therefore, the pharmacokinetic models developed on the basis of healthy volunteers data may be applied in patients with anemia due renal insufficiency, but it may have limited predictive value when applied to patients receiving chemotherapy. The possible involvement of EPOR in the elimination of rHuEPO could have important consequences for the pharmacokinetics of rHuEPO in this setting as the effect of chemotherapy on progenitor cells would reduce the number of EPOR available to clear. As a consequence, the rHuEPO clearance might be decreased and the rHuEPO elimination process would become slower than the absorption process. Therefore, the flip-flop phenomena observed in healthy subjects might disappear, leading to a completely different rHuEPO concentration time profile [44]. To explore this hypothesis fully, a more mechanistic model considering the complex interaction between the erythropoietic system, EPOR turnover, ESA pharmacokinetics and pharmacodynamics and the effect of chemotherapeutic agents would be necessary. However, the nonlinear disposition of rHuEPO has been mainly characterized with the most common model in dealing with saturation kinetics, Michaelis-Menten model, with a possible linear elimination pathway. Although such a model has been successfully applied to describe the nonlinear pharmacokinetic of rHuEPO in various species, it does not well represent underlying molecular events such as receptor binding, internalization, and degradation of rHuEPO. A general model for drugs exhibiting target-mediated drug disposition (TMDD) has been previously introduced [159]. It uses receptor binding and receptor-mediated endocytosis as a primary mechanism of nonlinear drug disposition. Furthermore, if it is assumed that free drug, free receptor, and drug-receptor complex are at quasi- or rapid equilibrium, the TMDD model can be simplified to its quasi-equilibrium form removing the need to estimate drug-binding constants such as k_{on} and k_{off} [160]. This approach has been suggested and applied to rat, monkey and human data [161]. The target mediated drug disposition model and the Michaelis-Menten model become identical when

$$\frac{R_{tot} \cdot k_D}{(k_D + C)^2} <<< 1 ,$$

where R_{tot}, k_D and C are the total number of drug receptors, the receptor affinity and the drug concentration, respectively [162].

A further indication of the possible involvement of receptor binding in the disposition of rHuEPO can be found when studying the intravenous data at lower doses and the pharmacokinetics after multiple dosing. In the first case, a correction factor was needed to explain about a 60% reduction in rHuEPO exposure for doses <20 IU/kg [44]. As the first measurements of rHuEPO concentration were usually obtained 30 min following dose administration, it may not have been possible to observe the presence of any early rapid distribution. Such distribution could be related to rapid binding, possibly to the EPOR. At higher doses, such binding could be saturated, leading to higher serum concentrations.

Some studies have reported a time-dependent clearance. For example, McMahon et al. [66] compared the single- and multiple-dose pharmacokinetics of epoetin alfa at 150 and 300 IU/kg intravenous and found that, at both dose levels, the mean clearance increased by 25% after five doses. Sans et al. found a 57–90% decrease in AUC and a 47–76% decrease in C_{max} after three subcutaneous doses of rHuEPO beta at 30, 60, or 100 IU/kg [65]. Similarly, most studies in nephrology patients have found that clearance increases and/or terminal half-life decreases after multiple dosing [107, 109, 163, 164]. Additionally, some investigators have reported that after the initial change (i.e., after 2–3 weeks), no subsequent changes occurred [163]. Conversely, other investigators reported no major change in pharmacokinetic properties from single-dose values, either after reaching a new steady-state hemoglobin value [165] or after 56 days [166]. Jensen et al. [64] found no differences in the pharmacokinetic properties of epoetin beta after long-term treatment except for a reduction in the volume of distribution. Nevertheless, upon multiple dosing, most data suggest that after several weeks of treatment, rHuEPO clearance increases about 10–30%, with an accompanying decrease in terminal half-life [61, 90]. It has been demonstrated that the PEGylated epoetin beta clearance in patients diagnosed with chronic kidney disease (0.8 L/day) increased 10–15% after 2 months of treatment [167]. These findings are consistent with a target mediated drug disposition [159] and the involvement of the EPOR-mediated pathway in the elimination of rHuEPO. As a consequence, once rHuEPO treatment starts, erythropoiesis is stimulated, the number of precursor cells is increased and more EPOR would be available to eliminate rHuEPO, which might result in a clearance increase and/or volume of distribution decrease, leading to a reduction in terminal half-life.

Summary

We have reviewed the clinical pharmacokinetics of protein-based ESA, including epoetin alfa, epoetin beta, PEGylated epoetin beta, and darbepoetin alfa. After subcutaneous administration, ESA enter into the systemic circulation through the blood capillaries or the lymphatic system. Absolute subcutaneous bioavailability is incomplete and about 40–70% for the dosing regimens used in the clinic. In addition, the rate of absorption is slower than the rate of elimination, the observed half-life after subcutaneous administration reflects the absorption rather than elimination rate. As expected from its large molecular weight, ESA's volume of distribution was similar to the plasma volume (40–60 mL/kg), suggesting confinement of EPO within the plasma circulation. Despite the long clinical experience with ESAs, the mechanism(s) of their clearance have not been fully elucidated and there is a paucity of information regarding which organ(s) and tissue(s) are important in the metabolism and elimination of these drugs. Clearly our understanding of the nature of the clearance pathways is incomplete and predicting their impact on clinical use of ESAs is unclear.

Acknowledgements
Parts of this review are updated version of previously published portions of three other book chapters:

1. *Elliot S, Heatherington AC, Foote MA. Erythropoietic Factors: Clinical Pharmacology and Pharmacokinetics. In: Morstyn G, Foote MA, and Lieschke GJ, eds., Hematopoietic Growths Factors in Oncology. Humana Press, NJ, USA, 2004; pp. 97–123.*
2. *Heatherington AC. Clinical pharmacokinetic properties of rHuEPO: a review. In: Molineux G, Foote MA, Elliott S, eds., Erythropoietins and Erythropoiesis: Molecular, Cellular, Preclinical, and Clinical Biology. Birkhauser, Basel, Switzerland, 2003; pp. 87–112.*
3. *Doshi S, Perez-Ruixo JJ, Jang GR, Chow A, Elliot S. Pharmacocinétique de les agents stimulant l'érythropoïèse. In: Rossert J, Casadevall N, Gisselbrecht C, eds., Les agents stimulant l'érythropoïèse. Paris, France, 2008.*

References

1. Halstenson CE, Macres M, Katz SA et al. Comparative pharmacokinetics and pharmacodynamics of epoetin alfa and epoetin beta. *Clin Pharmacol Ther* 1991;50;702–712.
2. Egrie JC, Strickland TW, Lane J et al. Characterization and biological effects of recombinant human erythropoietin. *Immunobiology* 1986;72:213–224.
3. Imai N, Kawamura A, Higuchi M et al. Physicochemical and biological comparison of recombinant human erythropoietin with human urinary erythropoietin. *J Biochem* 1990;107:352–359.
4. Rency MA, Scoble HA, Kim Y. Structural characterization of natural human urinary and recombinant DNA-derived erythropoietin. Identification of des-arginine 166 erythropoietin. *J Biol Chem* 1987;262:17156–17163.
5. Jelkmann W. Use of recombinant human erythropoietin as an antianemic and performance enhancing drug. *Curr Pharmaceut Biotechnol* 2000;1:11–31.
6. Macdougall IC, Gray SJ, Elston O et al. Pharmacokinetics of novel erythropoiesis stimulating protein compared with epoetin alfa in dialysis patients. *J Am Soc Nephrol* 1999;10:2392–2395.
7. Elliott S, Lorenzini T, Asher S et al. Enhancement of therapeutic protein *in vivo* activities through glycoengineering. *Nat Biotechnol* 2003;21:414–421.
8. Sinclair AM, Elliott S. Glycoengineering: The effect of glycosylation on the properties of thera-

peutic proteins. *J Pharm Sci* 2005;94:1626–1635.
9. Egrie JC, Dwyer E, Browne JK et al. Darbepoetin alfa has a longer circulating half-life and greater *in vivo* potency than recombinant human erythropoietin. *Exp Hematol* 2003;31:290–299.
10. Nissenson AR, Swan SK, Lindberg JS et al. Randomized, controlled trial of darbepoetin alfa for the treatment of anemia in hemodialysis patients. *Am J Kidney Dis* 2002;40:110–118.
11. Vansteenkiste J, Pirker R, Massuti B et al. Double-blind, placebo-controlled, randomized phase III trial of darbepoetin alfa in lung cancer patients receiving chemotherapy. *J Natl Cancer Inst* 2002;94:1211–1220.
12. Macdougall IC. CERA (Continuous Erythropoietin Receptor Activator): a new erythropoiesis-stimulating agent for the treatment of anemia. *Curr Hematol Rep* 2005;4:436–440.
13. Erslev AJ. *In vitro* production of erythropoietin by kidneys perfused with a serum-free solution. *Blood* 1974;44:77–85.
14. Syed RS, Reid SW, Li C et al. Efficiency of signalling through cytokine receptors depends critically on receptor orientation. *Nature* 1998;395:511–516.
15. Elliott S, Lorenzini T, Chang D et al. Mapping of the active site of recombinant human erythropoietin. *Blood* 1997;8:493–502.
16. Youssoufian H, Longmore G, Neumann D et al. Structure, function, and activation of the erythropoietin receptor. *Blood* 1993;81:2223–2236.
17. Fraser JK, Lin FK, Berridge MV. Expression of high affinity receptors for erythropoietin on human bone marrow cells and on the human erythroleukemic cell line, HEL. *Exp Hematol* 1988;16:836–842.
18. Sawyer ST, Krantz SB, Goldwasser E. Binding and receptor-mediated endocytosis of erythropoietin in Friend virus-infected erythroid cells. *J Biol Chem* 1987;262:5554–5562.
19. Matthews DJ, Topping RS, Cass RT et al. A sequential dimerization mechanism for erythropoietin receptor activation. *Proc Natl Acad Sci USA* 1996;93:9471–9476.
20. Philo JS, Aoki KH, Arakawa T et al. Dimerization of the extracellular domain of the erythropoietin (EPO) receptor by EPO: one high-affinity and one low-affinity interaction. *Biochemistry* 1996;35:1681–1691.
21. Sawyer ST, Krantz SB, Sawada K. Receptors for erythropoietin in mouse and human erythroid cells and placenta. *Blood* 1989;74:103–109.
22. Walrafen P, Verdier F, Kadri Z et al. Both proteasomes and lysosomes degrade the activated erythropoietin receptor. *Blood* 2005;105:600–608.
23. Gross AW, Lodish HF. Cellular trafficking and degradation of erythropoietin and novel erythropoiesis stimulating protein (NESP). *J Biol Chem* 2006;281:2024–2032.
24. Jelkmann W. Erythropoietin: structure, control of production, and function. *Physiol Rev* 1992;72:449–489.
25. Fisher JW. Erythropoietin: physiology and pharmacology update. *Exp Biol Med* 2003;228:1–14.
26. Koury MJ, Bondurant MC. Maintenance by erythropoietin of viability and maturation of murine erythroid precursor cells. *J Cell Physiol* 1988;137:65–74.
27. Gregoli PA, Bondurant MC. The roles of Bcl-XL and apopain in the control of erythropoiesis by erythropoietin. *Blood* 1997;89:630–640.
28. Chong ZZ, Kang JQ, Maiese K. Hematopoietic factor erythropoietin fosters neuroprotection through novel signal transduction cascades. *J Cereb Blood Flow Metab* 2002;22:503–514.
29. Sawyer T, Jacobs-Helber SM. Unraveling distinct intracellular signals that promote survival and proliferation: study of erythropoietin, stem cell factor, and constitutive signaling in leukemic cells. *J Hemather Stem Cell Res* 2000;9:21–29.
30. Kashii Y, Uchida M, Kirito K et al. A member of Forkhead family transcription factors FKHRL, is one of the downstream molecules of phosphatidylinositol 3-kinase-AKT activation pathway in erythropoietin signal transduction. *Blood* 2000;96:941–949.
31. Ghaffari S, Jagani Z, Kitidis C et al. Cytokines and BCR-ABL mediate suppression of TRAIL-induced apoptosis through inhibition of forkhead FOXO3a transcription factor. *Proc Natl Acad Sci USA* 2003;100:6523–6528.
32. Zamai L, Secchiero P, Pierpaoli S et al. TNF-related apoptosis-inducing ligand (TRAIL) as a negative regulator of normal human erythropoiesis. *Blood* 2000;95:3716–3724.
33. Jarsch M, Brandt M, Lanzendörfer M et al. Comparative erythropoietin receptor binding kinetics of C.E.R.A. and epoetin-beta determined by surface plasmon resonance and competition binding assay. *Pharmacology* 2008;81:63–69.
34. Canaud B, Mingardi G, Braun J et al. Intravenous C.E.R.A. maintains stable haemoglobin levels

in patients on dialysis previously treated with darbepoetin alfa: results from STRIATA, a randomized phase III study. *Nephrol Dial Transplant* 2008) Epub Jun 27.
35. Spivak JL. The biology and clinical applications of recombinant erythropoietin. *Semin Oncol* 1998;25:7–11.
36. Buemi M, Cavallaro E, Floccari F et al. The pleiotropic effect of erythropoietin in the central nervous system. *J Neuropathol Exp Neurol* 2003;62:228–236.
37. Spivak JL. Changing paradigms in anemia management. *Adv Studies Med* 2002;22:610–619.
38. Besarab A. Physiological and pharmacodynamic considerations for route of EPO administration. *Semin Nephrol* 2000;20:364–374.
39. Soulliard A, Audran M, Bressolle F et al. Pharmacokinetics and pharmacodynamics of recombinant human erythropoietin in athletes. Blood sampling and doping control. *Br J Clin Pharmacol* 1996;42:355–364.
40. Koopman MG, Koomen GC, Krediet RT et al. Circadian rhythm of glomerular filtration rate in normal individuals. *Clin Sci* 1989;77:105–111.
41. Wide L, Bengtsson C, Birgegard G. Circadian rhythm of erythropoietin in human serum. *Br J Haematol* 1989;72:85–90.
42. Pasqualetti P, Collacciani A, Casale R. Circadian rhythm of serum erythropoietin in myelodysplastic syndromes. *Eur Rev Med Pharmacol Sci* 2000;4:111–115.
43. Hayashi N, Kinoshita H, Yukawa E et al. Pharmacokinetic analysis of subcutaneous erythropoietin administration with nonlinear mixed effect model including endogenous production. *Br J Clin Pharmacol* 1998;46:11–19.
44. Olsson-Gisleskog P, Jacqmin P, Perez-Ruixo JJ. Population pharmacokinetics meta-analysis of recombinant human erythropoietin in healthy subjects. *Clin Pharmacokinet* 2007;46:159–173.
45. Kendall RG. Erythropoietin. *Clin Lab Haematol* 2001;23:71–80.
46. Erslev AJ. Erythropoietin. *N Engl J Med* 1991;324:1339–1344.
47. Jungers PY, Robino C, Choukroun G et al. Incidence of anaemia, and use of Epoetin therapy in pre-dialysis patients: a prospective study in 403 patients. *Nephrol Dial Transplant* 2002;17:1621–1627.
48. Nissenson AR, Nimer SD, Wolcott DL. Recombinant human erythropoietin and renal anemia: molecular biology, clinical efficacy, and nervous system effects. *Ann Intern Med* 1991; 114:402–416.
49. Adamson JW, Eschbach JW. Treatment of the anemia of chronic renal failure with recombinant human erythropoietin. *Ann Rev Med* 1990;41:349–360.
50. Macdougall IC. Meeting the challenges of a new millennium: optimizing the use of recombinant human erythropoietin. *Nephrol Dial Transplant* 1998;13:23–27.
51. Kato A, Hishida A, Kumagai H et al. Erythropoietin production in patients with chronic renal failure. *Ren Fail* 1994;16:645–651.
52. Ross RP, McCrea JB, Besarab A. Erythropoiein response to blood loss in hemodialysis patients is blunted but preserved. *ASAIO J* 1994;40:M880–885.
53. Miller CB, Jones RJ, Piantadosi S et al. Decreased erythropoietin response in patients with the anemia of cancer. *N Engl J Med* 1990;322:1689–1692.
54. Bron D, Meuleman N, Mascaux C. Biological basis of anemia. *Semin Oncol* 2001;28:1–6.
55. Endo Y, Nagai H, Watanabe Y et al. Heat-induced aggregation of recombinant erythropoietin in the intact and deglycosylated states as monitored by gel permeation chromatography combined with a low-angle laser light scattering technique. *J Biochem* 1992;112:700–706.
56. Pedain C, Herrero J, Kunzel W. Serum erythropoietin levels in ovarian cancer patients receiving chemotherapy. *Eur J Obstet Gynecol Reprod Biol* 2001;98:224–230.
57. Glaspy J, Henry D, Patel R et al. The effects of chemotherapy on endogenous erythropoietin levels and the pharmacokinetics and erythropoietic response of darbepoetin alfa: a randomised clinical trial of synchronous *versus* asynchronous dosing of darbepoetin alfa. *Eur J Cancer* 2005;41:1140–1149.
58. Schmalbach TK, Borch RF. Diethyldithiocarbamate modulation of murine bone marrow toxicity induced by cis-diammine (cyclobutanedicarboxylato) platinum (II). *Cancer Res* 1989; 49:6629–6633.
59. Boxenbaum H. Pharmacokinetic tricks and traps: flip-flop models. *J Pharm Pharmaceut Sci* 1998;1:90–91.
60. Varlet-Marie E, Gaudard A, Audran M et al. Pharmacokinetics-pharmacodynamics of recombinant human erythropoietins in doping control. *Sport Med* 2003;33:301–315.

61. Heatherington AC. Clinical pharmacokinetic properties of rHuEPO: a review. In: Molineux G, Foote MA, Elliott S, eds., Erythropoietins and Erythropoiesis: Molecular, Cellular, Preclinical, and Clinical Biology. Birkhauser, Basel, Switzerland, 2003; pp 87–112.
62. Cheung WK, Goon BL, Guilfoyle MC et al. Pharmacokinetics and pharmacodynamics of recombinant human erythropoietin after single and multiple subcutaneous doses to healthy subjects. *Clin Pharmacol Ther* 1998;64:412–423.
63. Markham A, Bryson HM. Epoetin alfa. A review of its pharmacodynamic and pharmacokinetic properties and therapeutic use in nonrenal applications. *Drug Eval* 1995;49:232–254.
64. Jensen JD, Madsen JK, Jensen LW et al. Reduced production, absorption, and elimination of erythropoietin in uremia compared with healthy volunteers. *J Am Soc Nephrol* 1994;5:177–185.
65. Sans T, Joven J, Vilella E et al. Pharmacokinetics of several subcutaneous doses of erythropoietin: potential implications for blood transfusions. *Clin Exp Pharmacol Physiol* 2000;27:179–184.
66. McMahon FG, Vargas R, Ryan M et al. Pharmacokinetics and effects of recombinant human erythropoietin after intravenous and subcutaneous injections in healthy volunteers. *Blood* 1990;76:1718–1722.
67. Supersaxo A, Hein WR, Steffen H. Effect of molecular weight on the lymphatic absorption of water-soluble compounds following subcutaneous administration. *Pharmacol Res* 1990;7:167–169.
68. Porter CJ, Charman SA. Lymphatic transport of proteins after subcutaneous administration. *J Pharm Sci* 2000;89:297–310.
69. McLennan DN, Porter CJ, Edwards GA et al. Pharmacokinetic model to describe the lymphatic absorption of r-metHu-leptin after subcutaneous injection to sheep. *Pharmacol Res* 2003;20:1156–1162.
70. Leak LV. The structure of lymphatic capillaries in lymph formation. *Fed Proc* 1976; 35:1863–1871.
71. McLennan DN, Porter CJ, Edwards GA et al. Lymphatic absorption is the primary contributor to the systemic availability of epoetin alfa following subcutaneous administration to sheep. *J Pharmacol Exp Ther* 2005;313:345–351.
72. McLennan DN, Porter CJ, Edwards GA et al. The absorption of darbepoetin alfa occurs predominantly via the lymphatics following subcutaneous administration to sheep. *Pharmacol Res* 2006;23:2060–2066.
73. Kagan L, Gershkovich P, Mendelman A et al. The role of the lymphatic system in subcutaneous absorption of macromolecules in the rat model. *Eur J Pharm Biopharm* 2007;67:759–765.
74. Bendayan M. Morphological and cytochemical aspects of capillary permeability. *Microsc Res Tech* 2002;57:327–349.
75. Kota J, Machavaram KK, McLennan DN et al. Lymphatic absorption of subcutaneously administered proteins: influence of different injection sites on the absorption of darbepoetin alfa using a sheep model. *Drug Metab Dispos* 2007;35:2211–2217.
76. Woo S, Krzyzanski W, Jusko WJ. Pharmacokinetic and pharmacodynamic modeling of recombinant human erythropoietin after intravenous and subcutaneous administration in rats. *J Pharmacol Exp Ther* 2006;319:1297–1306.
77. Ramakrishnan R, Cheung WK, Farrell F et al. Pharmacokinetic and pharmacodynamic modeling of recombinant human erythropoietin after intravenous and subcutaneous dose administration in cynomolgus monkeys. *J Pharmacol Exp Ther* 2003;306:324–331.
78. Krzyzanski W, Jusko WJ, Wacholtz MC et al. Pharmacokinetic and pharmacodynamic modeling of recombinant human erythropoietin after multiple subcutaneous doses in healthy subjects. *Eur J Pharm Sci* 2005;26:295–306.
79. Glaspy J, Henry D, Patel R et al. The effects of chemotherapy on endogenous erythropoietin levels and the pharmacokinetics and erythropoietic response of darbepoetin alfa: a randomised clinical trial of synchronous *versus* asynchronous dosing of darbepoetin alfa. *Eur J Cancer* 2005;41:1140–1149.
80. Agoram B, Sutjandra L, Sullivan JT. Population pharmacokinetics of darbepoetin alfa in healthy subjects. *Br J Clin Pharmacol* 2007;63:41–52.
81. Caliceti P, Veronese FM. Pharmacokinetic and biodistribution properties of poly(ethylene glycol)-protein conjugates. *Adv Drug Deliv Rev* 2003;55:1261–1277.
82. European Medicines Association. Scientific discussion: summary of product characteristics: MIRCERA (methoxy polyethyl-ene glycol-epoetin beta). Available at http://www.emea.europa.eu/humandocs/PDFs/EPAR/mircera/H-739-en6.pdf [Accessed 2008 May 20]

83. Fishbane S, Pannier A, Liogier X et al. Pharmacokinetic and pharmacodynamic properties of methoxy-polyethylene glycol-epoetin beta are unaffected by the site of subcutaneous administration. *J Clin Pharmacol* 2007;47:1390–1397.
84. Jensen JD, Jensen LW, Madsen JK. The pharmacokinetics of recombinant human erythropoietin after subcutaneous injection at different sites. *Eur J Clin Pharmacol* 1994;46:333–337.
85. Macdougall IC, Jones JM, Robinson MI et al. Subcutaneous erythropoietin therapy: comparison of three different sites of injection. *Contrib Nephrol* 1991;88:152–156.
86. Salmonson T, Danielson BG, Wikstrom B. The pharmacokinetics of recombinant human erythropoietin after intravenous and subcutaneous administration to healthy subjects. *Br J Clin Pharmacol* 1990;29:709–713.
87. Jensen JD, Madsen JK, Jensen LW et al. Reduced production, absorption, and elimination of erythropoietin in uremia compared with healthy volunteers. *J Am Soc Nephrol* 1994;5:177–185.
88. Deicher R, Horl WH. Differentiating factors between erythropoiesis-stimulating agents: a guide to selection for anemia of chronic kidney disease. *Drugs* 2004;64:499–509.
89. Karlsson MO, Sheiner LB. Estimating bioavailability when clearance varies with time. *Clin Pharmacol Ther* 1994;55:623–637.
90. Elliot S, Heatherington AC, Foote MA. Erythropoietic Factors: Clinical Pharmacology and Pharmacokinetics. In: Morstyn G, Foote M and Lieschke eds. Hematopoietic Growths Factors in Oncology. Humana Press (NJ) pp 97–123.
91. Cheung WK, Goon BL, Guilfoyle MC et al. Pharmacokinetics and pharmacodynamics of recombinant human erythropoietin after single and multiple subcutaneous doses to healthy subjects. *Clin Pharmacol Ther* 1998;64:412–423.
92. Cheung WK, Minton N, Gunawardena K. Pharmacokinetics and pharmacodynamics of epoetin alfa once weekly and three times weekly. *Eur J Clin Pharmacol* 2001;57:411–418.
93. Ramakrishnan R, Cheung WK, Wacholtz MC et al. Pharmacokinetic and pharmacodynamic modeling of recombinant human erythropoietin after single and multiple doses in healthy volunteers. *J Clin Pharmacol* 2004;44:991–1002.
94. Emanuele RM, Fareed J. The effect of molecular weight on the bioavailability of heparin. *Thromb Res* 1987;48:591–596.
95. Macdougall IC, Roberts DE, Coles GA et al. Clinical pharmacokinetics of epoetin (recombinant human erythropoietin). *Clin Pharmacokinet* 1991;20:99–113.
96. Kampf D, Kahl A, Passlick J et al. Single-dose kinetics of recombinant human erythropoietin after intravenous, subcutaneous and intraperitoneal administration. Preliminary results. *Contrib Nephrol* 1989;76:106–111.
97. Bommer J, Barth HP, Zeier M et al. Efficacy comparison of intravenous and subcutaneous recombinant human erythropoietin administration in hemodialysis patients. *Contrib Nephrol* 1991;88:136–143.
98. Flaharty KK, Caro J, Erslev A et al. Pharmacokinetics and erythropoietic response to human recombinant erythropoietin in healthy men. *Clin Pharmacol Ther* 1990;47:557–564.
99. Braeckman R. Pharmacokinetics and pharmacodynamics of protein therapeutics. In: *Peptide and protein drug delivery*. New York, Marcel Dekker, 1991; 391–384.
100. Kompella A, Lee VHL. Pharmacokinetics of peptide and protein drugs. In: *Peptides and protein drug analysis*. Reid ER (editor). New York, Marcel Dekker, 2000; 633–669.
101. Uji Y, Hirashima K, Hirasawa Y et al. Recombinant erythropoietin Phase I clinical study-single intravenous administration. *Med Consult N Rem* 1989;26:1–28.
102. Uji Y, Hirashima K, Hirasawa Y et al. Recombinant erythropoietin Phase I clinical study-multiple intravenous administration. *Med Consult N Rem* 1989;26:29–44.
103. Woo S, Jusko WJ. Interspecies comparison of pharmacokinetics and pharmacodynamics of recombinant human erythropoietin (rH-EPO). *Drug Metab Dispos* 2007;35:1672–1678.
104. Allon M, Kleinman K, Walczyk M et al. Pharmacokinetics and pharmacodynamics of darbepoetin alfa and epoetin in patients undergoing dialysis. *Clin Pharmacol Ther* 2002;72:546–555.
105. Heatherington AC, Schuller J, Mercer AJ. Pharmacokinetics of novel erythropoiesis stimulating protein (NESP) in cancer patients: preliminary report. *Br J Cancer* 2001;84:11–16.
106. Agoram B, Heatherington AC, Gastonguay MR. Development and evaluation of a population pharmacokinetic-pharmacodynamic model of darbepoetin alfa in patients with nonmyeloid malignancies undergoing multicycle chemotherapy. *AAPS J* 2006;8:E552–E563.
107. Brockmöller J, Kochling J, Weber W et al. The pharmacokinetics and pharmacodynamics of recombinant human erythropoietin in haemodialysis patients. *Br J Clin Pharmacol*

1992;34:499–508.
108. Kindler J, Eckardt KU, Ehmer B et al. Single-dose pharmacokinetics of recombinant human erythropoietin in patients with various degrees of renal failure. *Nephrol Dial Transplant* 1989;4:345–349.
109. Neumayer HH, Brockmöller J, Fritschka E et al. Pharmacokinetics of recombinant human erythropoietin after SC administration and in long-term IV treatment in patients on maintenance hemodialysis. *Contrib Nephrol* 1989;76:131–141.
110. Nielsen OJ. Pharmacokinetics of recombinant human erythropoietin in chronic haemodialysis patients. *Pharmacol Toxicol* 1990;66:83–86.
111. Xenocostas A, Cheung WK, Farrell F et al. The pharmacokinetics of erythropoietin in the cerebrospinal fluid after intravenous administration of recombinant human erythropoietin. *Eur J Clin Pharmacol* 2005;61:189–195.
112. Farrel FX, Juul S, Elliot C et al. Erythropoietin crosses the blood–brain barrier: an analysis in a nonhuman primate model. *Blood* 2001;98:148b.
113. Jumbe NL. Erythropoietic agents as neurotherapeutic agents: what barriers exist? *Oncology* 2002;16:S91–S107.
114. Grasso G, Buemi M, Alafaci C et al. Beneficial effects of systemic administration of recombinant human erythropoietin in rabbits subjected to subarachnoid hemorrhage. *Proc Natl Acad Sci USA* 2002;99:5627–5631.
115. Ehrenreich H, Hasselblatt M, Dembowski C et al. Erythropoietin therapy for acute stroke is both safe and beneficial. *Mol Med* 2002;8:495–505.
116. Katavetin P, Tungsanga K, Eiam-Ong S et al. Antioxidative effects of erythropoietin. *Kidney Internat* 2007;72:S10–S15.
117. Veng-Pedersen P, Widness JA, Pereira LM et al. Kinetic evaluation of nonlinear drug elimination by a disposition decomposition analysis: application to the analysis of the nonlinear elimination kinetics of erythropoietin in adult humans. *J Pharm Sci* 1995;84:760–767.
118. Widness JA, Veng-Pedersen P, Peters C et al. Erythropoietin pharmacokinetics in premature infants: developmental nonlinearity, treatment effects. *J Appl Physiol* 1996;80:140–148.
119. Veng-Pedersen P, Widness JA, Pereira LM et al. A comparison of nonlinear pharmacokinetics of erythropoietin in sheep and humans. *Biopharm Drug Dispos* 1999;20:217–223.
120. Kato M, Kamiyama H, Okazaki A et al. Mechanism for the nonlinear pharmacokinetics of erythropoietin in rats. *J Pharmacol Exp Ther* 1997;283:520–527.
121. Stohlman F. Observations on the physiology of erythropoietin and its role in the regulation of red cell production. *Ann N Y Acad Sci* 1959;77:710–724.
122. Jelkmann W, Wiedemann G. Serum erythropoietin level: relationships to blood hemoglobin concentration erythrocytic activity of the bone marrow. *Klin Wochensch* 1990;68:403–407.
123. Davies SV, Fegan CD, Kendall R et al. Serum erythropoietin during autologous bone marrow transplantion: relationship to measures of erythroid activity. *Clin Lab Haematol* 1995;17:139–144.
124. Grace RJ, Kendall RG, Hartley AE et al. Changes in serum erythropoietin levels during allogeneic bone marrow transplantation. *Eur J Haematol* 1991;47:81–85.
125. Pavlovic-Kentera V, Milenkovic P, Ruvidic R et al. Erythropoietin in aplastic anemia. *Blut* 1979;39:345–350.
126. Hammond D, Shore N, Movassaghi N. Production utilization excretion of erythropoietin. I. Chronic anemias. II. Aplastic crisis. III. Erythropoietic effects of normal plasma. *Ann N Y Acad Sci* 1968;149:516–527.
127. de Klerk G, Rosengarten P, Vet R et al. Serum erythropoietin titers in anemia. *Blood* 1981; 58:1164–1170.
128. Alexanian R. Erythropoietin excretion in bone marrow failure and hemolytic anemia. *J Lab Clin Med* 1973;82:438–445.
129. Beguin Y, Clemons GK, Pootrakul P et al. Quantitative assessment of erythropoiesis functional classification of anemia based on measurements of serum transferrin receptor erythropoietin. *Blood* 1993;81:1067–1076.
130. Embury SH, Garcia JF, Mohandas N et al. Effects of oxygen inhalation on endogenous erythropoietin, kinetics erythropoiesis properties of blood cells in sickle-cell anemia. *N Engl J Med* 1984;311:291–295.
131. Eckard KU, Dittmer J, Neumann R et al. Decline of erythropoietin formation at continuous hypoxia is not due to feedback inhibition. *Am J Physiol* 1990;258:F1432–F1437.

132. Kinoshita H, Ohishi N, Kato M et al. Pharmacokinetics distribution of recombinant erythropoietin in rats. *Drug Res* 1992;42:174–178.
133. Spivak JL, Hogans BB. The *in vivo* metabolism of recombinant human erythropoietin in the rat. *Blood* 1989:73:90–99.
134. Kato M, Kato Y, Sugiyama Y. Mechanism of the upregulation of erythropoietin-induced uptake clearance by the spleen. *Am J Physiol Endocrinol Metab* 1999;276:E887–E895.
135. Chapel S, Veng-Pedersen P, Hohl RJ et al. Changes in erythropoietin pharmacokinetics following busulfan-induced bone marrow ablation in sheep: evidence for bone marrow as a major erythropoietin elimination pathway. *J Pharmacol Exp Ther* 2001;298:820–824.
136. Walrafen P, Verdier F, Kadri Z et al. Both proteasomes and lysosomes degrade the activated erythropoietin receptor. *Blood* 2005;105:600–608.
137. Widness JA, Schmidt RL, Hohl RJ et al. Change in erythropoietin pharmacokinetics following hematopoietic transplantation. *Clin Pharmacol Ther* 2007;81:873–879.
138. Heatherington AC, Dittrich C, Sullivan JT et al. Pharmacokinetics of darbepoetin alfa after intravenous or subcutaneous administration in patients with non-myeloid malignancies undergoing chemotherapy. *Clin Pharmacokinet* 2006;45:199–211.
139. Piroso E, Erslev A, Flaharty K et al. Erythropoietin life span in rats with hypoplastic hyperplastic bone marrows. *Am J Hematol* 1991;36:105–110.
140. Agoram B, Molineux G, Jang G et al. Effects of altered receptor binding activity on the clearance of erythropoiesis-stimulating proteins: a minor role of erythropoietin receptor-mediated pathways. *Nephrol Dial Transplant* 2006;21:iv303 [abstract MP013].
141. Briddell R, Stoney G, Sutherland SG et al. An *in vitro* investigation of the mode of clearance of erythropoietin by human bone marrow cells. *Exp Hematol* 2003;31:214 [abstract 46].
142. Elliott S, Lorenzini T, Chang D et al. Mapping of the active site of recombinant human erythropoietin. *Blood* 1997;89:493–502.
143. Yoon WH, Park SJ, Kim IC et al. Pharmacokinetics of recombinant human erythropoietin in rabbits 3/4 nephrectomized rats. *Res Commun Mol Pathol Pharmacol* 1997;96:227–240.
144. Schuster SJ, Caro J. Erythropoietin: Physiologic basis for clinical applications. *Vox Sang* 1993;65:169–179.
145. Jang G. Darbepoetin alfa (Aranesp) pharmacokinetics is comparable in chronic kidney disease (CKD) patients receiving and not receiving dialysis, in pediatric patients and in healthy adults. Paper presented at: ASN 38th Annual Renal Week Meeting; Philadelphia, PA, 8–13 November 2005.
146. Fukuda MN, Sasaki H, Lopez L et al. Survival of recombinant erythropoietin in the circulation: the role of carbohydrates. *Blood* 1989;73:84–89.
147. Nielsen OJ, Egfjord M, Hirth P. The metabolism of recombinant erythropoietin in the isolated perfused rat liver. *Liver* 1990;10:343–349.
148. Widness JA, Veng-Pedersen P, Schmidt RL et al. *In vivo* ^{125}I-erythropoietin pharmacokinetics are unchanged after anesthesia, nephrectomy and hepatectomy in sheep. *J Pharmacol Exp Ther* 1996;279:1205–1210.
149. Jensen JD, Jensen LW, Madsen JK et al. The metabolism of erythropoietin in liver cirrhosis patients compared with healthy volunteers. *Eur J Haematol* 1995:54:111–116.
150. Elliott S, Egrie J, Browne J et al. Control of rHuEPO biological activity: the role of carbohydrate. *Exp Hematol* 2004;32:1146–1155.
151. Harris JM, Chess RB. Effect of pegylation on pharmaceuticals. *Nat Rev Drug Discov* 2003;2:214–221.
152. Molineux G. Pegylation: engineering improved biopharmaceuticals for oncology. *Pharmacotherapy* 2003;23:3S–8S.
153. Harris JM, Martin NE, Modi M. Pegylation: A novel process for modifying pharmacokinetics. *Clin Pharmacokinet* 2001:40:539–551.
154. Jolling K, Ruixo JJ, Hemeryck A et al. Population pharmacokinetic analysis of pegylated human erythropoietin in rats. *J Pharm Sci* 2004;93:3027–3038.
155. Jolling K, Perez Ruixo JJ, Hemeryck A et al. Mixed-effects modelling of the interspecies pharmacokinetic scaling of pegylated human erythropoietin. *Eur J Pharm Sci* 2005;24:465–475.
156. Jarsch M, Brandt M, Lanzendörfer M et al. Comparative erythropoietin receptor binding kinetics of C.E.R.A. and epoetin-beta determined by surface plasmon resonance and competition binding assay. *Pharmacology* 2008;81:63–69.
157. Veng-Pedersen P, Chapel S, Al-Huniti NH et al. Pharmacokinetic tracer kinetics analysis of

changes in erythropoietin receptor population in phlebotomy-induced anemia and bone marrow ablation. *Biopharm Drug Dispos* 2004;25:149–156.
158. Veng-Pedersen P, Widness JA, Pereira LM et al. A comparison of nonlinear pharmacokinetics of erythropoietin in sheep and humans. *Biopharm Drug Dispos* 1999;20:217–223.
159. Mager DE, Jusko WJ. General pharmacokinetic model for drugs exhibiting target-mediated drug disposition. *J Pharmacokinet Pharmacodyn* 2001;28:507–532.
160. Mager DE, Krzyzanski W. Quasi-equilibrium pharmacokinetic model for drugs exhibiting target-mediated drug disposition. *Pharmacol Res* 2005;22:1589–1596.
161. Woo S, Krzyzanski W, Jusko WJ. Target-mediated pharmacokinetics and pharmacodynamics of recombinant human erythropoietin (rHuEPO). *J Pharmacokinet Pharmacodyn* 2007;34:849–868.
162. Wagner JG. Biopharmaceutics and relevant pharmacokinetics. Hamilton, Ill., Drug Intelligence Publications, 1971.
163. Nielsen OJ. Pharmacokinetics of recombinant human erythropoietin in chronic haemodialysis patients. *Pharmacol Toxicol* 1990;66:83–86.
164. Lim VS, DeGowin RL, Zavala D et al. Recombinant human erythropoietin treatment in predialysis patients: a double-blind placebo-controlled trial. *Ann Intern Med* 1989;110:108–114.
165. Cotes JM, Pippard MJ, Reid CD et al. Characterization of the anemia of chronic renal failure and the mode of its correction by a preparation of human erythropoietin (r-HuEPO). An investigation of the pharmacokinetics of intravenous erythropoietin and its effects on erythrokinetics. *Q J Med* 1989;70:113–137.
166. Yamazaki C, Watanabe Y, Sakamoto N. Pharmacokinetic study of recombinant human erythropoietin treatment in pre-dialysis end stage renal disease patients. *Nippon Jinzo Gakkai Shi* 1993;35:1233–1242.
167. Locatelli F, Reigner B. C.E.R.A. Pharmacodynamics, pharmacokinetics and efficacy in patients with chronic kidney disease. Exp Opin Investig Drugs 2007;16:1649–1661.

Erythropoietins, Erythropoietic Factors and Erythropoiesis
Edited by S.G. Elliott, M.A. Foote, and G. Molineux

Use of erythropoietic stimulating agents in the setting of renal disease

David A. Goodkin

Goodkin Biopharma Consulting, LLC, Bellevue, WA, USA; and Arbor Research Collaborative for Health, Ann Arbor, MI, USA

Introduction

As kidneys fail, their capacity to produce erythropoietin (EPO) typically diminishes. This deficiency of EPO is the primary etiology of the progressive anemia of chronic kidney disease (CKD). Hemoglobin concentration decreases in association with the increase in blood urea concentration (Fig. 1) and the decline in creatinine clearance (Fig. 2) of progressive renal failure [1]. The anemic state is exacerbated by a shortened red blood cell lifespan and bone marrow resistance to EPO in the setting of advanced uremia. Bleeding time is prolonged, as well, and gastrointestinal blood loss is not uncommon. In addition, dialytic therapy may consume red cells, due to repeated diagnostic phlebotomy, hemolysis caused by the hemodialysis pump, bleeding associated with hemodialysis needle insertion and removal, recurrent anticoagulation, and incomplete return of blood from the hemodialysis filter and bloodlines at

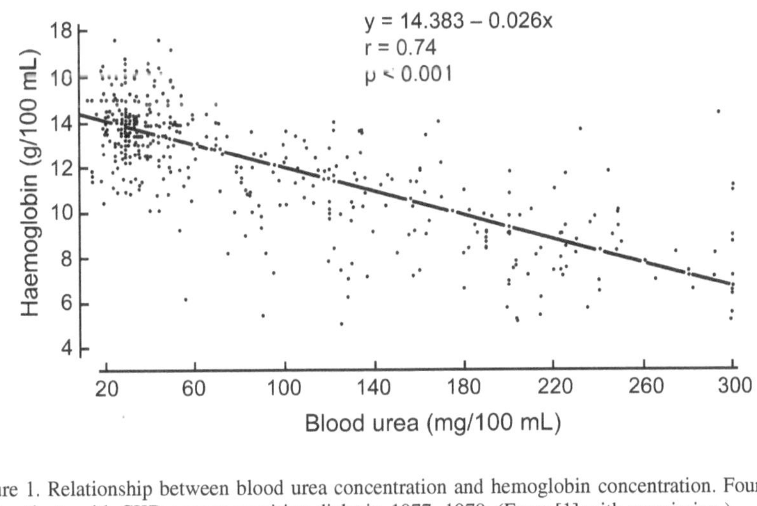

Figure 1. Relationship between blood urea concentration and hemoglobin concentration. Four hundred patients with CKD not yet requiring dialysis, 1977–1979. (From [1] with permission.)

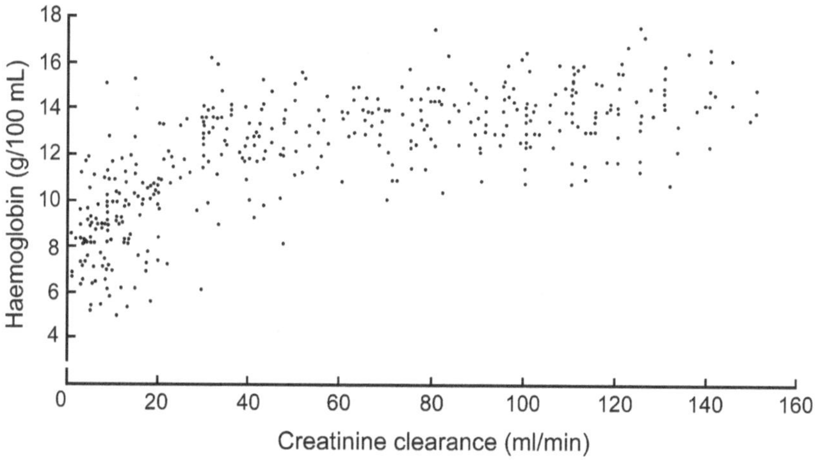

Figure 2. Relationship between creatinine clearance and hemoglobin concentration. Four hundred patients with CKD not yet requiring dialysis, 1977–1979. (From [1] with permission.)

the end of a treatment. Inflammatory cytokines and nutritional deficiencies may worsen the anemia of CKD.

If one visited a hemodialysis facility in the 1970s or 1980s, on any given day it was usual to see bags of blood suspended at multiple patient stations, with transfusions in progress. Many dialysis patients received 1–3 units of red blood cells per month [2], although some individuals had considerably higher transfusion requirements. Despite recurrent transfusions for approximately 25% of the patients, hematocrit values often oscillated within ranges <25%, causing persistent fatigue, limited capacity for exertion, and numerous other symptoms. In 1989, the first recombinant human erythropoietins (rHuEPO) were approved for commercial use. Most patients undergoing chronic dialysis treatments in developed nations and many patients with earlier stages of CKD now receive erythropoiesis-stimulating agent (ESA) therapy. Consequently, there has been a dramatic reduction in the rate of red blood cell transfusions and hematocrit values are considerably higher. In fact, dialysis professionals of more recent vintage have only seen blood transfusion as an occasionally-needed intervention for dialysis patients and not the common intradialytic process that it was in decades past.

Benefits of ESA treatment of anemia in patients with CKD

Reduction of red blood cell transfusions

A number of risks are associated with the transfusion of blood. Concern about the potential for transmission of viral infections such as hepatitis B, hepatitis C, and HIV was particularly worrisome for hemodialysis patients with chronic transfusion burdens before the ESA era. Improved screening techniques have greatly reduced the risk of transfusion-transmitted HIV-1 and hepatitis C virus to approximately 1 in 2×10^6 blood units from repeated donors in the United States [3]. Of course, most individuals prefer to avoid transfusion whenever feasible, and the possibility of transmitting other unscreened infectious diseases remains (e.g., bacteremia, Creutzfeld-Jakob disease, Chagas' disease, other hepatitis viruses). Recurrent red blood cell transfusions were historically associated with iron overload, sometimes severe and requiring chelation therapy. The incidence of cytotoxic antibody formation was increased by blood transfusion and this complication limited the prospects for kidney transplantation for some patients. Table 1 shows the estimated risks of various adverse transfusion reactions, as reported in 2005 [4].

Transfusion-related acute lung injury (TRALI) has emerged as the leading cause of death due to transfusion, and its incidence may be as high as between 1 in 5,000 to 1 in 1,323 transfusions [5]. Patients with TRALI develop tachypnea, dyspnea, cyanosis, temperature elevation, cyanosis, pulmonary infiltrates on radiographic exam, and acute hypoxemia with normal central venous pressure, in a clinical constellation that appears similar to acute respiratory distress syndrome. In addition, an immunosuppressive effect referred to as transfusion-related immune modulation (TRIM) is believed to result from a decrease in cell-mediated immunity after transfusion [4].

The most striking benefit of ESA therapy in the setting of renal failure has been the reduction in need for red blood cell transfusions. In the pivotal US Phase III trial of epoetin alfa, for example, erythrocyte transfusions were elim-

Table 1. Risk of adverse transfusion events. (From [4] with friendly permission by Wiley-Blackwell Publishing.)

Serious	Estimated risk
Mistransfusion	1:14,000 to 1:19,000
ABO-incompatible transfusion	1:38,000
Death due to ABO-incompatible transfusion	1:1.8 million
Acute hemolytic transfusion reaction	1:12,000
Delayed hemolytic transfusion reaction	1:4000 to 1:12,000
Transfusion-related acute lung injury	1:20,000 to 1:5000 (5% to 10% fatal)
Anaphylaxis	1:20,000 to 1:47,000
	1:150,000
	1:1600 (platelets)
	1:23,000 (erythrocytes)
Graft-*versus*-host disease	1:1 million (Canada)
Post-transfusion purpura	1:143,000 to 1:294,000
Fluid overload	1:708 to 1:3200
	1:7000 to 1:15,000
Less serious	
Febrile nonhemolytic transfusion reaction	1:500
Allergic (urticaria)	1:250

inated within 2 months of the onset of therapy among the 333 enrollees, who had required a total of 1,030 transfusions during the 6 months before treatment [6]. In clinical practice, the need for transfusion has not been eliminated, but it has been markedly reduced. In 2005, the incidence of transfusion for dialysis patients in the US had decreased to 31 units per 100 patient-years, from about 150 units per 100 patient-years in 1989 before the introduction of epoetin alfa [7]. The cost of transfusion of a single unit of packed red blood cells in an outpatient setting, including laboratory and personnel costs, has been cited as US$469 [8]. In the year before the availability of ESA therapy, more than 600,000 blood transfusions were estimated to have been given to dialysis patients in the United States [9]. The dialysis population has grown since then. ESA therapy thus relieves a serious burden on the blood-banking system and the reduction in transfusions offsets at least a small portion of the considerable cost of ESA medications.

Relief from anemic symptoms and complications

Patients with CKD experience many manifestations of anemia (Tab. 2). A number of the symptoms listed historically had been attributed to uremia. For this reason, the extent of symptomatic relief attained after ESA treatment of anemia came as a surprise to investigators early in the development of epoetin therapy.

Table 2. Manifestations of anemia

- Fatigue
- Weakness
- Malaise, depression
- Dyspnea
- Exertional chest pain
- Cardiac enlargement
- Impaired concentration/cognition
- Impaired sleep
- Impaired libido, erectile dysfunctional
- Impaired immune system
- Bleeding tendency
- Amenorrhea
- Anorexia
- Headache
- Intolerance to cold
- Pale skin

Cognitive function

Beneficial effects on brain function among hemodialysis patients after ESA correction of anemia have been demonstrated by both EEG waveform improvement and neuropsychiatric testing (attention, scanning, learning, memory functions) [10]. Cognitive function, including IQ scores, has been shown to improve after rHuEPO therapy in the setting of chronic ambulatory peritoneal dialysis [11].

Cardiac morphology and function

Left ventricular hypertrophy is common among CKD patients and is associated with increased mortality [12, 13]. Multiple studies of CKD patients have shown that increasing hematocrit values from ≤25% to 29–35% leads to marked reduction (11–35%) in left ventricular mass [14–18]. Study of patients on hemodialysis with asymptomatic cardiomyopathy and hemoglobin concentrations between 9–11 g/dL at entry, however, found that further increasing hemoglobin concentrations to the normal range with epoetin alfa (patients were randomly assigned to treatment to attain a hemoglobin concentration of 13.5 g/dL *versus* a control hemoglobin concentration of 10 g/dL) did not lead to further regression of left ventricular hypertrophy or dilation [19]. It may have prevented the development of *de novo* dilation.

The increase in oxygen delivery that accompanies correction of anemia improves exercise performance on treadmill testing and significantly reduces ST segment depression at maximal exercise, among hemodialysis patients with significant coronary artery disease [20–22]. Another cardiac benefit of ESA treatment is reduction of the chronically increased cardiac output associated with renal anemia [23], which has also been demonstrated after transfusional treatment [24].

It is important to note, however, that these cardiac effects have not been proven to result in improved patient survival. In fact, full correction of hemoglobin/hematocrit values to the normal range has resulted in increased mortality in two large prospective trials which are discussed in later sections.

Activity

Progressive renal failure leads to worsening exercise capabilities, likely due to several causes including uremic myopathy, disuse, and cardiovascular disease. Oxygen delivery is critical to enable activity and ESA therapy of anemia improves this capacity. Peak oxygen consumption ($\dot{V}O_2$ peak), usually measured at maximal exertion on a treadmill or stationary bicycle in studies of dialysis patients, is a surrogate marker for the cardiorespiratory capability to perform various tasks. As shown in Figure 3, $\dot{V}O_2$ peak, and hence exercise capacity, is far below normal in markedly anemic hemodialysis patients [25]. Each triangle in the figure represents average values from a separate study of dialysis patients. Increasing hematocrit to levels between 30–35% is associated with increases in $\dot{V}O_2$ peak of approximately 20–25%. Such increases enable

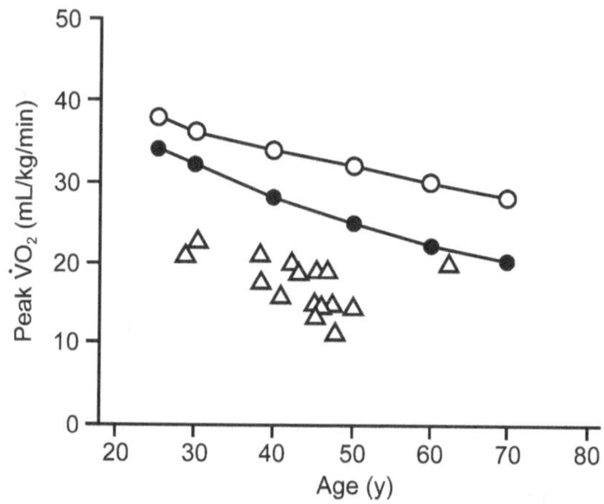

Figure 3. Average peak exercise capacity of patients receiving hemodialysis (open triangles) compared with expected levels for healthy sedentary men (open circles) and women (black circles). $\dot{V}O_2$ = oxygen consumption (From [25] with permission.)

more activities of daily living. For example, sedentary office work typically requires 5.7 mL O_2/kg/min, walking (at 4.9 km/h) requires 11.8 mL O_2/kg/min, and cooking requires 6.7–13.7 mL O_2/kg/min [26]. In contrast, vigorous activities such as playing soccer require >23 mL O_2/kg/min and are untenable for most anemic dialysis patients. Typically, a person is only able to sustain activities that consume oxygen at rates half of one's $\dot{V}O_2$ peak. For anemic dialysis patients, activities such as cooking require 40–50% of $\dot{V}O_2$ peak (moderate work), whereas walking slowly represents heavy work. $\dot{V}O_2$ peak values attained with partial correction of anemia may approach the lower limit of the normal range and such improvements ease functional limitations. In a study of eight patients, epoetin alfa therapy was used to increase hematocrit from the 18-to-24% range up to the 30-to-36% range, with significant improvements in $\dot{V}O_2$ peak and maximum aerobic exercise work capacity (watts) [27]. A larger study (n = 118) found that treatment of anemia significantly improved distance walked during stress testing, though not during a 6-min walk test [28].

Quality of life

Significant improvements were demonstrated in most quality-of-life parameters after 10 months of epoetin alfa treatment, resulting in an average hematocrit of 34%, during the open-label, pivotal, US Phase III study [29]. These measures included energy, activity level, functional ability, sleep, eating, satisfaction with health, sex life, well-being, psychological affect, life satisfaction, and happiness. A second open-label study included 484 dialysis patients starting epoetin alfa treatment [30]. Scores on the Medical Outcomes Study 36-item Short Form

Health Survey (SF-36) were compared between study entry and approximately 100 days later. Significant improvements were observed in vitality, physical functioning, social functioning, mental health, looking after the home, social life, hobbies, and satisfaction with sexual activity. The magnitudes of quality-of-life benefits were deemed clinically meaningful compared with studies of other disease states, although scores after treatment remained below the normal range for healthy subjects. A control group of patients, already receiving epoetin alfa at enrollment and whose mean hematocrit did not increase, exhibited no changes in SF-36 scores. Interestingly, regression analysis suggested that differences in hematocrit level explained only part of the change scores and the authors speculated that there could be an additional pharmacologic effect of treatment. It should be noted that the unblinded, uncontrolled nature of many of the quality-of-life studies of epoetin therapy is a less than ideal study design and the instruments used have not always been validated and rigorous.

In contrast, a double-blind prospective trial randomized patients to three groups: placebo *versus* treatment to a hemoglobin concentration of 9.5–11.0 g/dL *versus* treatment to 11.5–13.0 g/dL. This study demonstrated improvement in scores for fatigue, physical symptoms, relationships, and depression when the patients on epoetin were compared with those on placebo, though there was no significant difference in the improvement between the two epoetin doses [28]. Multiple aspects of sleep quality improve after ESA treatment [31]. In addition, epoetin therapy has been reported to decrease plasma histamine concentration and to relieve pruritus among hemodialysis patients [32].

Sexual and endocrine function

Sexual function may improve after ESA therapy due to an increase in energy, an increase in low testosterone concentration [33], and/or a decrease in elevated prolactin concentrations [34]. The latter effect has been reported to restore menses in female hemodialysis patients [35]. Other investigators have reported a lack of change in serum testosterone or prolactin concentrations, however [36, 37]. Additional reported hormonal changes include increased free T3 and free T4 [35], as well as ACTH and FSH [38].

Immune function

Patients with CKD are immunosuppressed and predisposed to infection. Partial correction of anemia repairs leukocyte respiratory burst activity, which is a critical component of phagocytosis [39], improves cell-mediated immunity [40], benefits humoral immunity/vaccine responses [41, 42], and increases expression of complement receptor 1 (CR1) on the red blood cell, which should aid bacterial clearance [43]. Normalization of hemoglobin concentration in hemodialysis patients with epoetin alfa further boosts cutaneous reactivity, increases CD8 cell counts, increases erythrocyte CR1 levels, and decreases anergy, compared with partial correction of anemia [44]. Also, decreased transfusion exposure enabled by ESA therapy may lead to diminished panel reactive antibodies against foreign HLA antigens and thereby improve a patient's candidacy for renal transplantation [45].

ESA benefits not mediated by an increase in red blood cell mass

Erythropoietin receptors (EPOR) have been reported on numerous cell types other than red blood cell precursors, including neurons, endothelial cells, retina, and cardiac myocytes. It should be briefly noted that there is exciting ongoing research into the potential direct benefits of EPO, including protection against ischemia/reperfusion injury (stroke, myocardial infarction), spinal cord injury, neuropathy (including diabetic neuropathy), and cognition [46, 47]. Although the EPO glycoprotein is >30 kDa, it has been shown to gain entry into the central nervous system. As described, quality-of-life benefits associated with ESA treatment of renal anemia have not been fully reconciled statistically to increases in hematocrit; the explanation may be that there are direct central nervous system benefits of EPO, too. It is possible that EPO may have direct detrimental effects, for example on diabetic retinopathy [48] or by causing vasoconstriction and hypertension at very high concentrations.

Adverse effects of ESA therapy in CKD

Hypertension

Blood pressure elevation accompanies the initiation of ESA therapy in as many as 35% of patients [6, 49]. Several mechanisms could theoretically underlie this tendency to hypertension and have been discussed in the literature: loss of the vasodilation induced by poor oxygen delivery to tissues, increased blood volume, increased blood viscosity, or direct epoetin-induced vasoconstriction. Increased cardiac output associated with anemia fails to decline appropriately after treatment in some patients who develop hypertension [50]. Markedly increasing epoetin dose to increase hemoglobin concentration from partial correction to normal did not have notable effects on blood pressure in two trials [51, 52], although there were significantly more reports of hypertension (30%) in the normal-hemoglobin group than in the anemic group (20%) in a third trial [53]. In clinical practice, attention to appropriate dialytic fluid removal, sodium and fluid intake, and adjustment of medication dosage has readily enabled management of blood pressure on ESA. Instances of malignant hypertension and seizures were reported during early trials, but these complications have occurred very rarely in later trials or general practice.

Thrombosis

Epoetin therapy affects hemostatic function. Many studies have been done, unfortunately often yielding conflicting results. One thorough literature review concludes that on balance the data indicate that therapy: 1) returns the prolonged bleeding time of CKD toward normal due to improved platelet aggregation and increased hematocrit (red cells are believed to displace platelets toward vessel walls); 2) transiently increases the number of circulating platelets, within the normal range; and 3) has minimal effects on the coagulation cascade (e.g., prothrombin time and partial activated thromboplastin time unchanged despite a variety of alterations in clotting proteins) [54]. While the decrease in bleeding time might beneficially diminish gastrointestinal or other bleeding, increased hemostatic activity could be detrimental to patients in other ways. Cardiovascular disease is the most common cause of death among patients with CKD and worsening of coronary or cerebral thrombosis would clearly be undesirable. For hemodialysis patients, clotting of the hemoaccess is a distressing, and expensive, complication. Whether or not ESA therapy increases access thrombosis after partial correction of anemia has been debated [55, 56], but a large, prospective study of normal *versus* low hematocrit values among hemodialysis patients with cardiac disease clearly demonstrated increased thrombosis rates of both native arteriovenous fistulae and synthetic grafts among patients randomized to the higher hematocrit target [51].

Dialytic clearance

Increases in red cell volume accompanied by decreased plasma volume theoretically could diminish the efficiency of solute removal during dialysis. Statistically significant, but small, increases in serum creatinine, phosphorus, and potassium concentrations were noted in a Phase III trial of epoetin alfa [6]. In general practice, hyperkalemia or worsening azotemia have not been issues and dialysis clearances are not significantly altered [57]. Regular attention to chemistry values, Kt/V results, and dialysis prescription have proven sufficient to avoid underdialysis during ESA therapy.

Pure red cell aplasia

A very rare, but serious, complication of treating anemia with recombinant proteins is pure red cell aplasia (PRCA). PRCA arises when patients develop antibodies to an ESA that also neutralize endogenous EPO. A severe, transfusion-requiring anemia results. The ESA must be discontinued and immunosuppressive therapy has been recommended [58]. PRCA has complicated therapy with epoetin alfa, epoetin beta, and darbepoetin alfa. There was a surge of cases in Europe between 1998 and 2002, associated with subcutaneous administration of epoetin alfa prepared by a single manufacturer, after a change in formulation. It is possible that rubber leachates from syringes acted as immune adjuvants [58]. A novel peptide currently in late stage development, Hematide, activates EPOR, but has no structural homology with EPO. Hematide rescued 10 patients with CKD (predialysis, hemodialysis, and peritoneal dialysis) who had PRCA [59].

Tumor progression

Some tumors have EPOR. ESA labeling information includes warnings that ESA therapy of anemia targeting hemoglobin concentrations above 12 g/dL has resulted in shorter times to tumor progression or death in the setting of several malignancies including metastatic breast, head and neck, nonsmall-cell lung, lymphoid, and cervical cancers.

Residual renal function

Animal studies and preliminary human trials raised uncertainty as to whether ESA therapy of anemia in CKD patients might slow the progression to end-stage renal disease (improved oxygen delivery to the kidney) or hasten deterioration of renal function (thrombosis, vasoconstriction). Larger clinical trials have not shown significant changes in the rate of CKD progression consequent to standard ESA therapy [60, 61].

Target hemoglobin/hematocrit

The initial target hemoglobin/hematocrit ranges for ESA therapy were selected empirically. Trials of epoetin alfa in the United States aimed for hematocrit values of 30–35% for several reasons: untreated hematocrit values often approximated 24% and it was felt that an increase in hematocrit roughly equivalent to a two-unit transfusion of red blood cells (i.e., an increase of 6 points to 30%) was a minimal desired benefit; there was an impression based on subjective patient responses that a hematocrit of 30% would ultimately be an appropriate point to start treatment during the course of CKD to avoid symptoms [62], and enthusiasm to correct anemia more aggressively was tempered by early cases of hypertension and seizures with rapid rises in hematocrit. Consultant hematologists contributing to study designs actually favored normal-hematocrit targets [63]. The original US product package insert for epoetin alfa specified a target range of 30–36%. Over the ensuing 19 years, some prospective trials have revealed deleterious outcomes at higher hematocrit values. Whether fortuitous or prescient, the original guidance regarding the maximal advisable hematocrit now appears appropriate.

Oxygen-carrying capacity of the blood increases linearly as the hematocrit increases. Blood viscosity, however, increases exponentially, so that a point of diminishing returns is reached for oxygen delivery (Fig. 4) [64]. These considerations suggest that a hematocrit of approximately 40% might be optimal, but they are based on viscosity measurements within glass capillary tubes, not within the dynamic environment of living blood vessels. A study of 27 patients with ischemic cerebrovascular disease indicated that oxygen delivery to the brain reaches a maximum at a hematocrit range of 40–45% [65], also suggesting that normalization of hematocrit might prove beneficial. Furthermore, observational epidemiologic studies of large databases found that higher hemoglobin/hematocrit values were associated with better survival rates among dialysis patients [66, 67] and small interventional studies of hematocrit normalization reported functional benefits and safety [31, 68–70].

Thus, it was an unexpected result of substantial import when the Normal Hematocrit Cardiac Trial (NHCT) demonstrated adverse outcomes among hemodialysis patients with congestive heart failure or ischemic heart disease targeted to normal hematocrit values (42%), compared with partial correction (hematocrit 30%) [51]. There were 183 deaths and 19 first nonfatal myocardial infarctions among 618 patients in the normal-hematocrit group *versus* 150 deaths and 14 myocardial infarctions among 615 patients in the low-hemat-

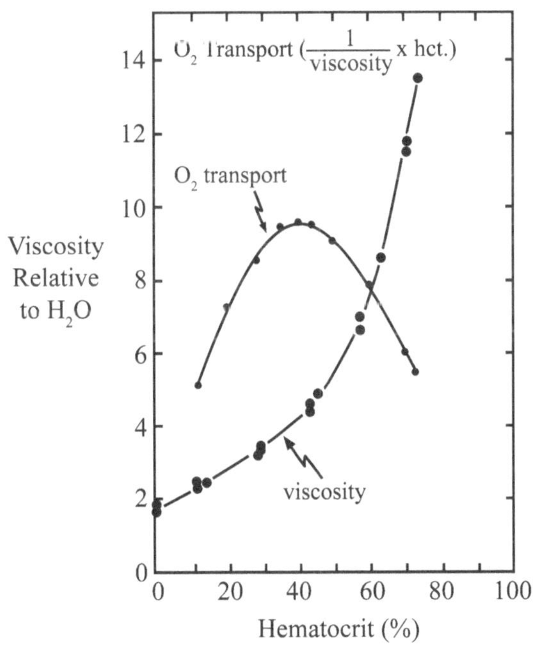

Figure 4. Effects of increasing hematocrit values on blood viscosity and oxygen delivery *in vitro*. Oxygen transport as calculated from the oxygen-carrying capacity of blood (hemoglobin concentration) and blood flow (reciprocal of viscosity). (From [64] with permission.)

ocrit group. The study was halted at the recommendation of an independent data monitoring committee after an interim analysis. The primary endpoint (time to death or myocardial infarction) results did not reach the prespecified stopping rule and were not statistically significant when adjusted for previous interim analyses [71], although the nominal P-value by log-rank test was 0.01. The latter statistic was included in the original study manuscript submitted (and was reported to the Food and Drug Administration [FDA]), but it was removed before publication after editorial input from *The New England Journal of Medicine* staff, which prompted an additional independent expert biostatistician review. The causes of death were similar between groups and death was most often due to cardiac disease. Vascular access thrombosis was more common in the normal group (39%) than in the low group (29%; $P = 0.001$). Significantly fewer patients required blood transfusion in the normal group (21%) than in the low group (31%; $P < 0.001$). Quality-of-life scores did not differ between the groups, though significantly better physical function scores were associated with higher hematocrit values. The mechanism underlying the increased mortality remains unclear. Mortality was not associated with dose of epoetin alfa and within treatment groups higher hematocrit values were not associated with higher mortality. Patients in the normal-hematocrit group received significantly greater amounts of intravenous iron, and among these patients, iron administration was associated with higher mortality, in *post hoc* analysis. The publication concluded that it is not recommended to target a hematocrit of 42% among hemodialysis patients with congestive heart failure or ischemic heart disease.

A subsequent study of 416 Scandinavian patients (predialysis, hemodialysis, and peritoneal dialysis) examined a normal-hemoglobin target range of 13.5–16.0 g/dL *versus* partial correction to 9.0–12.0 g/dL, using epoetin alfa [72]. Quality-of-life testing was administered to 253 dialysis patients. Significantly superior quality-of-life scores were reported for the normal-hemoglobin group, in physical symptoms, fatigue, depression, and frustration. Safety outcomes, including thrombotic events, were comparable between the groups.

An analysis of data from 2,858 US dialysis units (899,000 patient-years) between 1999 and 2002 found significantly lower mortality among facilities with greater percentages of patients with a hematocrit of at least 33%, after adjusting for case mix and patient characteristics [73].

Eight years after the publication of the NHCT results, two large interventional studies (CREATE [53] and CHOIR [52]) reported findings for normal *versus* lower hemoglobin concentrations among patients with CKD who did not require dialysis. The CREATE trial of 603 patients found no significant difference in cardiovascular outcomes between patients randomized to a hemoglobin target range of 10.5–11.5 g/dL and those targeted to 13.0–15.0 g/dL, prescribing epoetin beta. SF-36 quality-of-life scores were significantly better for the normal-hemoglobin group with regard to general health, mental health, physical function, physical role, social function, and vitality. No difference was seen between groups in calculated glomerular filtration rate (GFR), although time to initiation of dialysis was statistically significantly longer for the lower hemoglobin group (P = 0.03). In the normal-hemoglobin group, 52% of patients received intravenous iron, compared with 42% of those in the low-hemoglobin group (P-value not provided). The CHOIR trial randomly assigned 1,432 patients to either a hemoglobin target of 13.5 g/dL or 11.3 g/dL, prescribing epoetin alfa. The primary endpoint, a composite of cardiovascular events (death, myocardial infarction, congestive heart failure requiring hospitalization, and stroke), occurred more often among the normal-hemoglobin group than among the low-hemoglobin group (125 and 97 events, respectively; P = 0.03). Fewer than 3% of patients received intravenous iron. The groups did not differ in their quality-of-life scores or in progression to renal replacement therapy. The authors state that the study did not provide a mechanistic explanation for the poorer cardiovascular outcome with the use of the higher target hemoglobin concentration.

Currently, the labeling approved by regulatory authorities for each of the ESAs states that maintenance hemoglobin concentration should be 10–12 g/dL. The most recent anemia practice guidelines from the Kidney Disease Outcomes Quality Initiative (KDOQI) recommend a target range of 11–12 g/dL [74]. The European Best Practice Guidelines (EBPG), issued before the recent large studies of normal hemoglobin concentration, recommend a target of >11 g/dL, not to exceed 14 g/dL, with the exact concentration being guided by individual patient characteristics (sex, age, ethnicity, activity, and comorbid conditions) [75].

The Dialysis Outcomes and Practice Patterns Study (DOPPS) is a prospective, observational study [76], which tracks representative random samples of hemodialysis patients from 12 nations and collects extensive information on prescriptions, comorbid conditions, and laboratory data. Table 3 shows the average hemoglobin concentrations by country among prevalent patients at

Table 3. Average hemoglobin concentration, by nation

Country	No. patients	Mean Hgb ± SD (g/dL)
Australia/New Zealand	489	11.82 ± 1.53
Belgium	481	11.86 ± 1.33
Canada	511	11.76 ± 1.32
France	552	11.73 ± 1.43
Germany	553	11.65 ± 1.38
Italy	498	11.50 ± 1.46
Japan	1,803	10.44 ± 1.24
Spain	550	11.95 ± 1.50
Sweden	524	11.86 ± 1.37
United Kingdom	358	11.61 ± 1.53
United States	1,733	11.99 ± 1.36

Prevalent cross-section of hemodialysis patients (n = 8,052) at entry into the Dialysis Outcomes and Practice Patterns Study, 2005–2007. Hgb = hemoglobin concentration; SD = standard deviation.

study entry between 2005 and 2007 (unpublished data). Consistent with the guidance, the average values are between 11 and 12 g/dL, with the exception of the mean hemoglobin value of 10.44 g/dL in Japan, where nephrologists seek lower targets.

ESA prescription

As shown in Figure 5, epoetin alfa, epoetin beta, or darbepoetin alfa were prescribed to more than 90% of hemodialysis patients in North America, Europe, and Japan between 2005 and 2007 (DOPPS data, unpublished). The proportion of patients receiving each of the three products varies widely by nation. The reader is referred to the respective product labeling instructions, plus the KDOQI [77] and EBPG [75] guidelines, for details of dosage recommendations for the specific agents. Briefly, when initiating epoetin alfa therapy for an anemic patient, the recommended starting dose is 50–100 U/kg intravenously or subcutaneously thrice weekly; for epoetin beta, 20 U/kg subcutaneously thrice weekly or 40 U/kg intravenously thrice weekly; and for darbepoetin alfa, 0.45 μg/kg intravenously or subcutaneously once weekly. Dose should be titrated to avoid increases in hemoglobin concentration >1 g/dL during a 2-week period. If the hemoglobin concentration increases too quickly, it is generally preferable to reduce dose, perhaps 25%, rather than to withhold dosing, to avoid repeated cycles of undershooting and overshooting the desired hemoglobin range.

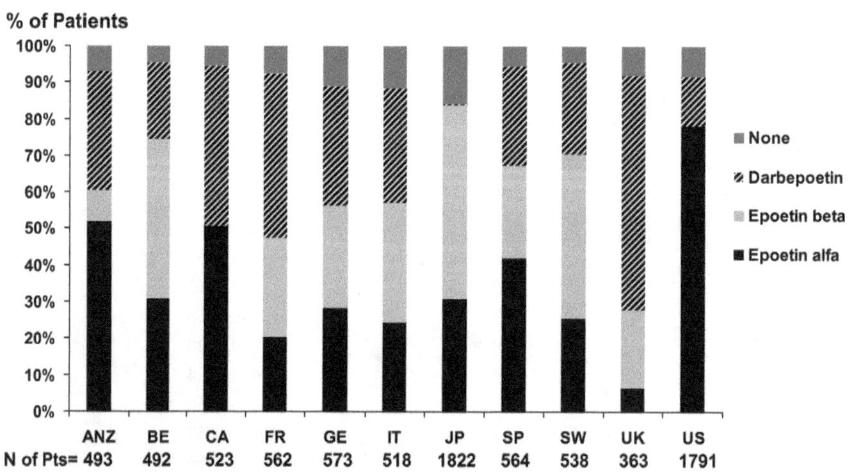

Figure 5. Distribution of ESA usage by nation. ESA prescriptions for prevalent hemodialysis patients at entry into the Dialysis Outcomes and Practice Patterns Study, 2005 to 2007 (n = 8,239). Abbreviations correspond to the nations listed in Table 3.

Route of administration

Despite lower bioavailability, subcutaneous administration of epoetin alfa allows lower doses than does intravenous dosing. There may be a threshold circulating drug concentration necessary to stimulate erythropoiesis. Drug concentration might decrease below this minimum level for a greater proportion of the week with bolus intravenous administrations than with the continued absorption that follows subcutaneous injections [78]. Another consideration is that lower nadir concentrations associated with intermittent intravenous administration might result in neocytolysis, a phenomenon wherein very low EPO concentration triggers reticuloendothelial hemolysis of young circulating red cells [79]. Two sizable, prospective trials comparing epoetin alfa dose requirements between the two routes of administration among hemodialysis patients both found that after 21 to 26 weeks the average dose is 27% lower with subcutaneous administration [80, 81].

ESAs are painless when administered intravenously during hemodialysis, whereas subcutaneous injection causes variable patient discomfort. The primary motivation for prescribing ESAs to hemodialysis patients subcutaneously is economics, as ESA therapy is expensive. (Paradoxically, in the US there are reimbursement incentives to dialysis centers that reward higher ESA consumption, and prescription via the intravenous route predominates there.) However, notwithstanding the average reduction in dose when a group of patients is converted from the intravenous to the subcutaneous route of administration, as many as 39% of individuals actually require the same or more epoetin subcutaneously than intravenously. For example, adipose tissue impedes subcutaneous drug absorption and seven of eight patients who weighed >82 kg did not have dose reductions after converting from intravenous to subcutaneous administration [82]. Hence, the greatest cost-savings might be attained by attempting both routes of administration for a given patient and then selecting the route with the lower dose requirement. In most cases this will prove to be the subcutaneous route. The half-life of darbepoetin alfa is considerably longer than the half-life of epoetin, and darbepoetin alfa dose requirements are the same for either route of administration.

Subcutaneous administration of epoetin alfa is associated with greater serial variability in hemoglobin concentration (standard deviation of 4 weekly hematocrit values = 2.02 volume %) than occurs with intravenous administration (1.37 volume %; $P < 0.01$) [82]. Many factors (e.g., body site of injection, adiposity, edema, temperature, and prior activity) alter the absorption of drugs such as epoetin, insulin, heparin, and growth hormone after subcutaneous injection, so it is not surprising that hemoglobin concentration will vary more with subcutaneous than with intravenous administration. Greater hemoglobin variability has been associated with higher mortality rates [83, 84], but it is unknown whether the magnitude of increased variability associated with subcutaneous administration of epoetin alfa is clinically significant. Predialysis and peritoneal dialysis patients lack chronic hemoaccess and are routinely treated subcutaneously.

Iron supplementation

Before the use of ESA therapy, transfusion-related iron overload was a serious complication. Stimulation of erythropoiesis with ESA therapy often leads to the opposite problem, iron deficiency, particularly for hemodialysis patients, who experience ongoing blood/iron loss associated with the procedure. Oral iron supplementation is generally believed to yield an inadequate rate of iron absorption and iron is typically prescribed intravenously to hemodialysis patients. Iron status should be assayed regularly and this is often attempted by measuring transferrin saturation and serum ferritin concentration every 1–3 months. Prescribers may wait until deficiency is detected and then infuse up to 1,000 mg in divided doses over 1–3 weeks, depending on the severity of the deficit and the iron formulation selected. A single, total-dose infusion has been described as a safe, convenient, and less expensive option [85]. Alternatively, ongoing infusions of smaller quantities of iron, e.g., with each session or weekly, are used at many centers in a strategy to prevent deficiency. Administration is prescribed more episodically for iron deficient predialysis and peritoneal dialysis patients, who lack permanent hemoaccess.

The aggregate evidence that intravenous iron supplementation decreases ESA dose requirement and aids hemoglobin maintenance is compelling. The treatment is not free of risk, however. First, intravenous iron dextran was historically associated with a low incidence of potentially lethal, anaphylactic-like reactions. Fortunately, there are alternate formulations that are safer (iron sucrose, ferric gluconate, and low molecular weight iron dextran); it has been recommended that high molecular weight iron dextran should be strictly avoided [86]. Second, patients with CKD have diminished immune function and are predisposed to infection and sepsis. Associations have been reported between increases of serum ferritin concentration >500 ng/mL and increased rates of infection [87, 88], and IV iron infusions could predispose to microbial proliferation. However, a 6-month observational study of 988 hemodialysis patients to determine risk factors for bacteremia showed no significant difference in serum ferritin concentration or receipt of intravenous iron within the past 6 months, between the 50 patients with bacteremia and the 935 without [89]. Third, the recurrent intravenous infusion of iron boluses could contribute to undesirable oxidative cardiovascular consequences, although the clinical validity of this concern has not been established. Cardiovascular disease is widely prevalent and highly deleterious in patients with CKD. Emerging data from studies in the general population suggest that binding of iron protects the heart. For example, infusion of desferoxamine during coronary artery bypass surgery, where iron exposure may result from hemolysis and cardioplegia solutions, improves postoperative cardiac function for at least 1 year [90]. Patients with type 2 diabetes and impaired coronary artery vasodilation respond favorably to desferoxamine [91]. The authors note that free iron ions available to stimulate deleterious oxygen radical reactions are typically very low (<5 micro mol/L) [91]. Experimentally, long-term iron chelation treatment of rats with streptozotocin-induced diabetes prevents impaired relaxation of aortic rings, suggesting that iron-catalyzed oxygen radical formation contributes to the development of diabetic endothelial dysfunction [92]. A historical-cohort study of 58,058 hemodialysis patients found that doses of intravenous iron up to 400 mg/month were associated with improved survival compared with no intravenous iron, but doses >400 mg/month were associated with higher death rates [93]. Although adjustments were made for case-mix, it is not certain that these observational associations are causal – the potential for confounding by indication still exists.

Notwithstanding the above considerations, the current state of the art, KDOQI guidance, and EBPG guidance all support the use of intravenous iron to improve ESA treatment of anemia. The DRIVE study found that intravenous ferric gluconate treatment (n = 66), *versus* none (n = 66), improved hemoglobin response without causing an increase in infectious or cardiac adverse events among hemodialysis patients with serum ferritin concentration ≥500 ng/mL and transferrin saturation ≤25% [94]. Although the 6-week observation period and sample size are grossly inadequate to permit conclusions about long-term safety, the results do not signal early dangers. Large, randomized, prospective, chronic, survival studies of various doses of intravenous iron therapy would better inform optimal anemia management and are clearly lacking.

Hyporesponse to ESA therapy

It has long been recognized that there is wide variety in dose-requirements for epoetin treatment. For example, in the pivotal US Phase III trial of epoetin alfa in the setting of hemodialysis, the intravenous dose needed to maintain hematocrit values of 35% varied 42-fold (from 12.5–525 U/kg thrice weekly) among 333 iron-replete patients who at entry were deemed medically stable, without blood loss, and expected to live at least 6 months [6]. Nonetheless, patients requiring unusually high doses of ESA, or, in particular, those who previously maintained a desirable hemoglobin concentration but develop increasing dose requirements must be evaluated. Table 4 delineates causes of

Table 4. Causes of hyporesponse to ESA therapy

- Iron deficiency
- Inflammation/infection
- Bleeding
- Hemolysis
- Malignancy (particularly hematologic)
- Osteitis fibrosa (hyperparathyroidism)
- Hemoglobinopathy (e.g., thalassemia)
- Angiotensin converting enzyme inhibition
- Nutritional/vitamin deficiency
- Inadequate dialysis
- Antibody-mediated pure red cell aplasia (PRCA)

hyporesponse. For predialysis or peritoneal dialysis patients, the clinician should first verify that the patient is complying with the prescribed subcutaneous injections of the ESA. Iron deficiency, inflammation, and bleeding will account for most cases of hyporesponse and should be sought carefully. Vitamin deficiency (particularly of vitamin B_{12} or folate) may cause resistance to therapy. These deficiencies should not occur when a daily oral multivitamin supplement is taken to offset the loss of water-soluble vitamins during dialysis, but surprisingly the prescription rate of multivitamins ranges only from 4–72% across five European nations, Japan, and the United States [95]. Data are lacking to prove that angiotensin-converting enzyme inhibition or inadequate dialysis commonly or markedly inhibits ESA response, but such etiologies likely contribute occasionally. Aluminum toxicity was historically a cause of worsening anemia in CKD, but this is rarely a problem now that aluminum-based phosphate binder use has been abandoned. Some publications suggest that carnitine deficiency also contributes to ESA hyporesponse, but the data are not convincing.

Of interest, a recent retrospective case-control study suggests that hepatitis C infection *decreases* ESA dose requirements [96]. The authors speculate that viral inflammation may stimulate EPO release from the liver, but they provide no data comparing EPO levels in the absence of exogenous therapy. This new observation related to hepatitis C infection remains to be confirmed.

Novel ESA therapy: what's next?

Epoetin alfa and epoetin beta, 165-amino-acid rHuEPO, were the original ESA treatments and have established records of safety and efficacy. 'Biosimilar' or 'follow-on' recombinant versions have reached the market in some regions, as patent protections for the original compounds have expired. The miniepidemic of PRCA, which followed a subtle change in the epoetin alfa manufacturing process by one company, shows that close surveillance will be required for biologic follow-on products to assure equivalence to the original formulations. A publication describes multiple variations in the characteristics of biosimilar epoetin products procured from around the world that might impact immunogenicity/safety, including differences in pH, osmolality, protein aggregates, and even endotoxin [97].

Darbepoetin alfa, available since 2001, is a second-generation ESA, with two additional carbohydrate chains carrying increased sialic acid residues linked to the protein backbone, to prolong serum half-life and allow less-frequent dosing (weekly, every other week, and in some cases monthly). Five amino acids in the backbone were substituted to enable the carbohydrate engineering.

PEGylated-epoetin beta has entered the market in Europe and has been approved by the FDA in the United States. 'PEGylation' refers to the addition of poly[ethylene] glycol moieties, to slow clearance. The new compound has a longer serum half-life than does darbepoetin alfa. A large, open-label, trial compared two intravenous dosing intervals (once every 2 weeks and once every 4 weeks) of PEGylated-epoetin beta *versus* standard epoetin in dialysis patients and found noninferior control of hemoglobin concentration and similar adverse event profiles across groups [98]. Monthly dosing could prove an advantage, particularly to patients who require subcutaneous injections.

Hematide is an engineered peptide, also PEGylated, that activates EPOR, although it lacks homology with EPO. The molecule has been shown to control anemia in Phase II studies of hemodialysis and predialysis patients, given either intravenously or subcutaneously every 4 weeks [99]. The incidence, if any, of antibody formation against the synthetic peptide was not reported. Hematide has proven effective in reversing ESA-induced PRCA [59].

Also in development is a very novel oral medication approach to anemia therapy. These agents competitively inhibit the enzymes that inactivate hypoxia inducible transcription factors (HIF). As a result, increased quantities of HIF are transferred to the nuclei of cells, mimicking a hypoxic state. Gene expression of EPO is stimulated, along with expression of other proteins that enhance iron availability and otherwise aid erythropoiesis. Preliminary clinical investigation has shown that EPO and hemoglobin concentrations increase after treatment with either of two HIF stabilization agents in patients with CKD [100, 101]. HIF stabilization induces the production of a number of nonerythropoietic proteins, as well; the safety profiles of the agents need to be determined with larger studies. Such oral ESA therapy might prove particularly advantageous to expand the treatment of anemia among predialysis patients and to overcome inflammatory resistance to erythropoiesis.

Asialo and carbamylated EPO analogs, which do not stimulate the bone marrow, are under development for nonanemic indications, such as ischemia-reperfusion injury states.

Also, novel approaches are being investigated to improve iron availability in the setting of inflammation, which could aid erythropoiesis in renal failure. Hepcidin, increased during inflammatory states, is a protein that diminishes iron absorption from the gastrointestinal tract and impedes release of iron from the reticuloendothelial system. Commercial efforts are underway to develop antibodies and antagonists to hepcidin and to manipulate hemojuvelin, which in turn affects hepcidin concentrations.

References

1. Andreucci VE. Chronic renal failure. In: Castro JE (ed.): *The Treatment of Renal Failure*, Appleton-Century-Crofts, New York. 1982;33–167
2. Blagg CR. In memory of Joseph Wetherill Eschbach. *Kidney Int* 2008;73:665–666
3. Dodd RY, Notari EP, Stramer SL. Current prevalence and incidence of infectious disease markers and estimated window-period risk in the American Red Cross blood donor population. *Transfusion* 2002;42:975–979
4. Luban NLC. Transfusion safety: where are we today? *Ann N Y Acad Sci* 2005;1054:325–341
5. Silliman CC, Ambruso DR, Boshkov LK. Transfusion-related acute lung injury. *Blood* 2005;

105:2266–2273
6. Eschbach JW, Abdulhadi MH, Browne JK et al. Recombinant human erythropoietin in anemic patients with end-stage renal disease: results of a phase III multicenter clinical trial. *Ann Intern Med* 1989;111:992–1000
7. Mayne TJ, Gitlin M, Krishnan M. Variation in hemoglobin distribution and transfusions in patients with chronic renal failure (CRF) undergoing dialysis. *J Am Soc Nephrol* 2007;18:702A (abstract)
8. Cremieux P-Y, Barrett B, Anderson K et al. Cost of outpatient blood transfusion in cancer patients. *J Clin Oncol* 2000;18:2755–2761
9. Goodnough LT, Strasburg D, Riddel J et al. Has recombinant human erythropoietin therapy minimized red-cell transfusions in hemodialysis patients? *Clin Nephrol* 1994;41:303–307
10. Marsh JT, Brown WS, Wolcott D et al. rHuEPO treatment improves brain and cognitive function of anemic dialysis patients. *Kidney Int* 1991;39:155–163
11. Temple RM, Deary IJ, Winney RJ. Recombinant erythropoietin improves cognitive function in patients maintained on chronic ambulatory peritoneal dialysis. *Nephrol Dial Transplant* 1995;10:1733–1738
12. Silberberg JS, Rahal DP, Patton DR et al. Role of anemia in the pathogenesis of left ventricular hypertrophy in end-stage renal disease. *Am J Cardiol* 1989;64:222–224
13. Foley RN, Parfrey PS, Harnett JD et al. Clinical and echocardiographic disease in patients starting end-stage renal disease therapy. *Kidney Int* 1995;47:186–192
14. Silberberg J, Racine N, Barre P et al. Regression of left ventricular hypertrophy in dialysis patients following correction of anemia with recombinant human erythropoietin. *Can J Cardiol* 1990;6:1–4
15. Colan S, Jabs K, Harmon W. Treatment with epoetin alfa is associated with a reduction in ventricular mass and an increase in exercise tolerance in young hemodialysis patients. *J Am Soc Nephrol* 1991;2:374 (abstract)
16. Low-Friedrich I, Grutzmacher P, Marz W et al. Therapy with recombinant human erythropoietin reduces cardiac size and improves heart function in chronic hemodialysis patients. *Am J Nephrol* 1991;11:54–60
17. Pascual J, Teruel JL, Moya JL et al. Regression of left ventricular hypertrophy after partial correction of anemia with erythropoietin in patients on hemodialysis: a prospective study. *Clin Nephrol* 1991;35:280–287
18. Goldberg N, Lundin AP, Delano B et al. Changes in left ventricular size, wall thickness, and function in anemic patients treated with recombinant human erythropoietin. *Am Heart J* 1992; 124:424–427
19. Foley RN, Parfrey PS, Morgan J et al. Effect of hemoglobin levels in hemodialysis patients with asymptomatic cardiomyopathy. *Kidney Int* 2000;58:1325–1335
20. Macdougall IC, Lewis NP, Saunders MJ et al. Long-term cardiorespiratory effects of amelioration of renal anaemia by erythropoietin. *Lancet* 1990;335:489–493
21. Hase H, Imamura Y, Nakamura R et al. Effects of rHuEPO therapy on exercise capacity in hemodialysis patients with coronary artery disease. *Jpn Circ J* 1993;57:131–137
22. Wizemann V, Kaufmann J, Kramer W. Effect of erythropoietin on ischemia tolerance in anemic hemodialysis patients with confirmed coronary artery disease. *Nephron* 1992;62:161–165
23. Nonnast-Daniel D, Frei U, Brabant EG et al. Atrial natriuretic peptide and central haemodynamics during correction of renal anaemia by recombinant human erythropoietin treatment in regular dialysis treatment patients. *Nephrol Dial Transplant* 1989;4:478 (abstract)
24. Neff MS, Kim KE, Persoff M et al. Hemodynamics of uremic anemia. *Circulation* 1971;43:876–883
25. Painter P. The importance of exercise training in rehabilitation of patients with end-stage renal disease. *Am J Kidney Dis* 1994;24:S2–S9
26. Lentner C (ed.): *Geigy Scientific Tables. 8th ed.* New Jersey, Ciba-Geigy Corporation, 1981;228–231
27. Thompson JR, Stray-Gunderson J. Cardiovascular adaptations in maximal exercise among patients with end stage renal disease (ESRD) after graded increases in hemoglobin (Hgb). *J Am Soc Nephrol* 1991;2:389 (abstract)
28. Canadian Erythropoietin Study Group. Association between recombinant human erythropoietin and quality of life and exercise capacity of patients receiving haemodialysis. *Brit Med J* 1990;300:573–578
29. Evans RW, Rader B, Manninen DL, Cooperative Multicenter EPO Clinical Trial Group. The quali-

ty of life of hemodialysis recipients treated with recombinant human erythropoietin. *JAMA* 1990;263:825–830

30. Beusterien KM, Nissenson AR, Port FK et al. The effects of recombinant human erythropoietin on functional health and well-being in chronic dialysis patients. *J Am Soc Nephrol* 1996;7:763–773
31. Benz RL, Pressman MR, Hovick ET et al. A preliminary study of the effects of correction of anemia with recombinant human erythropoietin therapy on sleep, sleep disorders, and daytime sleepiness in hemodialysis patients (The SLEEPO study). *Am J Kidney Dis* 1999;34:1089–1095
32. De Marchi S, Cecchin E, Villalta D et al. Relief of pruritus and decreases in plasma histamine concentrations during erythropoietin therapy in patients with uremia. *N Engl J Med* 1992;326:969–974
33. Haley NR, Matsumoto AM, Eschbach JW et al. Low testosterone levels increase in male hemodialysis patients treated with recombinant human erythropoietin. *Kidney Int* 1989;35:193 (abstract)
34. Schaefer RM, Kokot F, Wernze H et al. Improved sexual function in hemodialysis patients on recombinant erythropoietin: a possible role for prolactin. *Clin Nephrol* 1989;31:1–5
35. Yeksan M, Tamer N, Cirit M et al. Effect of recombinant human erythropoietin (rHuEpo) therapy on plasma FT3, FT4, TSH, FSH, LH, free testosterone and prolactin levels in hemodialysis patients. *Int J Artif Organs* 1992;15:585–589
36. Bommer J, Kugel M, Schwobel B et al. Improved sexual function during recombinant human erythropoietin therapy. *Nephrol Dial Transplant* 1990;5:204–207
37. Foresta C, Mioni R, Bordon P et al. Erythropoietin stimulates testosterone production in man. *J Clin Endocrinol Metab* 1994;78:753–756
38. Watschinger B, Watzinger U, Templ H et al. Effect of recombinant human erythropoietin on anterior pituitary function in patients on chronic hemodialysis. *Horm Res* 1991;36:22–26
39. Veys N, Vanholder R, Ringoir S. Correction of deficient phagocytosis during erythropoietin treatment in maintenance hemodialysis patients. *Am J Kidney Dis* 1992;19:358–363
40. Singh AB, Singh M, Palekar S et al. The effects of recombinant human erythropoietin on the cell mediated immune response of renal failure patients. *J Med* 1992;23:289–302
41. Sennesael JJ, Van der Niepen P, Verbeelen DL. Treatment with recombinant human erythropoietin increases antibody titers after hepatitis B vaccination in dialysis patients. *Kidney Int* 1991;40:121–128
42. Schaefer RM, Paczek L, Berthold G et al. Improved immunoglobulin production in dialysis patients treated with recombinant human erythropoietin. *Int J Artif Organs* 1992;15:204–208
43. Hebert LA, Birmingham DJ, Dillon JJ et al. Erythropoietin therapy in humans increases erythrocyte expression of complement receptor type 1 (CD35). *J Am Soc Nephrol* 1994;4:1786–1791
44. Roman RMB, Lobo PI, Taylor RP et al. Prospective study of the immune effects of normalizing the hemoglobin concentration in hemodialysis patients who receive recombinant human erythropoietin. *J Am Soc Nephrol* 2004;15:1339–1346
45. Grimm PC, Sinai-Trieman L, Sekiya NM et al. Effects of recombinant human erythropoietin on HLA sensitization and cell mediated immunity. *Kidney Int* 1990;38:12–18
46. Bogoyevitch MA. An update on the cardiac effects of erythropoietin cardioprotection by erythropoietin and the lessons learnt from studies in neuroprotection. *Cardiovasc Res* 2004;63:208–216
47. Lipton SA. Erythropoietin for neurologic protection and diabetic neuropathy. *N Engl J Med* 2004;350:2516–2517
48. Watanabe D, Suzuma K, Matsui S et al. Erythropoietin as a retinal angiogenic factor in proliferative diabetic retinopathy. *N Engl J Med* 2005;353:782–792
49. Sundal E, Kaeser U. Correction of anaemia of chronic renal failure with recombinant human erythropoietin: safety and efficacy of one year's treatment in a European multicentre study of 150 haemodialysis-dependent patients. *Nephrol Dial Transplant* 1989;4:978–987
50. Raine AEG, Roger SD. Effects of erythropoietin on blood pressure. *Am J Kidney Dis* 1991;18:76–83
51. Besarab A, Bolton WK, Browne JK et al. The effects of normal as compared with low hematocrit values in patients with cardiac disease who are receiving hemodialysis and epoetin. *N Engl J Med* 1998;339:584–590
52. Singh AK, Szczech L, Tang KL et al. Correction of anemia with epoetin alfa in chronic kidney disease. *N Engl J Med* 2006;355:2085–2098
53. Drueke TB, Locatelli F, Clyne N et al. Normalization of hemoglobin level in patients with chron-

ic kidney disease and anemia. *N Engl J Med* 2006;355:2071–2084
54. Tang WW, Stead RA, Goodkin DA. Effects of epoetin alfa on hemostasis in chronic renal failure. *Am J Nephrol* 1998;18:263–273
55. Eschbach JW. Erythropoietin is not a cause of access thrombosis. *Semin Dial* 1993;6:180–184
56. Muirhead N. Erythropoietin is a cause of access thrombosis. *Semin Dial* 1993;6:184–188
57. Delano BG, Lundin AP, Galonsky R et al. Dialyzer urea and creatinine clearances are not significantly altered in erythropoietin treated maintenance hemodialysis patients. *ASAIO Trans* 1990;36:36–39
58. Pollock C, Johnson DW, Horl WH et al. Pure red cell aplasia induced by erythropoiesis-stimulating agents. *Clin J Am Soc Nephrol* 2008;3:193–199
59. Macdougall IC, Casadevall N, Stead RB et al. Successful treatment of anti-erythropoietin antibody-mediated pure red cell aplasia with Hematide, a novel, synthetic, peptide-based erythropoietin receptor agonist. *J Am Soc Nephrol* 2007;18:80A (abstract)
60. The US Recombinant Human Erythropoietin Predialysis Study Group. Double-blind, placebo-controlled study of the therapeutic use of recombinant human erythropoietin for anemia associated with chronic renal failure in predialysis patients. *Am J Kidney Dis* 1991;18:50–59
61. Austrian Multicenter Study Group of r-HuEPO in Predialysis Patients. Effectiveness and safety of recombinant human erythropoietin in predialysis patients. *Nephron* 1992;61:399–403
62. Eschbach JE. The anemia of chronic renal failure: pathophysiology and the effects of recombinant erythropoietin. *Kidney Int* 1989;35:134–148
63. Nissenson AR, Besarab A, Bolton WK et al. Target haematocrit during erythropoietin therapy. *Nephrol Dial Transplant* 1997;12:1813–1816
64. Erslev AJ, Caro J, Schuster SJ. Is there an optimal hemoglobin level? *Transfus Med Rev* 1989;3:237–242
65. Kusunoki M, Kimura K, Nakamura M et al. Effects of hematocrit variations on cerebral blood flow and oxygen transport in ischemic cerebrovascular disease. *J Cereb Blood Flow Metab* 1981;1:413–417
66. Ma JZ, Ebben J, Xia H et al. Hematocrit level and associated mortality in hemodialysis patients. *J Am Soc Nephrol* 1999;10:610–619
67. Locatelli F, Pisoni RL, Akizawa T et al. Anemia management for hemodialysis patients: Kidney Disease Outcomes Quality Initiative (K/DOQI) guidelines and Dialysis Outcomes and Practice Patterns Study (DOPPS) findings. *Am J Kidney Dis* 2004;44:S27–S33
68. Eschbach JW, Glenny R, Robertson T et al. Normalizing the hematocrit (HCT) in hemodialysis patients (HDP) with EPO improves quality of life (Q/L) and is safe. *J Am Soc Nephrol* 1993;4:425 (abstract)
69. Barany P, Svendenhag J, Katzarski K et al. Physiological effects of correcting anemia in hemodialysis patients (HD pts) to a normal HB. *J Am Soc Nephrol* 1996;7:1472 (abstract)
70. Nissenson AR, Pickett JL, Theberge DC et al. Brain function is better in hemodialysis (HD) patients (PTS) when hematocrit is normalized with erythropoietin (rHuEPO). *J Am Soc Nephrol* 1996;7:1459 (abstract)
71. Besarab A, Goodkin DA, Nissenson AR. The normal hematocrit study – follow-up. *N Engl J Med* 2008;358:4
72. Furuland H, Linde T, Ahlmen J et al. A randomized controlled trial of haemoglobin normalization with epoetin alfa in pre-dialysis and dialysis patients. *Nephrol Dial Transplant* 2003;18:353–361
73. Wolfe RA, Hulbert-Shearon TE, Ashby VB et al. Improvements in dialysis patient mortality are associated with improvements in urea reduction ratio and hematocrit, 1999 to 2002. *Am J Kidney Dis* 2005;45:127–135
74. National Kidney Foundation. K/DOQI clinical practice guidelines and clinical practice recommendations for anemia in chronic kidney disease: 2007 update of hemoglobin target. *Am J Kidney Dis* 2007;50:474–530
75. Locatelli F, Alijama P, Barany P et al. European best practice guidelines (EBPG) for the management of renal anaemia in patients with chronic renal failure. *Nephrol Dial Transplant* 2004;19:ii1–ii47
76. Young EW, Goodkin DA, Mapes DL et al. The Dialysis Outcomes and Practice Patterns Study (DOPPS): an international hemodialysis study. *Kidney Int* 2000;57:S74–S81
77. National Kidney Foundation. K/DOQI clinical practice guidelines and clinical practice recommendations for anemia in chronic kidney disease. *Am J Kidney Dis* 2006;47:S1–S145

78. Besarab A, Flaharty KK, Erslev AJ et al. Clinical pharmacology and economics of recombinant human erythropoietin in end-stage renal disease: the case for subcutaneous administration. *J Am Soc Nephrol* 1992;2:1405–1416
79. Alfrey CP, Fishbane S. Implications of neocytolysis for optimal management of anaemia in chronic kidney disease. *Nephron Clin Pract* 2007;106:c149–c156
80. Paganini EP, Eschbach JW, Lazarus JM et al. Intravenous *versus* subcutaneous dosing of epoetin alfa in hemodialysis patients. *Am J Kidney Dis* 1995;26:331–340.
81. Kaufman JS, Reda DJ, Fye CL et al. Subcutaneous compared with intravenous epoetin in patients receiving hemodialysis. *N Engl J Med* 1998;339:578–583
82. Goodkin DA, Gimenez LF, Graber SE et al. Hematocrit stability following intravenous *versus* subcutaneous administration of epoetin alfa to dialysis patients: a *post hoc* analysis. *Clin Nephrol* 1999;51:367–372
83. Gilbertson DT, Ebben JP, Foley RN et al. Hemoglobin level variability: associations with mortality. *Clin J Am Soc Nephrol* 2008;3:133–138
84. Brunelli SM, Joffe MM, Israni RK et al. History-adjusted marginal structural analysis of the association between hemoglobin variability and mortality among chronic hemodialysis patients. *Clin J Am Soc Nephrol* 2008;3:777–782
85. Auerbach MA, Winchester J, Wahab A et al. A randomized trial of three iron dextran infusion methods for anemia in EPO-treated dialysis patients. *Am J Kidney Dis* 1998;31:81–86
86. Rodgers GM, Auerbach M, Cella D et al. High-molecular weight iron dextran: a wolf in sheep's clothing? *J Am Soc Nephrol* 2008;19:833–840
87. Tielemans CL, Lenclud CM, Wens R et al. Critical role of iron overload in the increased susceptibility of haemodialysis patients to bacterial infections: beneficial effects of desferrioxamine. *Nephrol Dial Transplant* 1989;4:883–887
88. Hoen B, Kessler M, Hestin D et al. Risk factors for bacterial infections in chronic haemodialysis adult patients: a multicentre prospective survey. *Nephrol Dial Transplant* 1995;10:377–381
89. Hoen B, Paul-Dauphin A, Hestin D et al. EPIBACDIAL: a multicenter prospective study of risk factors for bacteremia in chronic hemodialysis patients. *J Am Soc Nephrol* 1998;9:869–876
90. Paraskevaidis IA, Iliodromitis EK, Vlahakos D et al. Deferoxamine infusion during coronary artery bypass grafting ameliorates lipid peroxidation and protects the myocardium against reperfusion injury: immediate and long-term significance. *Eur Heart J* 2005;26:263–270
91. Nitenberg A, Paycha F, Ledoux S et al. Coronary artery responses to physiological stimuli are improved by deferoxamine but not by L-arginine in non-insulin-dependent diabetic patients with angiographically normal coronary arteries and no other risk factors. *Circulation* 1998;97:736–743
92. Pieper GM, Siebeneich W. Diabetes-induced endothelial dysfunction is prevented by long-term treatment with the modified iron chelator, hydroxyethyl starch conjugated-deferoxamine. *J Cardiovasc Pharmacol* 1997;30:734–738
93. Kalantar-Zadeh K, Regidor DL, McAllister CJ et al. Time-dependent associations between iron and mortality in hemodialysis patients. *J Am Soc Nephrol* 2005;16:3070–3080
94. Coyne DW, Kapoian T, Suki W et al. Ferric gluconate is highly efficacious in anemic hemodialysis patients with high serum ferritin and low transferrin saturation: results of the Dialysis Patients' Response to IV Iron with Elevated Ferritin (DRIVE) Study. *J Am Soc Nephrol* 2007;18:975–984
95. Fissell RB, Bragg-Gresham JL, Gillespie BW et al. International variation in vitamin prescription and association with mortality in the dialysis outcomes and practice patterns study (DOPPS). *Am J Kidney Dis* 2004;44:293–299
96. Khurana A, Nickel AE, Narayanan M et al. Effect of hepatitis C infection on anemia in hemodialysis patients. *Hemodial Int* 2008;12:94–99
97. Singh AK. Biosimilar epoetins: potential for variation reinforces need for regulation. *Nephrology Times* April 2008:2
98. Levin NW, Fishbane S, Canedo FV et al. Intravenous methoxy polyethylene glycol-epoetin beta for haemoglobin control in patients with chronic kidney disease who are on dialysis: a randomised non-inferiority trial (MAXIMA). *Lancet* 2007;370:1415–1421
99. Macdougall IC, Leong R, Wiecek A et al. Use of Hematide, a novel, synthetic peptide-based erythropoiesis stimulating agent, in the management of hemoglobin (Hb) in non-dialysis and hemodialysis chronic kidney disease (CKD) patients. World Congress of Nephrology, April 21–25. 2007 (abstract)

100. Provenzano R, Fadda G, Bernardo M et al. FG2216, a novel oral HIF-PHI, stimulates erythropoiesis and increases hemoglobin concentration in patients with non-dialysis CKD. NKF Spring Clinical Meetings Abstracts 2008: A80 (abstract)
101. Frohna PA, Milwee S, Pinkett J et al. Preliminary results from a randomized, single-blind, placebo-controlled trial of FG-4592, a novel hypoxia inducible factor prolyl hydroxylase inhibitor, in CKD anemia. *J Am Soc Nephrol* 2007;18:763A (abstract)

Erythropoietins, Erythropoietic Factors and Erythropoiesis
Edited by S.G. Elliott, M.A. Foote, and G. Molineux

Abuse of recombinant erythropoietins and blood products by athletes

Don H. Catlin and Caroline K. Hatton

Anti-Doping Research, Inc., 3873 Grand View Blvd, Los Angeles, CA 90066, USA

History

History of doping

Ever since the beginnings of recorded time, there has been evidence that men experiment with performance enhancement. In ancient times, medicinal plants were used for doping. Now sport is coping with drugs developed with recombinant-DNA technology and genetic manipulation is being discussed.

According to Philostratus, a Greek physiologist in the third century AD, "[athletes should be]... freed from the use of clay and mud and irksome medicine" [1]. He describes Olympians raising their achievement level by eating bread soaked in opium. Centuries ago, the Incas chewed coca leaves to sustain strenuous work and Berserkers ate mushrooms containing muscarine before battle. Some of the earliest pharmaceuticals, amphetamine, strychnine, and ephedrine rapidly became interesting to athletes in search of speed. Canal swimmers are reported to have used strychnine to win races in Amsterdam. The first documented doping fatality at the Olympics shocked the 1960 Games [2] and it involved amphetamine abuse.

"There can be no doubt that stimulants are today widely used by athletes participating in competitions; the record-breaking craze and the desire to satisfy an exacting public play a more and more prominent role, and take higher rank than the health of the competitors itself" [3]. Written in 1939 by a physiologist, this quote is still apropos today and it shows how long ago the 'win at all cost' mentality was already apparent. In the 1970s, Congressional hearings in the United States documented significant abuse of stimulants and anabolic steroids [4], medical organizations took anti-doping stands, and widespread amphetamine abuse was described in professional football.

Characteristics of doping

Not all athletes dope – fortunately most do not – but those who do have a profound influence on sport. Ben Johnson set a world record in the 100-meter dash at the 1988 Olympics in Seoul, but he was doped with stanozolol, an anabolic steroid. Until this Games-stopping event, the hope had been that doping was confined to strength sports. The Johnson affair led to vociferous cries from clean athletes for more testing and stricter penalties, and spurred sport to devote more resources to solving the problem. In retrospect, the resources were woefully inadequate but it was a start.

An axiom of drugs and sport is that some athletes will experiment with any new drug that has the potential to enhance performance. The trials are conducted secretly and of course never published. Sometimes information is shared with close partners, but generally not. Often members of the athletes' entourage are deeply involved in doping.

Doping has occurred in virtually all sports but certain ones have predilections [5, 6]. Endurance events (such as long-distance running, cycling, swimming, and skiing events) experience difficulties with various types of blood doping, including the use of erythropoietic proteins. Short-distance speed events have always had stimulant abuse. Weightlifting, throwing events, and several other sports have been plagued with anabolic androgenic steroid doping. Even shooting events have difficulties as beta-blockers improve shooting scores [7].

Development of testing

By the mid-1960s, international sport officials were growing very concerned about drugs and were actively exploring several approaches to the issue. Until then the main approach was education, but this was unlikely to deter elite athletes who were beginning to receive substantial financial rewards for success. In 1967, Prince de Mérode presented a survey of the problem to the International Olympic Committee (IOC); shortly he became chairman of the IOC Medical Commission and plans were laid to implement urine testing [8]. After experimenting with urine testing at the Winter Olympics of Grenoble in 1968, the IOC implemented widespread testing at the Summer Olympic Games of Munich in 1972. In 1984, there were only five IOC-accredited laboratories. Now there are 34 such laboratories operating in 31 countries. In the early years, the testing was mostly confined to the Olympic Games, then it spread to other competitions, and now athletes are at risk to be tested all year-around. Out-of-competition testing was implemented when it became clear that the ergogenic effects of drugs, such as anabolic steroids and erythropoietic proteins, lasted for many days after discontinuation. Otherwise simply by discontinuing the drugs before announced in-competition tests, one could evade detection. More than 200,000 samples are tested per year [9], and of these, about half are collected out of competition.

The Court of Arbitration for Sport (CAS)

The Court of Arbitration for Sport (CAS) was created by the IOC in 1983 to resolve sport-specific disputes [10]. It provides a forum for judicial resolution of disputes among athletes, national and international sport federations, national Olympic Committees, games organizers, sponsors, and others. Disputes are typically presided over by a panel of three arbitrators selected from a pool of CAS arbitrators. All CAS decisions are final with almost no possibility of appeal.

Doping cases are subject to the strict liability rule which states that "doping is… the presence in the athlete's body of a Prohibited Substance…" [11]. Thus, if a prohibited substance is found in an athlete's body, a doping violation has occurred irrespective of any other factor [10]. At times this rule results in seemingly harsh consequences as was demonstrated by the case of the young Romanian gymnast who placed first in the Women's Individual All-Round Event at the Olympic Games of Sydney. The athlete tested positive for pseudoephedrine, which was allegedly given to her by the team doctor.

The two most common subjects that are argued in CAS cases are the validity of the analytical methods used to establish that a prohibited substance has been detected in the athlete's body fluid, and the chain-of-custody of the sample. The principles of the analytical methods that may be applied, and the chain-of-custody are specified in the Medical Code of the Olympic Movement [11]. During the adjudication procedure, the laboratory director typically defends the analytical results and the sample collection officials defend the custody issues.

The World Anti-Doping Agency (WADA)

Faced with many positive test results, mounting numbers of drugs and methods to consider banning, laboratories to accredit and re-accredit, and complex issues involving new fields of science, the IOC fostered the development of a new agency, the World Anti-Doping Agency (WADA) to take over the drug-control effort. This move coincided with revelations in 1998 that a great many cyclists competing in the Tour de France were doped with recombinant human erythropoietin (rHuEPO). Since the IOC was sometimes criticized for being slow to develop adequate anti-doping programs, the WADA was structured to operate outside the direct control of the IOC.

WADA was established in 1999 and began operations in 2000. In 2004, the IOC transferred the management of the prohibited list to WADA. In addition to updating the list and the regulatory requirements of the International Standard for Laboratories annually, WADA has published a few "Technical Documents" containing recommendations and criteria on specific laboratory testing topics.

Doping by expanding the red cell mass

In the early 2000s, induced erythrocythemia was the term used to describe expansion of the blood volume by transfusion of whole blood, transfusion of packed red cells, or by administration of erythropoietic proteins. The term 'blood doping' has been used to describe any of these three methods, although there are obvious differences in the techniques and the infrastructure required to use them.

Erythropoietic proteins, other than Epogen, the first product, are referred to as follow-on biologics (FOBs) to reflect the fact that manufacturers cannot exactly duplicate the Epogen profile. The new term 'erythropoiesis-stimulating agents (ESAs)' includes Epogen and FOBs and reflects the remarkable advances in the field, and the fact that drugs other than rHuEPO can stimulate erythropoiesis. A recent comprehensive and authoritative review of ESAs and other aspects of doping discusses other compounds such as hypoxia inducible factors, desferrioxamine, cobalt, perfluorocarbons, that are beyond the scope of this review [12]. Other recent reviews cover strategies for detecting rHuEPO [13] and biosimilar rHuEPO [14]. In addition a children's book that is widely read by adults focuses on testing for rHuEPO at the Winter Olympic Games of Salt Lake City [15].

Autologous and homologous blood doping in sport

Both homologous and autologous blood products have been used by sportsmen. The typical technique for autologous transfusion involves phlebotomy, freezing the red blood cells under glycerol, allowing 8–12 weeks for hemoglobin recovery while continuing training, and the red blood cell reinfusion 1–2 days before the event. In other words, blood from one person is stored and reinfused to the same person. The use of homologous blood products avoids the regeneration period but is associated with multiple risks and adverse effects. The terms 'blood packing' and 'blood boosting' usually refer to doping with homologous and autologous blood products.

Many rumors and anecdotes suggest that a foreign airliner equipped with refrigeration devices was used to support doping at the 1976 Games of Montreal. The first documented case of blood doping occurred in the 1980 Olympics in Moscow where a Finnish distance runner freely admitted receiving two units shortly before he won medals in the 5- and 10-km [16]. At the 1984 Games in Los Angeles, seven cyclists from the US team blood doped [17]. One cyclist doped by autologous transfusion and performed his personal-best at the Olympic trials a month before the Games. By the time this news was circulated to team-mates, it was too late to use the autologous reinfusion paradigm, so with the aid of a physician, whole blood from relatives and unrelated donors was transfused in a motel room [18]. Four of the seven won medals, but so did others who did not blood dope. During the investigation, the athletes justified the act by pointing out that the IOC list of prohibited substances did not explicitly ban blood doping. Shortly thereafter, the IOC added to the list doping with blood products. Another documented case is that of a US skier who admitted to blood doping with autologous blood in 1987 [19]. To date, the only case detected by a laboratory blood test is that of a US cyclist at the 2004 Vuelta de Espana cycling road race.

Doping with recombinant human erythropoietin in sport

Given that blood doping with transfusions is a complicated matter that requires substantial infrastructure such as blood harvesting equipment, processing devices, and refrigeration, it is understandable that the availability of a drug, even one that had to be administered intravenously or subcutaneously, would be used. Exactly when doping with rHuEPO began is not known. The secrecy and privacy that surrounds the topic precludes detailed knowledge, but it is understood that athletes and their medical entourage follow pharmaceutical developments closely. Thus it is not surprising that rumors of rHuEPO's potential for abuse by athletes first surfaced before pharmaceutical rHuEPO was introduced (1987), and that rumors of actual use circulated before the 1988 Winter Games of Calgary.

In the late 1980s, there was a remarkable cluster of up to 20 deaths among elite cyclists in the Netherlands and Belgium. The story was widely reported and there was an official investigation, but no details ever emerged. It is speculated that rHuEPO may have been involved [20, 21]. We were unsuccessful in obtaining the details through contacts in the sport community. Because the deaths occurred after the release or rHuEPO, it was widely speculated that they were due to excessive doses of it. In 1996, 'erythropoietin' and analogues were added to the IOC list of prohibited substances.

Although it was widely accepted that rHuEPO was being abused throughout the 1990s, it was not until the 1998 Tour de France that use was documented. The discovery of rHuEPO and other drugs in the trunk of an automobile just before the race led to a widespread investigation that eventually resulted in recovery of many drug products, confessions of several athletes, and evidence of widespread involvement of cycling teams and coaches [22–24]. This expose differed from other sensational drugs in sport stories (such as Ben Johnson's in 1988) in that most of the competitors were involved, not just one athlete. It convinced cycling officials that rHuEPO doping was pervasive and that users and winners were highly correlated. Further, officials in other endurance sports understood that their sports were also at risk. Some of the steps taken collectively by sport were further implementation of blood screening methods (*vide infra*), monitoring athletes' red blood cell parameters, increased funding for research on detection, formation of expert committees, and more emphasis on the role of police authorities. A most unfortunate aspect of such developments is that success is equated with doping, thereby raising suspicions about any athlete who performs well, which has led drug-free athletes to speak out and to pressure authorities for more effective action. Another remarkable benefit of the affair is that it led the IOC to create the WADA.

Prevalence of doping with erythropoietic proteins

It is nearly impossible to estimate the prevalence of blood doping of any kind. Only one survey has attempted to estimate the prevalence of blood doping, and this study commingled transfusion doping with doping with rHuEPO. Scarpino et al. [25] interviewed 1,015 Italian athletes and reported that 7% regularly used red blood cell reinfusion or rHuEPO techniques and 25% were 'occasional' users. In the same study, coaches, managers, and team physicians also estimated the regular users at 7%. Sport hot-lines are another source of information. Throughout the 1990s, the United States Olympic Committee operated an anonymous hot line which athletes could call to request information about drugs and sport. Committee reports indicate that the number of calls regarding rHuEPO increased during the 1990s.

The lay sport literature reports widespread doping with erythropoietic proteins in endurance sports such as track, cycling, and cross country skiing. These reports are largely confined to elite athletes competing in major events. From the frequency of these sensational reports, one gains the impression that such doping is common, but prevalence cannot be estimated from such data. To date, neither college sport administrators nor the media have reported doping with erythropoietic proteins in the college sport population.

Now that a urine test for erythropoietic proteins (*vide infra*) has been deployed, some data have been forthcoming from sports organizations [9]. The WADA publishes laboratory statistics annually, including the total number of tests worldwide, and the number of A samples reported to contain rHuEPO or darbepoetin alfa (Tab. 1) [26].

Table 1. Number of worldwide doping control urine tests and A samples reported to contain recombinant erythropoietic proteins

Year	EPO	Darbepoetin alfa	N tests
2000	*	0	117,314
2001	*	0	125,701
2002	*	3	131,373
2003	51	7	151,210
2004	38	0	169,187
2005	15	1	183,337
2006	17	1	198,143
2007	22	2	225,000

Each time the United States Anti-Doping Agency (USADA) sanctions an athlete, it issues a press release that includes the athlete's name, sport, the date of the test, and the name of the event in the case of in-competition testing. USADA also publishes quarterly statistics. From 2000 to 2002, the media reported several cases of doping with erythropoietic proteins [27]. The media reports occur many months before the adjudication proceedings that could find that doping did not occur. In 2002, two cases of alleged doping with rHuEPO were heard by the CAS. In one case, the Court upheld the laboratory result and determined that the athlete had doped with rHuEPO [28]. In the second case, CAS concluded that rHuEPO doping had not been proven due to a flaw in the analysis [29]. Although no one really knows to what extent erythropoietic proteins are used by athletes, it is increasingly obvious from all sources of data that they are. Although the use currently appears to be confined to the highest echelons of sport, if the history of doping with other agents (notably steroids) is any indication, that use will spread.

The first confession of rHuEPO use was by the female winner of the 2004 Ironman Triathlon [30]. It has been followed with many more confessions by athletes, some of whom tested positive and some of whom never did.

Does expanding the red cell mass enhance performance?

The underlying theory is that increased oxygen-carrying capacity increases performance of muscle tissue by increasing oxygen supply. The ability to perform sustained aerobic exercise depends on both oxygen delivery to muscles and ability of the tissues to use it. Which of the two limits exercise capacity has been the subject of many investigations and debates [18, 31–33]. $VO_{2\,max}$, a widely accepted index of physical fitness [34], correlates with red cell mass [35], thereby, supporting the hypothesis that an expanded red cell mass enhances performance by increasing the amount of oxygen delivered to muscle. The increase in blood viscosity due to increased concentrations of hemoglobin does not limit delivery as long as the hemoglobin is approximately <20 g/L [36]. Other possible theories advanced to explain the increase exercise capacity are that the expanded blood volume leads to increased cardiac output, improved buffering capacity for the lactic acid accumulated during exercise, and enhanced heat dissipation.

A paradox exists in exercise physiology. Athlete fitness is negatively correlated with hematocrit, yet increasing the hematocrit by transfusion or by doping with erythropoietic proteins improves $VO_{2\,max}$ and performance [37, 38]. Exercise training expands plasma volume and decreases hematocrit and hemoglobin concentration, thus resulting in a negative correlation between hematocrit and fitness [38–40]. Values for hematocrit, hemoglobin, and red blood cell count are in the lower range of normal for athletes [37, 41]. Athletes with the lowest hematocrits had the highest aerobic working capacity and isometric adductor strength [38].

At first it was thought that the hemoconcentration that accompanies endurance competitions would lead to increased blood viscosity and that this would severely limit cardiac output and thereby limit the usefulness of rHuEPO. Hematocrit is highly correlated with blood viscosity [38]; however since hematocrits of 50% and more are often recorded among competing athletes, other factors must be operating. In one study of athletes competing in an ultramarathon, the immediate post-race hematocrits were not different from the baseline values, and on the day after the competition, the values were lower than baseline. Apparently replenishing lost fluid and electrolytes during very long events is sufficient to avoid serious increases in viscosity [37].

The idea that dehydration during long endurance events will lead to hemoconcentration and increased blood viscosity, and therefore result in athletes experiencing strokes and other complications of increased blood viscosity, is difficult to confirm. No media reports exist of athletes experiencing such catastrophes during endurance events. The cluster of deaths in cyclists in the late 1980s [42], could be related to increased blood viscosity, but the athletes were not competing at the time.

Autologous blood infusion increases performance

Many clinical experiments have shown that transfusions increase human performance [32, 43–46]. The most convincing study [32] used a double-blind, sham-infusion controlled, cross-over design. Highly trained elite athletes received 900 mL autologous red blood cells collected and frozen approximately 7 weeks earlier; 24 h later they experienced a 35% increase in run-time to exhaustion, a 5% increase in $VO_{2\,max}$, and a 7% increase in hemoglobin concentration. The authors concluded that the limit to aerobic activity was the transport of oxygen to muscle. In another study, this time using 10-km race-time as the outcome rather than physiological measurements, six highly-trained but not elite athletes improved their mean run-times from 33.3 to 32.1 min and their hematocrits increased 5% shortly after receiving a 400-mL autologous infusion of red cells collected 11 weeks earlier [46]. An interesting aspect of autologous infusions is that the magnitude of the improvement is related to the baseline level of fitness. Individuals who are moderately fit experience the greatest improvement while individuals who are poorly fit or extremely fit experience less improvement in maximal oxygen uptake [45]. Improvement in human performance is a consistent feature of blood transfusions provided that the amount of red cells infused is sufficient to increase serum hemoglobin concentration and hematocrit and thereby deliver more oxygen to working muscles. The improvements can be shown in race-times, fixed-run times, run-times to exhaustion, $VO_{2\,max}$, and other measures of performance. Accordingly, sport officials search for ways to deter those athletes who are determined to dope with transfusions or any other means to increase the delivery of oxygen to muscles.

rHuEPO enhances performance in healthy subjects

It is well known that when rHuEPO is administered to patients with anemia, they experience significant improvement in fatigue, physical symptoms, and physical performance [47]. In the present context, the question is whether rHuEPO enhances physical performance in healthy subjects and particularly in athletes. Since the release of rHuEPO almost 15 years ago, there have been few studies on athletes and most of these have involved its effect on red cell indices and other markers.

Two placebo-controlled studies have shown that rHuEPO increased $VO_{2\,max}$ by 6.0% to 7.7% after 3–4 weeks of subcutaneous rHuEPO (150 IU/kg/week) [48, 49]. Typically the weekly dose is administered subcutaneously in doses of 150 IU/kg/week. At a higher dose (180–210 IU/kg/week, subcutaneously), the $VO_{2\,max}$ increase was similar (7%) [50]. These studies, which used recreational athletes, establish that doses of rHuEPO that are sufficient to increase the hematocrit to nearly 50% also increase the $VO_{2\,max}$ up to 7%. Anecdotal reports, however, indicate that after a few weeks of modest doses (150 IU/kg/week), athletes lower the weekly dose to about 60 IU/kg/week. To determine the effect of such a regimen, after 3 weeks at 150 IU/kg/week, a study [49] decreased the dose to 60 IU/kg/week for an additional 5 weeks and found that the improvement in $VO_{2\,max}$ (4.7–9.7%) and increased hematocrits continued. This study and a similar low dose study [51] provide support for the anecdotal reports from athletes, and confirm that maintenance doses of 60 IU/kg/week are capable of enhancing aerobic performance for at least 3 weeks.

Adverse effects of rHuEPO in athletes

Unlike studies in patients with anemia, there are no surveys of the adverse effects of rHuEPO in athletes. Athletes tend to be healthy and that may offer some protection against the known side effects in anemic patients; however, athletes are deliberately increasing their hematocrits to levels associated with thromboembolic and other complications. Only one case report appears to link rHuEPO to a cerebral thrombosis in an athlete who admitted to using rHuEPO but was also taking other drugs [52]. Athletes who use rHuEPO are at risk to develop true iron deficiency, functional iron deficiency, or iron overload. Of course careful professional medical management could avoid these complications; however, Cazzola reports that an investigation by Italian Magistrates [53] reveals that some professional cyclists have evidence of iron overload with ferritin values in excess of 1,000 ng/mL [54]. The investigation also provided data consistent with a risk of post-treatment blunted production of endogenous EPO [54]. In addition, Berglund and Ekblom [51] have studied the effect of rHuEPO on the blood pressure of athletes. The systolic and diastolic blood pressure values at rest were unchanged after rHuEPO treatment; however, there was a marked increase in systolic blood pressure during submaximal exercise. The initial and final values were 177 mmHg and 191 mmHg. The propensity of athletes to titrate their hematocrits to high levels and to take rHuEPO without adequate medical supervision, together with their risk for iron disorders and exercise induced elevated systolic pressure, make it likely that more adverse effects are occurring than are reported in the medical literature. This under-reporting is inherent to the secretive nature of doping.

Risks and benefits

Shermer's risks and benefits analysis is that the choice to dope, especially with rHuEPO, makes sense because it is highly effective at enhancing performance, the risk of getting caught is small, the financial rewards are astronomical, and clean athletes may not be competitive [55]. Fortunately, he is able to propose ways to tip the balance of risks and benefits to favor the choice of staying drug-free.

Detecting erythropoietic proteins in body fluids

Direct and indirect tests

Sport classifies tests for doping substances as direct or indirect. A direct test identifies the substance by an unambiguous method such as gas chromatography-mass spectrometry (GC-MS), whereas indirect tests measure, for example, the serum concentration of markers that correlate with the use of a prohibited substance, without directly identifying the substance. Direct tests sufficed for many years, but now that doping has turned to endogenous steroids and glycoproteins, a variety of new strategies and indirect tests have developed [5, 13]. Indirect tests have not been used to declare that an athlete has used a substance, but it is hoped that with sufficient validation, they could become definitive.

An example of indirect test for a different class of doping agents is the detection of the use of pharmaceutical testosterone. It was a challenge with GC-MS because pharmaceutical and endogenous testosterone could not be distinguished. The problem is partly solved by performing longitudinal tests of the urinary steroid profile and by determining the $^{13}C/^{12}C$ ratio of urinary testosterone by isotope ratio mass spectrometry [56]. The glycoproteins present a special challenge because so far no MS-based method has been developed that is sensitive enough to detect either rHuEPO or recombinant human growth hormone (rHuGH) in human urine. In another example of indirect testing, Sonksen and colleagues set out to find a group of blood parameters that would indicate recent use of rHuGH. After collecting baseline blood samples, they administered rHuGH for several days and monitored serum levels of six substances known to be influenced by rHuGH [57]. A composite score indicative of recent use of rHuGH was developed. The results showed good separation between the scores of the placebo-treated and rHuGH-treated subjects. The investigators intend to expand the studies to a much larger number of subjects and to determine the variability of the markers in various ethnic groups. The hope is to find a composite score that is so convincing that the indirect test is considered definitive. Until 2000, there was no practical direct test for rHuEPO, therefore certain sports implemented indirect tests to identify potential users of rHuEPO.

Indirect tests for doping with erythropoietic proteins

The hematocrit 'health' test

The International Cycling Union (UCI), concerned that rHuEPO had pervaded elite competition, declared that male and female athletes could not compete with a hematocrit >50% and 47%, respectively [58, 59]. The International Ski Federation (FIS) implemented a similar rule based on hemoglobin values >185 g/L (men) or 165 g/L (women). One strategy was to determine the values immediately before an event and withhold the athletes from competition if the limits were exceeded. The hematocrit cutoff values have been changing but generally they are 50% for men and 47% for women. The hematocrit test has been dubbed a 'health test' because it is considered dangerous for an athlete to compete if the hematocrit is greater than the cutoff. The term 'health test' preempts legal action because the athlete is not declared a drug user, he/she is only kept from competing, which is lifted after 5–15 days provided that the hematocrit has decreased below the limit. For example, at the 2006 Winter Olympics in Torino, several athletes were kept from competing for 5 days [60].

Aside from the health issue, the argument in favor of a hematocrit test is that by imposing an upper limit, the test prevents excessive use of rHuEPO. An argument against the hematocrit test is that it would discriminate against individuals who have naturally high hematocrits [61, 62]. Partially countering this argument is the finding that of 334 hematocrits determined on 34 professional cyclists before rHuEPO was available, the values ranged from 39–48% (mean, 43%) [63]. In addition, a protocol was developed to determine if an athlete's hematocrit naturally exceeds 50%. Postural changes in hematocrit are avoided by taking samples after sitting for 15 min. Increased hematocrits due to dehydration have not been a significant problem [64]. Despite the controversial nature of the test, there is evidence that it is having the intended effect: hemoglobin values among elite cross-country skiers increased dramatically from 1994 to 1996, and declined after the test was implemented [65].

A disadvantage of the hematocrit test is that it draws attention to the benefits of an increased hematocrit, thus tempting athletes with a natural baseline <50% to find a way to increase it. Indeed, it is common lore that some athletes check their hematocrits using portable centrifuges and self-administer saline infusions and phlebotomy if the values are too high. They are also reported to take anticoagulants to prevent thromboembolic events.

Hypochromic macrocytes
Casoni et al. [66] studied red cell indices after administering rHuEPO every other day at an average dose of 15 IU/kg/day for up to 45 days. The parameters that changed the most were mean corpuscular volume (MCV) and mean corpuscular hemoglobin (MCH). The authors defined red cells with MCV >128 fl and MCH <28 pg as hypochromic macrocytes and proposed a cutoff of 0.6% to distinguish rHuEPO users from healthy controls subjects. No further work on this particular index has been published.

Serum transferrin receptor (sTfr) and ferritin
Serum concentrations of soluble transferring receptor (sTfR) and ferritin are regulated by cellular iron status, and cellular iron uptake is facilitated by transferrin receptor-mediated endocytosis. As a result of externalization of TfR during the endocytic cycle, a soluble form of TfR can be detected in serum. Thus, the major determinants of sTfR concentration are cellular iron demands and red cell proliferation rate. Since rHuEPO expands the red cell mass, it was logical to determine if sTfR and ferritin could serve as indirect markers of rHuEPO administration.

Several studies have shown that levels of sTfR increase, those of ferritin decrease, and the ratio of sTfR/ferritin increases when rHuEPO is administered [50, 67, 68]. Furthermore these changes are detectable in most subjects for up to 1 week after discontinuing rHuEPO [50]. The change in sTfR/ferritin ratio was less dramatic in subjects treated with supplemental iron [50] than in subjects who did not receive iron [67], thus the absence of supplemental iron may exaggerate the sTfR/ferritin ratio. Most users of rHuEPO are likely to take supplemental iron. The specificity of the changes in sTfR, ferritin, and sTfR/ferritin has not been evaluated in a large group of healthy normal control subjects of various ethnicities, subjects with disease, or under various conditions of iron supplementation, altitude, or training. Nevertheless these markers are relatively simple to measure and they may be useful as indirect markers of the use of erythropoietic proteins, either alone or in combination with other markers.

Multiple markers of erythropoietic activity – the Australian studies
In the year before the Sydney Olympic Games of 2000, Australian scientists performed an extensive series of studies designed to find a combination of blood markers that would indicate which athletes were using rHuEPO [48, 69]. The approach was to administer rHuEPO for 3 weeks, collect blood for analyses before, during, and for 2 weeks after administration was discontinued, and measure a variety of potential markers in serum and blood. The data were analyzed by various statistical models that provided the optimum combination of variables and the weight to attach to each. The main study compared placebo

with rHuEPO-treated subjects who received 50 IU/kg three times per week for 25 days [67]. The treatment group comprised 49 subjects from Australia and 24 subjects from Beijing, China.

The study found that five markers provided the most discrimination between the placebo and rHuEPO-treated groups. These were hematocrit, reticulocyte hematocrit (MCV of reticulocytes * number of reticulocytes), percent macrocytes, serum concentration of EPO, and serum concentration of sTfr. The total score was referred to as the 'on-score' [69]. By the end of the 3-week rHuEPO administration period the 'on-scores' of the rHuEPO-treated and placebo groups differed significantly. Serum EPO and sTfR were particularly increased compared with the control group. No differences were seen between the Australian and the Chinese subjects or between men and women.

The Australian investigators calculated the 'on-score' of about 1,200 elite athletes from 12 countries [69]. Based on these data, they calculated cutoffs for the 'on-score' values beyond which the risk of a false-positive was very low. At a meeting of experts 3 months before the Sydney Games [70], the 'on-score' was not approved as a standalone index of use of erythropoietic proteins largely due to legal and medical concerns. However, it was approved as a technique to indicate which urines should be tested by the recently developed, definitive urine test (*vide infra*). The rules stated that an athlete could not be declared 'positive' for rHuEPO unless the 'on-score' was greater than the cutoff and the urine test showed rHuEPO. Samples could be declared 'suspicious' if the 'off-score' (*vide infra*) was increased or if only one of the blood or urine tests was positive. No athletes were declared positive during the Games of Sydney but seven were reported as 'suspicious' [71].

Index of recent use of erythropoietic proteins

The Australian researchers also found that after rHuEPO was discontinued, the reticulocyte count and serum EPO were depressed while the hematocrit remained increased [69]. From these three parameters, they calculated an 'off-score' and proposed that an elevated 'off-score' be used to identify athletes who had recently discontinued rHuEPO. The 'off-score' was increased from the third to the twelfth post-administration day. Although the statistical certainty that the 'off-score' indicated recent use was quite high, the IOC expert committee did not accept the test largely because of the legal difficulty of proving that an athlete used rHuEPO on the basis of an indirect test. The 'off-score' is used today to indicate which athletes should be followed closely with longitudinal testing.

Urine EPO concentrations by immunoassay

Attempts to detect abuse of erythropoietic proteins by urine immunoassay have not been successful. Since rHuEPO and natural urinary EPO cannot be distinguished by the available antibodies [72], a diagnostic test would depend on finding urinary EPO concentrations that are far above the normal range. In addition, less than 5% of the dose is excreted in urine [73] and urinary EPO concentrations are affected by pH, specific gravity, and exercise [74, 75]. Small increases in urinary EPO were detected by a immunoradiometric assay after large doses (200 IU/kg every other day for 10 days) [72].

Monitoring an individual's hematologic variables

Until very recently, all doping control samples were collected using the cross-sectional model, which means that the analytical results are interpreted on their face value, not compared to previous samples collected from the same person; however, if the latter is done, a linked or trend analysis can be performed. This test is a much more powerful type of analysis because results will vary from day to day between persons, however within persons, the results are generally more stable. For example, if a person has a hematocrit of 39%, it will vary by only a few percent from day to day, but if that person's hematocrit increases to 46%, even though that is well within the normal range for the whole population, for that person it is an outlier and it needs to be investigated. The WADA's Athlete's Passport, UCI Biological Passport, and USADA Project Believe, all include trend analysis.

WADA Athlete's Passport

The WADA Athlete's Passport [76] is based on monitoring an individual's selected parameters, which reflect the effects of doping indirectly, as opposed to the traditional direct detection of doping. Biologic tracing throughout an athlete's sports career should make any doping agent far more difficult to use. From a medical point of view, biologic monitoring is also likely to reveal disease, whether or not it is due to doping, making early and effective treatment possible. The Athlete's Passport will eventually improve the effectiveness of the fight against doping and protect the health of athletes. In addition, it will support the intelligent targeting of athletes for doping control. Although the concept is simple, in that different models have already been validated for widespread use in other scientific fields, such as epidemiology, legal medicine, or even veterinary monitoring, several key considerations must be taken into account before its widespread use in the field of anti-doping. While examining legal and disciplinary issues, as well as feasibility, WADA started a pilot project with the UCI in 2008; however, because of a UCI lawsuit against WADA, WADA withdrew its support of the UCI in relation to WADA's pilot project of the Athlete's Passport in the sport of cycling and will continue the Athlete's Passport pilot project with another sport. WADA's view is that the Athlete's Passport will augment, not replace traditional anti-doping testing.

UCI Biological Passport

The UCI Biological Passport is designed to maintain an individual's electronic record including urine test results and the steroid profile, blood test results and the corresponding profile; establish individual limits; detect doping indirectly; assess significant variations; and potentially use those as a basis for opening disciplinary proceedings.

USADA's Project Believe

In April 2008, USADA created a voluntary, longitudinal pilot testing program and began testing American Olympic hopefuls [77]. The volunteers agreed to provide additional blood and urine samples beyond those required by the regular in- and out-of-competition testing program. The analytical results were to be used to establish individual baselines.

Private trend analysis programs

These programs are generally operated by teams within sports. They are designed to supplement the WADA- or National Anti-Doping Organization (NADO) directed testing. They are particularly useful for sports that have major doping issues. They are private in that the data are controlled by the program rather than WADA or the NADO, meaning that the program must have careful oversight so that any aberrant results or positive results are not ignored.

Results of monitoring individual data over time

The most comprehensive study monitored hemoglobin concentration and percent reticulocytes in samples obtained from 440 female and 634 male elite cross-country skiers for 6 years [78]. For the men, the hemoglobin concentrations decreased dramatically (16.2 to 14.8 g/dL) from 1997/1999 to 2002/2003. In the same time period, the percent reticulocytes also declined significantly. The results for the women were similar. These desirable and promising results were attributed to an enhanced blood testing program and to imposing cutoff values. Unfortunately hemoglobin values increased steadily from 2002/2003 to 2005/2006 while the reticulocyte count continued to decrease. No apparent reason has been found why the hemoglobin values have been increasing. The speculation includes changes in drugs or administration details of the doping agents used by these athletes.

Direct tests for doping with erythropoietic proteins

The isoelectric focusing (IEF) test for urinary rHuEPO

A method for detecting rHuEPO (epoetin alfa or epoetin beta) in urine by electrophoresis was first described in 1995 [79]. Although this test had practical limitations, it demonstrated conclusively that the isoform pattern of urinary endogenous EPO differs from the pattern of urinary rHuEPO, and it was the first successful attempt to develop a direct test for urinary rHuEPO. At that time the only rHuEPO was epoetin alfa, although we now know that the test works equally well on epoetin beta. Similarly in 2002 Skibeli et al. [80] isolated EPO from human serum and showed, using gel electrophoresis, that endogenous and recombinant EPO differed.

A significant improvement in practical detection occurred in 2000 when Lasne and de Ceaurriz [81] described a method for detecting rHuEPO in urine based on isoelectric focusing (IEF) with immunoblotting. The method included a novel and critical step: a second blot ('double-blot'). After the isoforms of rHuEPO are separated by IEF, the first blot is performed, then the membrane containing the transferred proteins is incubated with anti-EPO antibody [81, 82]. The second blot transfers only the anti-EPO antibodies to a second membrane, and the second membrane is incubated with a second antibody directed against the first antibody. This step markedly reduces nonspecific binding and yields clear isoform patterns [81, 82]. After the second antibody is incubated with streptavidin-horseradish peroxidase and substrate, the emitted chemiluminescence is captured with a charge-coupled device camera to produce an image of the gel. Figure 1 is an electropherogram showing the band pattern of darbepoetin alfa in a urine sample obtained from a patient who was treated with darbepoetin alfa.

Despite the fact that the IEF test is technically demanding and may take up to 3 days to complete, at this time there is a great deal of interest in it because it is a direct test that can yield conclusive results [81]. Further, the results of the test and the underlying IEF method have been accepted by the CAS [28].

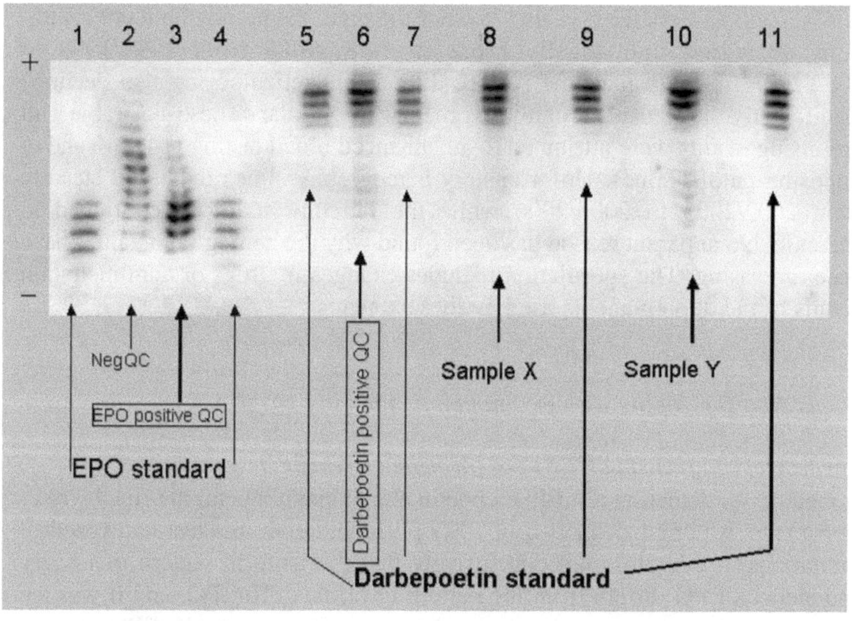

Figure 1. Electropherogram of rHuEPO and darbepoetin alfa standards; of extracts of urine obtained from drug-free control subjects and subjects treated with rHuEPO or darbepoetin alfa; and of extracts of urine from unknown individuals. The anode and cathode sides of the electropherogram are labeled + and –, respectively. The rHuEPO standard is in lane 1 and 4 and the darbepoetin standard is in lanes 5, 7, 9, and 11. The Neg QC (lane 2) is an extract of urine from drug-free subjects showing the normal pattern of endogenous hEPO. The EPO positive QC (lane 3) and darbepoetin positive QC (lane 6) are urine extracts obtained after the administration of rHuEPO or darbepoetin alfa, respectively. Samples X and Y are urine extracts from two unknown individuals (lanes 8 and 10). The bands of endogenous urinary EPO are faint or absent in lanes 6, 8, and 10.

Beginning in 2005, the WADA Technical Document TD2004EPO harmonized inter-laboratory criteria for the identification of epoetin alfa and epoetin beta, and darbepoetin alfa by IEF double-blotting and chemiluminescent detection [83]. The document was revised in 2007 [84].

Lasne and co-workers explored the interpretation of IEF data by a rational statistical approach, namely discriminant analysis [85]. Using a dataset of 116 rHuEPO-positive profiles from 26 subjects and 131 negative profiles from 131 subjects, they numbered the bands numbered according to a template of 16 possible positions. The band relative intensities were the 16 variables for the statistical analysis. The method was tested with data from an administration trial of low doses (6.7–10 IU/kg) after high-dose (265 IU/kg) injections; 71 profiles from one subject. With 10-fold validation, they had no false positives with 100,000 simulations. They concluded that they achieved improved detection of low doses.

Investigators are working on means to improve the IEF test, to expand its use to all WADA-accredited laboratories, and to further define criteria for determining if rHuEPO is present. In addition, investigators are working on ways to detect erythropoietic proteins with mass spectrometry.

Darbepoetin alfa detected at Winter Olympic Games of Salt Lake, 2002

As the IOC-accredited laboratory designated to perform the testing at the Games of Salt Lake, our laboratory began implementing the IEF technique for detecting erythropoietic proteins about 8 months before the Games. Since darbepoetin alfa was likely to be approved by the FDA before the Games, we were aware that athletes might have access to it at the time of the Games. Accordingly, we obtained a urine from a patient being treated with darbepoetin alfa. As shown in Figure 1, the electropherogram of this urine confirmed that the isoforms of darbepoetin alfa were more acidic than those of rHuEPO [86] because the former migrated to the more acidic region of the IEF gel. This sample was important because IOC laboratory regulations require that analyses include a reference standard, a known positive quality control urine, and a known negative quality control urine.

During the Games of Salt Lake, our laboratory analyzed 598 urine samples for the drugs on the IOC List of Prohibited Substances. In addition, 1,222 blood samples were obtained from athletes competing in the endurance sports of cross country skiing, biathlon, Nordic combined, long-track speed skating, and short-track speed skating. These blood samples were tested at the venues by international sport federations before competition for hemoglobin and reticulocytes.

If the reticulocytes exceeded 2.0% or if the hemoglobin concentration exceeded 16.5 gm% for women or 17.5 gm% for men, the athlete returned after competition and an additional blood and urine were obtained for EPO analysis at the laboratory. Of the 1,222 blood samples, 133 (10.6%) had reticulocytes >2%, and eight samples (0.6%) had increased hemoglobin concentrations [87]. In total, the laboratory received 77 combined blood and urine samples and of these, the samples from three athletes met our criteria for darbepoetin alfa. All three athletes were cross country skiers. Together they had won eight medals.

According to the procedure in effect at these Olympics, before announcing test results to the public, a committee reviews the laboratory findings and two hearings are held. The hearings, which are attended by the representatives of the athlete with or without the athlete, delve into the details of the testing. If the testing results are upheld, the IOC Executive Board reviews the case and makes the final decision to announce the findings to the public. In addition, the athlete has the right to appeal the IOC decision to the CAS. As there is little to lose, most athletes do appeal to the CAS. The darbepoetin alfa cases were adjudicated by the CAS in 2003 and 2004. As a result, the three skiers returned all eight medals, and the athletes who crossed the finish line behind them were moved up in the rankings.

Epoetin alfa and darbepoetin alfa detection studies

Athletes do not reveal their doses or dosing regimen, thus the only experimental approach is to administer likely doses and analyze the urine at various times after the last dose. The early work of Wide et al. [79] established that their gel electrophoresis assay detected urinary rHuEPO for up to 24 h after administering rHuEPO at 60 IU/kg/week for 7–9 weeks. We have used the IEF method [81] to determine the approximate time-course of detection of epoetin alfa in urine. We gave 150 IU/kg/week to 15 healthy young subjects for 3 weeks. If the hematocrit exceeded 48%, the next dose was withheld. After discontinuing the epoetin alfa, all urines were positive by IEF for 2 days and about 50% were positive for 4 days.

We have also determined that darbepoetin alfa is detectable in urine for up to 12 days. A modest single dose (0.40 μg/kg) was detected in eight of nine subjects for 2–4 days post-administration and in some subjects for up to 12 days. Thus, the detectability of darbepoetin alfa appears to be similar or longer than that of rHuEPO, which is consistent with darbepoetin alfa's longer serum half-life [88].

One limitation of detection-time investigations is that they do not necessarily mimic the dosage regimens used by athletes. Typically athletes become aware of the retrospectivity of a test and adjust their doping schedules so as to evade detection. In the case of a rapidly evolving new test, such as the IEF test for urinary rHuEPO, athletes are undoubtedly searching for test-beating options as diligently as laboratory scientists are working on improvements. Perhaps the athletes who got caught using darbepoetin alfa at the 2002 Games of Salt Lake did not expect the test to detect it.

The effects of microdose recombinant human erythropoietin regimens in athletes

Ashenden and co-workers assessed the plausibility of anecdotal reports that microdosing with rHuEPO can allow athletes to escape detection by doping control tests [89]. Two men first received high initial doses of rHuEPO (six injections of approximately 260 IU/kg injections between days 0 and 11) with a single treatment of iron (100 mg), aiming for 170 g/L hemoglobin. For the next 3 weeks, microdoses (<10% of initial dose) were injected every 2–3 days and adjusted based only on hematologic parameters, e.g., reticulocyte counts, hemoglobin values remained increased. Most urine samples collected >24 h after a microdose tested negative, as well as some samples collected 12–18 h after a microdose. The urine test showed the reappearance of endogenous EPO during microdosing. The authors recommend that authorities supplement the urine test with an approach with greater retrospectivity and confirm that tests must be done out-of-competition to be effective.

Lamon and co-workers estimated the detection window of darbepoetin alfa after one subcutaneous injection (4,769 IU or 40 μg or 70 IU/kg) in three men as close to 7 days, approximately twice as long as for rHuEPO (epoetin beta) (4,000 IU or approximately 70 IU/kg) [90]. The detection window was shortened in subjects with a naturally increased or stimulated EPO production rate. The authors recommend revising WADA criteria for darbepoetin alfa from IEF band intensity ratio to the position and specific distribution of the bands in the most acidic area of the gel, a less conservative set of criteria.

Retrospectivity of the urinary IEF test

The ability of the urinary IEF test to control the abuse of erythropoietic proteins among athletes is a complex function of dose, dosing regimen, detection times, urine collection time relative to last dose, and whether the test is announced in advance or a surprise (short-notice or out-of-competition testing). Sport administrators control the testing time and it is generally agreed that unannounced short-notice testing is the most effective deterrent. The athlete controls the dose and dosing interval, and the ability of the test to detect erythropoietic proteins is related to the urinary pharmacokinetics of the protein and the inherent sensitivity of the test. A ‘good’ test is one that can detect the erythropoietic proteins for as long as the beneficial effects of the drug are present. The interplay of these factors is only beginning to be explored.

The pharmacodynamic effects of erythropoietic proteins that are particularly pertinent in the context of detecting users and that have been studied in healthy subjects, as opposed to patients with anemia, are the effects on hematocrit, hemoglobin, and $VO_{2\,max}$. The effect on $VO_{2\,max}$ is the most relevant because it is a direct measure of performance. rHuEPO administered at doses of 150–230 IU/kg/week for approximately 3 weeks produces an increase of $VO_{2\,max}$ of 6–8%, as expected [48–50]. Only two studies have monitored $VO_{2\,max}$ after rHuEPO was discontinued. In the first, rHuEPO was administered for 25 days (150 IU/kg/week) and 4 weeks after the last dose the $VO_{2\,max}$ determined was not distinguishable from the $VO_{2\,max}$ of the placebo group [91]. It has been suggested that athletes may use a low-dose maintenance regimen that is sufficient to maintain the hematocrit just beneath the 50% ‘health’ test threshold [49, 81]. To explore the effects of a low-dose regimen on $VO_{2\,max}$, Russell et al. [49] administered rHuEPO for 3 weeks at a dose of

150 IU/kg/week, then lowered the dose to 60 IU/kg/week for an additional 5 weeks. They determined that the $VO_{2\,max}$ in week 12, 4 weeks after the last dose, was still increased (3.1% to 4.5%). The hematocrit remained increased for about 17 days after the last dose. Although these results need to be confirmed, the data on low doses of rHuEPO suggest that for a urine test for rHuEPO to be effective it should be capable of detecting rHuEPO for 3–4 weeks after discontinuing a dose of 60 IU/kg/week. Such data are not available.

Another approach to the retrospectivity issue is to assume that the effect on $VO_{2\,max}$ lasts as long as the hematocrit or hemoglobin remain increased. While this assumption has not been directly investigated, it is reasonable to assume that the duration of action on $VO_{2\,max}$ will correlate with hematocrit and hemoglobin. In the three studies [49, 50, 91] that measured hematocrits after rHuEPO was discontinued, a fixed dose of rHuEPO (80–230 IU/kg/week) was administered for 3–5 weeks. During the rHuEPO administration phase, the hematocrit steadily increased until a plateau of approximately 50% was reached at 12–14 days. The hematocrit-time graphs of these studies show that the hematocrit is unchanged (approximately 50%) for 12–20 days after rHuEPO is discontinued. The hematocrit was last measured on days 24, 28, and 30 post administration day; for each study, the hematocrit was still greater than that of the placebo group. Thus, performance as measured by $VO_{2\,max}$ and the hematocrit appear to follow more or less parallel time courses, again consistent with 3–4 weeks of enhancement after the last dose. An athlete could be enhanced and yet have a negative urinary IEF test.

Between lab comparisons of IEF results

Lundby et al. [92] administered rHuEPO to eight subjects, split each urine into two aliquots, and sent them to two WADA-accredited laboratories for EPO analysis by the WADA-approved IEF method. The subjects received rHuEPO every other day for 2 weeks (boosting period), followed by 1 weekly treatment over the next 2 weeks (maintenance period). Both laboratories reported negative results as expected for the eight baseline urine samples, however Laboratory B reported no positive results in the 52 samples collected during boosting and maintenance period and for the 2 weeks after. The results from Laboratory A were positive in all eight samples collected during the 2 weeks of boosting period, in six of the 16 samples collected during the maintenance period, and in two of the 20 samples collected after discontinuation of rHuEPO. Although none of the samples analyzed in Laboratory B fulfilled WADA criteria for reporting positive, five of 18 samples collected during the maintenance period were classified as 'suspicious'. Unfortunately, no electropherograms were presented so the reader could not evaluate the quality of the analyses, the characteristics of the 'suspicious' samples, or compare the data from the two laboratories. This study is the first between-laboratory study that has been published thus there is no way to know how common these discrepant findings are. At the time that this chapter was written, considerable controversy had arisen over the veracity of the findings.

Other erythropoietins (follow-on biologics)

Until relatively recently, there was only one rHuEPO and only one manufacturer. Now the numbers of both rHuEPOs and manufacturers are proliferating. The term 'follow-on biologic' or biosimilars indicates that they are not identical to each other due to factors such as the cells of origin, and specific processes of production, purification, and formulation [93]. Athletes will experiment with these agents as soon as they become available. If laboratories are going to keep up with this fast-moving field, they need detailed information on the new products, a reference standard of the product, and at a minimum, a urine sample collected from a subject known to be receiving the substance. Ideally, the laboratories would administer the substance to subjects to obtain data on the isoform patterns and the pharmacokinetics of detection. These requirements are not easy to fulfill, particularly in the United States where ethics committees generally require that the drug be available as an FDA-approved product. If not, the investigator must file an application for an Investigational New Drug (IND), a lengthy, complicated, and expensive process. Alternatively the investigator may collaborate with colleagues who are working under an FDA-approved protocol.

rHuEPO (Epogen) has been and will continue to be one of the most important drugs that was ever developed. It has remarkable efficacy in the clinic and has improved the lives of a great many people with anemia and renal disease. Now that the original patents have expired a host of follow-on biologics with erythrocyte-stimulating properties are here or coming [93]. In addition, fusion proteins, second- and third-generation EPOs, EPO dimmers, and many more are in various stages of development. All show some degree of erythropoiesis stimulation, and therefore can be justified as legitimate pharmaceuticals being developed to help patients. The anti-doping movement has no way to know if someone is purposely developing these substances to aid and abet cheaters. We hope their use for legitimate medical purposes vastly overwhelms the uses for doping.

Epoetin alfa, epoetin beta, and darbepoetin alfa

The current IEF test [81] is able to detect administration of epoetin alfa, epoetin beta, and darbepoetin alfa, which enables sport to control these erythropoietic proteins provided that the urine is collected when sufficient amounts of the material are present. Much work is needed on the time-course of detection after various doses have been administered. A few such studies are underway.

Epoetin omega

This epoetin has been administered to hemodialysis subjects with anemia [94] and some information is available on its structure [95]. It is produced in baby hamster kidney cells and is less acidic than epoetin alfa or epoetin beta. In the current IEF assay, a reference standard of epoetin omega migrates to the most basic area of the electropherogram. Further, we recently encountered a urine sample with isoforms consistent with epoetin omega. The isoelectric profile of epoetin omega shows that, compared with epoetin alfa, it has several additional bands in the basic region [96].

Epoetin delta

Epoetin delta (Dynepo) was released in 2007 in Europe for the treatment of the anemia of chronic renal disease and for end-stage renal disease with dialysis. It is the first rHuEPO on the market that was produced in cultured human cells. Its electropherogram profile is readily distinguished from that of epoetin omega and the combination of epoetin alfa and epoetin beta [96].

PEGylated epoetin beta (CERA)

Continuous erythropoietin receptor activator (CERA) is a third-generation erythropoiesis stimulating agent [97, 98]. It is used to correct anemia and maintain hemoglobin concentrations in patients with renal failure. CERA is epoetin beta that has been PEGylated. Its detectability in urine has not yet been demonstrated and there is some concern in the anti-doping context that its high molecular weight may prevent it from being excreted into urine.

Hematide

Hematide is a PEGylated synthetic dimeric peptide that binds to and activates the EPO receptor (EPOR) and causes proliferation and differentiation of erythroid progenitor cells [99, 99–101]. Its chemical structure is unrelated to any of the erythropoietins. In a Phase I study, Hematide produced a dose-dependent increase in hemoglobin concentration and reticulocyte numbers. Currently, it is in a Phase III clinical trial [101]. No means to detect Hematide in humans has been described yet.

Future of detecting rHuEPO and follow-on biologics in body fluids

After less than 7 years of only having to contend with rHuPO and darbepoetin alfa, the WADA criteria for reporting the finding of a recombinant form of EPO have to be modified to cover the follow-on biologics. The delta and omega epoetins and many more that we have studied in the laboratory have more basic bands than epoetin alfa and epoetin beta. Although no published studies show electropherograms after administration to humans, we can be quite sure that the extra bands will appear in clinical samples as extra bands in the basic region. These electropherograms will not fit the current WADA criteria for reporting and they will have to be distinguished from exercise patterns. Already laboratories are observing such patterns. Currently, they must be classified as 'negative' even though there is compelling evidence that some are true positives. This frustrating situation may be overcome by revising the WADA TD on EPO criteria, however it is more likely that additional criteria such as longitudinal blood indices will be added to the definition of a positive.

Characterization by SDS-Page mobility

Kohler and co-workers took a new approach based on sodium dodecyl sulfate-polyacrilamide gel electrophoresis (SDS-PAGE) gels, using two internal standards (darbepoetin alfa and recombinant rat EPO) to calculate the relative mobility of endogenous EPO and recombinant analogs [102]. The analysis of urine samples from a reference group of 53 volunteers (30 men, 23 women) and from one volunteer who received epoetin delta (man, single subcutaneous injection, 50 IU/kg, 4,000 IU) led to significant discrimination between endogenous and rHuEPO

Characterization by mass spectrometry

Experiments conducted by Stubiger et al. [103] with reference standards of epoetin alfa, epoetin beta, and darbepoetin alfa show that they can be characterized and differentiated by mass spectrometry, based on exact molecular weight determination on intact glycoproteins by matrix-assisted laser desorption/ionisation (MALDI) and time-of-flight (TOF) mass spectrometry. Application to the testing of human samples would begin with enrichment from body fluids. MALDI-TOF analysis of enzymatically desialylated molecules improved mass spectral resolution by decreasing microheterogeneity. De-*N*-glycosylation also improved resolution, but not as much.

LC-MS-MS test for rHuEPO in horses

Because animal athletes are used by humans for sport, entertainment, or gain, it is no surprise that the abuse of recombinant EPO has surfaced in both equine and canine races [104–106]. Administration of rHuEPO to horses causes the production of antibodies that may cross-react with endogenous EPO and cause anemia and even death. Antibody detection is the simplest approach to a test. For confirmation purposes, Guan and co-workers have proposed a liquid chromatography coupled to tandem mass spectrometry (LC-MS-MS) to identify rHuEPO and darbepoetin alfa in equine plasma [107]. After immunoaffinity enrichment with anti-rHuEPO antibodies linked to magnetic beads and trypsin digestion, LC-MS-MS analysis allows identification of proteotypic peptides. The limit of detection was 0.1 ng/mL. One research horse received rHuEPO on two occasions and it could be detected for 24–48 h afterwards. Another research horse received darbepoetin alfa weekly for 7 weeks and it could be detected for 120 h after the last injection. The first test from this group could not differentiate rHuEPO from darbepoetin alfa [108]; however the second test could [107]. Both recombinant erythropoetins were found in the serum of four racehorses. The authors present the method as sensitive and reliable. They believe in its deterrence value. Applying tests and sanctions more often would improve effectiveness.

EPO gene manipulation

International sport organizations are mindful that gene manipulation might be used in the future to produce athletes with exceptional characteristics. At the first meeting of the IOC Gene Therapy Working Group, gene therapy was defined such that it would not enhance athletes' performance ("...transfer of genetic material to human somatic cells for the treatment or prevention of disease or disorders") [109]. Further, the group opined that gene manipulation is not on the immediate horizon, but that methods such as proteomic analysis might be developed so as to be prepared when gene-based doping becomes a reality.

Participants in the third Gene Doping Symposium organized by the WADA and Russian sport authorities in Saint Petersburg agreed that there was no evidence that gene doping was being used now (2008); however, some investigators have been approached by athletes. Further, it was noted that serious conceptual and technical problems continue to plague clinical investigators, and patients have experienced severe and unanticipated setbacks and adverse events, including death and the induction of leukemia. WADA is currently spending approximately 25% of its research budget on detection of gene doping.

One example of the research on 'gene doping' with EPO is a report that cDNA in primate (cynomolgus macaque) muscle is detectable by the IEF test [110]. The test was applied to serum instead of urine. It distinguished endogenous EPO from the product of *in vivo* gene transfer (transfer of homologous cDNA into skeletal muscle by injection of recombinant adeno-associated virus).

Detecting homologous blood transfusion

Nelson and co-workers proposed to prove homologous blood transfusion by quantitation of blood group antigens [111]. This approach exploits the likelihood of differences in minor blood group antigens probed with 12 antisera used with a secondary antibody directly conjugated with fluorescein to label IgG-coated red blood cells. The method was applied to blood samples from 25 patients who had received blood transfusions, to determine percentages of antigen-negative and positive red cells. Mixed red cell populations were detected in 22 of 25 transfused patients. The three patients who were 'negative' for transfusions turned out not to have received their scheduled transfusions. The technique can detect a second population represented by <5% of red cells. Apart from doping, mixed cell populations can arise as a result of disease, or after bone marrow or peripheral blood stem cell transplantation, or through blood transfusion. Rare occurrences are hemorrhage between women and fetus, intrauterine twin-twin transfusion, or in the rare tetragametic chimeras. Increasing the number of antisera used in the test would increase the test's effectiveness.

Voss and co-workers validated a flow cytometry approach to detecting homologous blood transfusion, using eight primary antibodies (an improvement over the 12 in Nelson's method) and two secondary antibodies [112]. For each primary antibody, Voss et al. established linearity, specificity, recovery, precision, robustness and interday-precision. Signal amplification resulted in better separation of mixed populations. Their method can detect a second population represented by 0.3–2.0% of red cells, an improvement over the 5% or less in the method of Nelson et al. [111].

Arndt and coauthors reviewed the detection of allogeneic (homologous) blood transfusions by flow cytometry and highlight the power of testing blood samples taken at different times because it provides an opportunity to detect a decrease in percentage of a minor red blood cell population as an indication that it arose from transfusion [113]. After the 2004 Vuelta de Espana cycling, stage road race, flow cytometry showed the presence of a mixed red blood cells population in one athlete, who was charged with a doping violation, namely for testing positive for transfusing another person's blood. The athlete appealed before the American Arbitration Association, who denied the appeal in 2005 [114]. The athlete's second appeal before sport's highest court, the International CAS), was denied in 2006 and a 2-year suspension imposed [115]. The CAS panel had considered the athlete's theoretical explanations including a disappearing twin, unsubstantiated problems with his positive test results, and even an extortion plot by a fan of another team.

Summary

The inherent complexity of the topic stems from interface between human behavior and the many disciplines that have something to offer – chemistry, pharmacology, medicine, law, ethics, and sociology, to name a few. Ultimately if sport continues to rely on testing as the solution, it is necessary to focus resources on physical and chemical methods for detecting drugs. At the same time sport hopes to change its culture through education and espousing ethical positions.

Blood doping, the most recalcitrant and threatening problem in recent years, goes through successive phases of coming under control and growing in a new direction. Threatening because the drugs are highly efficacious, and they enhance performance in endurance sports and perhaps even sports that rely on short bursts of energy. Recalcitrant because the methods to detect erythropoietic proteins in body fluids are complex, therefore global implementation is all the more difficult. This problem is in addition to sports usual vexing problem of tracking athletes in their travels and implementing collection procedures wherever they may be.

The current IEF test together with the current WADA reporting criteria are adequate from the perspective of unambiguous identification of epoetin alfa, epoetin beta, and darbepoetin alfa, and legal defensibility, however the test is probably not sensitive enough to detect those three recombinant erythropoietic proteins for as long as performance enhancement will last. Fortunately such a degree of sensitivity is not required to contain the use of these three products: the current test simply needs to be applied widely enough in short notice, out-of-competition testing. Meanwhile, the WADA is working on updating criteria and exploring new approaches to move ahead of the problem.

Acknowledgements
We are grateful for financial support from the United States Anti-Doping Agency and the International Olympic Committee for studies on erythropoietic proteins, and to Françoise Lasne and Jacques de Ceaurriz of the WADA-accredited laboratory in Paris for assistance and training in the IEF method. Amgen Inc. generously provided epoetin alfa and darbepoetin alfa, and Steve Elliott, PhD at Amgen provided invaluable advice on the interpretation of IEF analysis results and related matters.

References

1 Gordon B. Grecian athletic training in the third century (AD). *Ann Med History* 1935;6:513–519.
2 Beckett AH, Cowan DA. Misuse of drugs in sport. *Br J Sports Med* 1979;12:185–194.
3 Boje O. Doping: A study of the means employed to raise the level of performance in sport. *Bull of the Health Organisation, League Nations* 1939;8:439–469.
4 Committee on the Judiciary – United States Senate. Proper and improper use of drugs by athletes. U.S. Government Printing Office, Washington, DC, 1973.
5 Catlin DH, Hatton CK. Use and abuse of anabolic and other drugs for athletic enhancement. *Adv Intern Med* 1991;36:399–424.
6 Catlin DH, Murray TH. Performance-enhancing drugs, fair competition, and Olympic sport. *JAMA* 1996;276:231–237.
7 Kruse P, Ladefoged J, Nielsen U et al. Beta-blockade used in precision sports: effect on pistol shooting performance. *J Appl Physiol* 1986;61:417–420.
8 Dirix A. Medical Guide. Blackwell Scientific Publication, International Olympic committee, Lausanne, Switzerland, 1992.
9 Catlin DH, Fitch KD, Ljungqvist A. Medicine and Science in the Fight Against Doping is Sports. *J Intern Med* 2008;264:95–98.
10 McClaren RH. Introducing the Court of Arbitration for Sport: the *ad hoc* division at the Olympic Games. *Marquette Sports Law* 2001;12:514–542.
11 IOC. Olympic Movement Anti-Doping Code. http://www.olympic.org. Lausanne, Switzerland, International Olympic Committee. (Accessed 1 October 2002)
12 Elliott S. Erythropoiesis-stimulating agents and other methods to enhance oxygen transport. *Br J Pharmacol* 2008;154:529–541.
13 Segura J, Pascual JA, Gutierrez-Gallego R. Procedures for monitoring recombinant erythropoietin and analogues in doping control. *Anal Bioanal Chem* 2007;388:1521–1529.
14 Combe C, Tredree RL, Schellekens H. Biosimilar epoetins: an analysis based on recently implemented European medicines evaluation agency guidelines on comparability of biopharmaceutical proteins. *Pharmacotherapy* 2005;25:954–962.
15 Hatton C. *The night Olympic Team – Fighting to Keep Drugs out of the Games*. Boyds Mills Press, Honesdale, PA, 2008.
16 Higdon H. Blood-doping among endurance athlete. *Am Med News* 1985;39–41.
17 Cramer RB. Olympic cheating: the inside story of illicit doping and the US cycling team. *Rolling Stone* 1985;25–30.
18 Klein HG. Blood transfusion and athletics. Games people play. *N Engl J Med* 1985;312:854–856.
19 Gatlin LD. *Christian Science Monitor* 1998;22:18.
20 Adamson JW, Vapnek D. Recombinant erythropoietin to improve athletic performance. [letter] *N Engl J Med* 1991;324:698–699.
21 Fisher LM. Stamina-building drug linked to athletes' deaths. *New York Times*, 22. 5–19–1991.
22 Swift EM. Drug pedaling. *Sports Illustrated*, 60–65. 7–5–1999.
23 Jarvis CA. Tour de Farce. *Br J Sports Med* 1999;33:142–143.
24 A Sport in Shame. *Sports Illustrated*, 28–33. 7–27–1998.
25 Scarpino V, Arrigo A, Benzi G et al. Evaluation of prevalence of "doping" among Italian athletes. *Lancet* 1990;336:1048–1050.
26 World Anti-Doping Agency. www.WADA-AMA.org. 2008 (Accessed 16 June 2008)

27 Mohammad Mourit tests positive for EPO. *Chicago Tribune*; 2002.
28 Court of Arbitration for Sport. Roland Meier v. Swiss Cycling. CAS 2001/A/345 Meier v/Swiss Cycling, Lausanne, Switzerland, Court of Arbitration for Sport. (Accessed 27 January 2002)
29 Court of Arbitration for Sport. Union Cycliste Internationale (UCI) v. Bo Hamburger. CAS 2001/A/343 UCI v/Hamburger, Lausanne, Switzerland, Court of Arbitration for Sport. (Accessed 28 January 2002)
30 Kraft confesses. http://sports.espn.go.com/oly/news/story?id=1920706. 2004. (Accessed 16 June 2008)
31 Gledhill N. Blood doping and related issues. a brief review. *Med Sci Sport Exercise* 1982; 14:183–189.
32 Buick FJ, Gledhill N, Froese AB et al. Effect of induced erythrocythemia on aerobic work capacity. *J Appl Physiol* 1980;48:636–642.
33 Spriet LL, Gledhill N, Froese AB et al. Effect of graded erythrocythemia on cardiovascular and metabolic responses to exercise. *J Appl Physiol* 1986;61:1942–1948.
34 Hill AV, Long CNH, Lupton H. Muscular exercise, lactic acid and the supply and utilization of oxygen. Parts VII-VIII. *Proc Royal Soc Lond* 1924;97:155–176.
35 Kjellberg SR, Ruhde, U Sjostrand T. The amount of hemoglobin (blood volume) in relation to the pulse rate and heart volume during work. *Acta Physiol Scand* 1950;19:152–169.
36 Castle WB, Jandl JH. Blood viscosity and blood volume: opposing influences upon oxygen transport in polycythemia. *Semin Hematol* 1966;3:193–198.
37 Neumayr G, Pfister R, Mitterbauer G et al. Short-term effects of prolonged strenuous endurance exercise on the level of haematocrit in amateur cyclists. *Int J Sport Med* 2002;23:158–161.
38 Brun JF, Bouchahda C, Chaze D et al. The paradox of hematocrit in exercise physiology: which is the "normal" range from an hemorheologist's viewpoint? *Clin Hemorheol Microcirc* 2000;22:287–303.
39 Ernst E. Changes in blood rheology produced by exercise. *JAMA* 1985;253:2962–2963.
40 Ernst E, Matrai A. Hematocrit and plasma volume in runners. *Ann Intern Med* 1984;101:571.
41 Clement DB, Asmundson RC, Medhurst CW. Hemoglobin values: comparative survey of the 1976 Canadian Olympic team. *Can Med Assn J* 1977;117:614–616.
42 Adamson JW, Vapnek D. Recombinant erythropoietin to improve athletic performance. *N Engl J Med* 1991;324:698–699.
43 Gledhill N. The influence of altered blood volume and oxygen transport capacity on aerobic performance. *Exerc Sport Sci Rev* 1985;13:75–93.
44 Williams MH, Wesseldine S, Somma T et al. The effect of induced erythrocythemia upon 5-mile treadmill run time. *Med Sci Sport Exercise* 1981;13:169–175.
45 Sawka MN, Young AJ, Muza SR et al. Erythrocyte reinfusion and maximal aerobic power. An examination of modifying factors. *JAMA* 1987;257:1496–1499.
46 Brien AJ, Simon TL. The effects of red blood cell infusion on 10-km race time. *JAMA* 257:2761–2765.
47 Canadian Erythropoietin Study Group. Association between recombinant human erythropoietin and quality of life and exercise capacity of patients receiving haemodialysis. *BMJ* 1990;300:573–578.
48 Parisotto R, Gore CJ, Emslie KR et al. A novel method utilising markers of altered erythropoiesis for the detection of recombinant human erythropoietin abuse in athletes. *Haematologica* 2000;85:564–572.
49 Russell G, Gore CJ, Ashenden MJ et al. Effects of prolonged low doses of recombinant human erythropoietin during submaximal and maximal exercise. *Eur J Appl Physiol* 2002;86:442–449.
50 Birkeland KI, Stray-Gundersen J, Hemmersbach P et al. Effect of rhEPO administration on serum levels of sTfR and cycling performance. *Med Sci Sport Exercise* 2000;32:1238–1243.
51 Berglund B, Ekblom B. Effect of recombinant human erythropoietin treatment on blood pressure and some haematological parameters in healthy men. *J Intern Med* 1991;229:125–130.
52 Lage JM, Panizo C, Masdeu J et al. Cyclist's doping associated with cerebral sinus thrombosis. *Neurology* 2002;58:665.
53 Comitato Olimpico Nazionale Italiano. Commissione Scientifica Antidoping del CONI. "Progetto Io non rischil la salute!". 2000. Rome, Italy.
54 Cazzola M. A global strategy for prevention and detection of blood doping with erythropoietin and related drugs. *Haematologica* 2000;85:561–563.
55 Shermer M. The doping dilemma. *Sci Amer* 2008;298:82–89.

56 Aguilera R, Chapman TE, Catlin DH. Performance characteristics of a carbon isotope ratio method for detecting doping with testosterone based on urine diols: controls and athletes with elevated testosterone/epitestosterone ratios. *Clin Chem* 2001;47:292–300.

57 Wallace JD, Cuneo RC, Baxter R et al. Responses of the growth hormone (GH) and insulin-like growth factor axis to exercise, GH administration, and GH withdrawal in trained adult males: a potential test for GH abuse in sport. *J Clin Endocrinol Metab* 1999;84:3591–3601.

58 Union Cycliste Internationale. U.C.I. Lettre aux groupes sportifs. *Lettre a tous les coureurs*. 7–2–1997.

59 Vergouwen PC, Collee T, Marx JJ. Haematocrit in elite athletes. *Int J Sport Med* 1999; 20:538–541.

60 Lynn Zinser and John Eligon. Athletes Are Facing Increased Vigilance.http://www.nytimes.com/2006/02/12/sports/olympics/12drugs.html?_r=1&scp=3&sq=2006+olympics+health+test&st=nyt&oref=slogin.2008. (Accessed 12 February 2006)

61 Marx JJ, Vergouwen PC. Packed-cell volume in elite athletes. *Lancet* 1998;352:451.

62 Schumacher YO, Schmid A, Lenz T et al. Blood testing in sports: hematological profile of a convicted athlete. Clin J. *Sport Med* 2001;11:115–117.

63 Saris WH, Senden JM, Brouns F. What is a normal red-blood cell mass for professional cyclists? *Lancet* 1998;352:1758.

64 Neumayr G, Pfister R, Mitterbauer G et al. Short-term effects of prolonged strenuous endurance exercise on the level of haematocrit in amateur cyclists. *Int Sport Med* 2002;23:158–161.

65 Videman T, Lereim I, Hemmingsson P et al. Changes in hemoglobin values in elite cross-country skiers from 1987–1999. *Scand J Med* Sc Sports 2000;10:98–102.

66 Casoni I, Ricci G, Ballarin E et al. Hematological indices of erythropoietin administration in athletes. *Int J Sport Med* 1993;14:307–311.

67 Gareau R, Audran M, Baynes RD et al. Erythropoietin abuse in athletes. *Nature* 1996; 380:113.

68 Audran M, Gareau R, Matecki S et al. Effects of erythropoietin administration in training athletes and possible indirect detection in doping control. Effects of erythropoietin administration in training athletes and possible indirect detection in doping control. *Med Sci Sport Exercise* 1999;31:639–645.

69 Parisotto R, Wu M, Ashenden MJ et al. Detection of recombinant human erythropoietin abuse in athletes utilizing markers of altered erythropoiesis. *Haematologica* 2001;86:128–137.

70 International Olympic Committee. http://www.olympic.org/uk/news/publications/press_uk.asp?release=56.(Accessed 20 October 2002)

71 Kazlauskas R, Howe C, Trout G. Strategies for rhEPO detection in sport. *Clin J Sport Med* 2002;12:229–235.

72 Souillard A, Audran M, Bressolle F et al. Pharmacokinetics and pharmacodynamics of recombinant human erythropoietin in athletes. Blood sampling and doping control. *Br J Clin Pharmacol* 1996;42:355–364.

73 Flaharty KK, Caro J, Erslev A et al. Pharmacokinetics and erythropoietic response to human recombinant erythropoietin in healthy men. *Clin Pharmacol Ther* 1990;47:557–564.

74 Remacha AF, Ordonez J, Barcelo MJ et al. Evaluation of erythropoietin in endurance runners. *Haematologica* 1994;79:350–352.

75 Emslie KR, Howe C, Trout G. Measurement of urinary erythropoietin levels in athletes. *Recent advances in doping analysis*. 1999;7:291.

76 World Anti-Doping Agency. WADA Athlete's Passport Q & A. http://www.wada-ama.org/en/dynamic.ch2?pageCategory.id=754. 2008. (Accessed 16 June 2008)

77 USADA. USADA News Release. http://www.usantidoping.org/files/active/resources/press_releases/Enhanced%20Initiatives%20-%20Bejing.pdf. 2008. (Accessed 17 June 2008)

78 Morkeberg J, Saltin B, Belhage B et al. Blood profiles in elite cross-country skiers: a 6-year follow-up. *Scand J Med Sci Sports* 2008; Epub ahead of print

79 Wide L, Bengtsson C, Berglund B et al. Detection in blood and urine of recombinant erythropoietin administered to healthy men. *ExercMed Sci Sport Exercise 1995*;27:1569–1576.

80 Skibeli V, Nissen-Lie G, Torjesen P. Sugar profiling proves that human serum erythropoietin differs from recombinant human erythropoietin. *Blood* 2001;98:3626–3634.

81 Lasne F de Ceaurriz J. Recombinant erythropoietin in urine. *Nature* 2000;405:635.

82 Lasne F. Double-blotting: a solution to the problem of non-specific binding of secondary antibodies in immunoblotting procedures. *J Immunol Method* 2001;253:125–131.

83 WADA EPO committee. WADA Technical Document TD2004EPO. http://www.wada-ama.org/rtecontent/document/td2004epo_en.pdf 2004. (Accessed 4 June 2008)
84 WADA EPO committee. WADA Technical Document TD2007EPO. http://www.wada-ama.org/rtecontent/document/td2007epo_en.pdf. 2007. (Accessed 4 June 2008)
85 Lasne F, Thioulouse J, Martin L et al. Detection of recombinant human erythropoietin in urine for doping analysis: interpretation of isoelectric profiles by discriminant analysis. *Electrophoresis* 2007;28:1875–1881.
86 Egrie J, Browne J. Darbepoetin alfa is more potent *in vivo* and can be administered less frequently than rHuEPO. *Br J Cancer* 2002;87:476–477.
87 IOC Medical Commission. Post Games Report – Salt Lake City 2002, Phase 4. www.cafdis-antidoping.net. (Accessed 1 June 2002)
88 Macdougall IC. Optimizing the use of erythropoietic agents – pharmacokinetic and pharmacodynamic considerations. *Nephrol Dial Transplant* 2002;17 Suppl 5:66–70.
89 Ashenden MJ, Varlet-Marie E, Lasne F et al. The effect of microdose recombinant human erythropoietin regimens in athletes. *Haematologica* 2006;91:1143–1148.
90 Lamon S, Robinson N, Mangin P et al. Detection window of Darbepoetin-alfa following one single subcutaneous injection. *Clin Chim Acta* 2007;379:145–149.
91 Parisotto R, Gore CJ, Hahn AG et al. Reticulocyte parameters as potential discriminators of recombinant human erythropoietin abuse in elite athletes. *Int J Sport Med* 2000;21:471–479.
92 Lundby C, Achman-Anderson NJ, Thomsen JJ et al. Testing for recombinant human erythropoietin in urine: problems associated with current anti doping testing. *J Appl Physiol* 6/26/2008 (on line).
93 Schellekens H. Follow-on biologics: challenges of the "next generation". *Nephrol Dial Transplant* 2005;20 Suppl 4:iv31–iv36.
94 Bren A, Kandus A, Varl J et al. A comparison between epoetin omega and epoetin alfa in the correction of anemia in hemodialysis patients: a prospective, controlled crossover study. *Artif Organs* 2002;26:91–97.
95 Sikole A, Spasovski G, Zafirov D et al. Epoetin omega for treatment of anemia in maintenance hemodialysis patients. *Clin Nephrol* 2002;57:237–245.
96 Belalcazar V, Ventura R, Segura J et al. Clarification on the detection of epoetin delta and epoetin omega using isoelectric focusing. *Am J Hematol* 2008;83:754
97 Macdougall IC. Novel erythropoiesis-stimulating agents: a new era in anemia management. *Clin J Am Soc Nephrol* 2008;3:200–207.
98 McGahan L. Continuous erythropoietin receptor activator (Mircera) for renal anemia. *Issues Emerg Health Technol* 2008; vol:1–6.
99 Connolly PJ, Wetter SK, Murray WV et al. Synthesis and erythropoietin receptor binding affinities of N,N-disubstituted amino acids. *Bioorg Med Chem Lett* 2000;10:1995–1999.
100 Fan Q, Leuther KK, Holmes CP et al. Preclinical evaluation of Hematide, a novel erythropoiesis stimulating agent, for the treatment of anemia. *Exp Hematol* 2006;34:1303–1311.
101 DelVecchio L, Locatelli F. New erythropoiesis-stimulating agents: how innovative are they? *Contrib Nephrol* 2008;161:255–260.
102 Kohler M, Ayotte C, Desharnais P et al. Discrimination of recombinant and endogenous urinary erythropoietin by calculating relative mobility values from SDS gels. *Int J Sport Med* 2008;29:1–6.
103 Stubiger G, Marchetti M, Nagano M et al. Characterisation of intact recombinant human erythropoietins applied in doping by means of planar gel electrophoretic techniques and matrix-assisted laser desorption/ionisation linear time-of-flight mass spectrometry. *Rapid Commun Mass Spectrom* 2005;19:728–742.
104 Associated Press Horse Racing News Wire. Police arrest four on suspicion of drug possession, race fixing. http://sports.espn.go.com/espn/wire?section=horse&id=2395684.2008. (Accessed 18 June 2008)
105 Standardbred Canada Insider News. ORC Raids Drug Supplier. www.standardbredcanada.ca/news/iss0406/0420.html. (Accessed 18 June 2008)
106 Bartlett C, Clancy GJ, Cowan DA et al. Detection of the administration of human erythropoietin (HuEPO) to canines. *J Anal Toxicol* 2006;30:663–669.
107 Guan F, Uboh CE, Soma LR et al. Differentiation and identification of recombinant human erythropoietin and darbepoetin Alfa in equine plasma by LC-MS/MS for doping control. *Anal Chem* 2008;80:3811–3817.

108 Guan F, Uboh CE, Soma LR et al. LC-MS/MS method for confirmation of recombinant human erythropoietin and darbepoetin alpha in equine plasma. *Anal Chem* 2007;79:4627–4635.
109 Bulman E. IOC Prepares for possible future genetic cheating. Nando Media. (Accessed 2 June 2001)
110 Lasne F, Martin L, Larcher T et al. "Genetic Doping" with erythropoietin cDNA in primate muscle is detectable. *Mol Ther* 2004;10:409–410.
111 Nelson M, Popp H, Sharpe K et al. Proof of homologous blood transfusion through quantification of blood group antigens. *Haematologica* 2003;88:1284–1295.
112 Voss SC, Thevis M, Schinkothe T et al. Detection of homologous blood transfusion. *Int J Sport Med* 2007;28:633–637.
113 Arndt PA, Kumpel BM. Blood doping in athletes-Detection of allogeneic blood transfusions by flow cytofluorometry. *Am J Hematol* 2008;83:657–667
114 CAS. CAS 2005/A/884 Tyler Hamilton V/USADA & UCI. Http://www.usantidoping.org/files/active/arbitration_rulings/CAS%20Decision-Tyler%20Hamilton_Feb2006.pdf. 2005. (Accessed 23 May 2008)
115 USADA. USADA press release on Hamilton decision.http://www.usantidoping.org/files/active/resources/press_releases/usada%20press%20release-hamilton_february%202006.pdf. 2008. (Accessed 23 may 2008)

Erythropoietins, Erythropoietic Factors and Erythropoiesis
Edited by S.G. Elliott, M.A. Foote, and G. Molineux

Role and regulation of iron metabolism in erythropoiesis and disease

Tara L. Arvedson and Barbra J. Sasu

Amgen Inc., One Amgen Center Drive, M/S 15-2-A, Thousand Oaks, California 91320, USA

Introduction

Iron is an essential element for normal cellular metabolism and growth as an enzyme cofactor, heme constituent and oxygenation sensor. In excess however, free iron is toxic. Living organisms have, therefore, evolved sophisticated and tightly regulated mechanisms to control iron uptake, transport, and release. Defects in any part of this process can lead to disease. For example, excessive uptake can lead to systemic iron overload and associated toxicity. Inappropriate or inefficient transport can lead to iron maldistribution or deficiency. Red blood cells are the primary consumers of iron and the largest body iron pool (approximately 50% of the body's iron is incorporated in heme); hence, fluctuations in iron supply can have significant effects on red blood cell production and function. This chapter provides an overview of the proteins and pathways involved in iron metabolism as they relate to normal red blood cell biology and disorders of iron excess or deficiency.

Molecular coordination of iron metabolism

Iron uptake from the diet

Absorption of iron from the diet involves two discrete steps: uptake of iron by the intestinal epithelia from the lumen of the gut and transfer of iron into the blood. Dietary iron is found in the form of inorganic ferric (Fe^{3+}) or ferrous (Fe^{2+}) ions (primarily from nonanimal sources) or heme (from hemoglobin or myoglobin). Most inorganic iron is in the ferric form, which readily precipitates if the pH is raised above 3. Precipitation is prevented by association of the iron with other dietary components such as amino acids, citrate, or sucrose. Transport of ferrous iron across the intestinal mucosa is thought to be mediated primarily by the divalent metal transporter 1 (DMT1) in the duodenum [1]. Other proteins such as the zinc transporter Zip14 have been suggested to have a role in this transport [2]. Both DMT1 and Zip14 transport ferrous iron and reduction of ferric iron is thought to be achieved by the ferrireductase duodenal cytochrome b (DcytB) [3]. However, mice lacking DcytB do not show signs of iron deficiency or abnormal erythropoiesis, suggesting there may be other proteins or components capable of reducing ferric iron in the gut [4].

Heme is the main source of organic iron in the diet and elucidating the mechanism of heme uptake is an evolving area. As heme is hydrophobic, it may diffuse across the enterocyte membrane. Alternatively, uptake may be receptor mediated. Recently heme was shown to be taken up in an energy-dependent process by the heme carrier protein 1 (HCP-1), a protein that also transports folate [5, 6]. It is also possible that heme may enter the enterocyte by associating with a cell surface receptor that mediates endocytosis. Data suggest that heme enters the cell intact [7]. Heme degradation within the enterocyte may be mediated by heme oxygenase and the released iron likely joins the same intracellular pool as inorganic iron. Once inside the enterocyte, iron may be used immediately, stored, or transported to the basolateral surface for export. Figure 1 illustrates the key molecules governing iron movement through the duodenal enterocyte into the plasma iron pool.

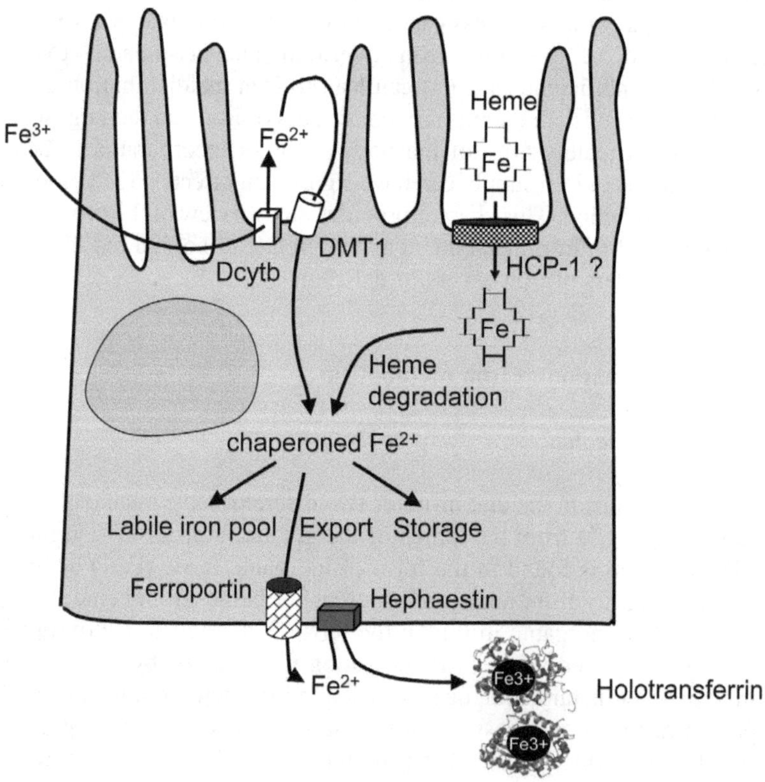

Figure 1. Molecular mechanisms of intestinal iron uptake. See text for further explanation.

Iron export

Iron is exported from the basolateral membrane of enterocytes through the multipass transmembrane protein ferroportin, which like DMT1, transports ferrous iron. In addition to a role in the enterocyte, ferroportin is also required for iron transport across the placenta and is important for iron export from macrophages [8]. Oxidation is required for iron release from ferroportin, suggesting an intimate relationship between ferroportin and an associated ferroxidase. In the enterocytes, this oxidation step is mediated largely by hephaestin, an enzyme with a copper-containing active site [9]. Oxidation in other tissues is mediated by ceruloplasmin, a copper-containing protein that is closely related to hephaestin [10]. Ceruloplasmin is present in the circulation and is also anchored to the cell surface of some cell types. Approximately 90% of the copper in serum is associated with ceruloplasmin. Both copper deficiency and aceruoplasminemia can cause anemia, emphasizing the importance of these ferroxidases in the iron metabolism pathway [11]. After oxidation of iron to the ferric form, it is loaded onto transferrin for transport to distal sites in the body.

Cellular iron uptake

Under normal conditions, most of the iron in the circulation is bound to transferrin. In this state, iron is not reactive; moreover, the association between iron and transferrin is of extremely high affinity and in practical terms, no dissociation occurs in serum ($K_D \sim 10^{-20}$ M) [12]. Uptake of iron from plasma is mediated by association of holotransferrin (fully iron saturated) with a membrane-bound transferrin receptor (TfR1). TfR1 is ubiquitously expressed and binds holotransferrin with high affinity (1–5 nM) [13]. After binding of holotransferrin to TfR1, the complex is internalized and as holotransferrin-TfR1-containing vesicles become acidified (pH ~ 5.5), iron is released [14]. The ferric iron is then reduced to ferrous iron by the ferrireductase Steap3 and transported out of the vesicle by DMT1 to be either stored or used by the cell [15]. After release of iron, the apotransferrin (no associated iron)-TfR1 complex is recycled to the cell surface where apotransferrin dissociates from the receptor and is reused. Figure 2 illustrates the molecules involved in cellular iron uptake. On average, each molecule of transferrin will make 100–200 circuits in its lifetime [16]. The transferrin compartment typically contains about 3 mg of iron at one time and turns over approximately 10 times a day [17].

Iron utilization and storage

Normal individuals have approximately 3–4 g of iron distributed in the following areas: approximately 2 g in red blood cells (hemoglobin), approximately 1 g in hepatocytes (storage), approximately 600 mg in macrophages,

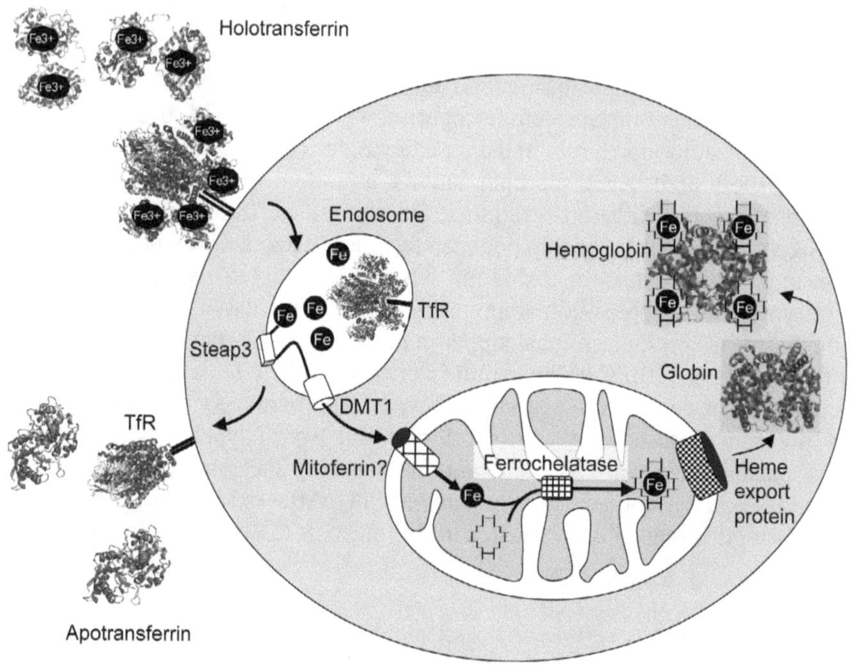

Figure 2. Iron uptake by red blood cell precursors. See text for further explanation.

approximately 300 mg in the bone marrow, approximately 300 mg in muscle (myoglobin), and approximately 3 mg associated with transferrin or other serum transport molecules [18].

Once inside the cell, iron may be incorporated into key enzymes involved in DNA synthesis, electron transport, and ATP production. The mechanisms of intracellular iron transport are not fully understood. Several theories exist to explain iron movement within the cell, such as vesicular-mediated transcytosis or small molecule- or protein chaperone-mediated transport [19]. Surplus intracellular iron within cells is stored in cytoplasmic ferritin, a spherical multimer with 24 subunits capable of storing up to 4,500 iron ions. Ferritin takes up the highly reactive ferrous iron and stores it as ferric iron in a nonreactive mineral core. The change in oxidation state is mediated by the ferritin heavy (H) chain, which has oxidative capability. The remainder of the ferritin molecule comprises the light (L) chain which appears to have primarily a structural function. The ratio of H:L chains in ferritin varies in different tissues. In those tissues where prevention of iron-mediated oxidative stress is critical, such as the heart or neurons, the ratio of H:L chains favors the H chain. This ratio is reversed in the liver, spleen, and microglia. Ferritin expression is increased in response to intracellular iron concentrations and oxidative stress [20]. Ferritin is also found in the serum and is believed to be the result of leakage from damaged or iron loaded cells or secretion in response to inflammation as an acute phase reactant. Serum ferritin is L-rich and differs from intracellular ferritin in that it is reported to contain less iron and be more heavily glycosylated [21, 22].

Iron in red blood cells

On average 200 billion new red blood cells are produced each day requiring approximately 20 mg of iron for hemoglobin synthesis. Reticulocytes acquire iron primarily through transferrin, although ferritin uptake is also possible [23]. TfRs are expressed on red blood cell precursors from the BFU–E stage through the late reticulocyte stage [24]. Hemoglobin accumulation is first apparent during the polychromatophilic erythroblast stage. Hemoglobin synthesis continues after the nucleus is ejected and stops when the mitochondria and ribosomes are degraded. Eventually each red blood cell will contain approximately 300 million hemoglobin molecules [25].

Normal adult hemoglobin is a tetramer of four globin chains (mainly $\alpha_2\beta_2$) each with an associated heme (iron and porphyrin) molecule. Heme synthesis occurs in the mitochondrial matrix. As a result, iron that is taken up from the serum by TfR1 must be transported from the endosome, through the outer and inner mitochondrial membranes, and into the mitochondrial matrix. This process is not well defined. Transport across one or more of the mitochondrial membranes may be mediated by mitoferrin [26]. In the mitochondrial matrix, iron can be incorporated into heme, used for iron sulfur cluster formation, or stored in mitochondrial ferritin [27]. Once heme is generated, it is exported out of the mitochondria for subsequent combination with globin. A number of candidate molecules may be involved in heme export, including FLVCR [28], ABCG2 (a xenobiotic transporter also known as the breast cancer [chemotherapy] resistance protein) [29], and the ABC-mitochondrial erythroid transporter (ABC-me) [30]. The process of iron uptake, transport and incorporation into heme is rapid. *In vitro* studies using reticulocytes indicate the process requires only 6–8 min at 37 °C [31]. *In vivo* studies indicate the process occurs in less than an hour [32]. Figure 2 shows the molecules involved in iron transit within the mitochondria and incorporation into hemoglobin.

Defects in mitochondrial iron uptake, transport, or metabolism may play a role in diseases such as myelodysplastic syndrome, some sideroblastic anemias, or lead poisoning [33, 34]. Although the etiology of these disorders is often unknown, they are characterized by the presence of sideroblasts in the bone marrow. Sideroblasts are abnormal erythroblasts containing iron-loaded mitochondria that cluster around the nucleus. Because of this unique morphology, they are often called ringed sideroblasts [35]. This abnormal accumulation of mitochondrial iron may be a consequence of impaired iron management or impaired porphyrin or globin coordination.

The biosynthesis of porphyrin and globin are tightly correlated with iron homeostasis in erythroblasts. The first and rate limiting step of porphyrin biosynthesis is mediated by erythroid-specific α-aminolevulinate synthase (ALAS-E). Expression of this enzyme is increased in response to iron levels [36]. Ferrochelatase, the enzyme, which inserts iron into porphyrin, is also upregulated by the presence of an iron surplus [37]. Together this tight regulation ensures that neither free iron nor free porphyrin accumulates, since both are toxic. In addition, both alpha [38] and beta globin [39] synthesis are regulated by heme availability.

Disease states develop in conditions where the carefully orchestrated process of hemoglobin production is disrupted. ALAS-E mutation results in X-linked hereditary sideroblastic anemia due to insufficient production of porphyrin and concurrent iron accumulation [40], which is exacerbated by systemic iron overload [41]. Ferrochelatase mutation causes erythropoietic protoporphyria due to buildup of excessive protoporphorphyrin and unused iron [42]. Defective globin synthesis is exemplified by thalassemia major. Clinical symptoms of thalassemia major include anemia, liver and spleen enlargement due to excessive red blood cell destruction and iron overload due to transfusions and increased absorption. The major cause of mortality in thalassemia major patients is iron overload-related organ damage [43].

Iron recycling from red blood cells

Red blood cells circulate for an average of 120 days. As each red blood cell nears the end of its lifespan, its membrane characteristics change such that it is recognized and phagocytosed by macrophages of the reticuloendothelial system. Changes leading to recognition may include flipping or deterioration of the membrane to expose antigens (e.g., phosphatidylserine, desialylated β-galactosyl residues) [44]. Once phagocytosed, the red blood cells are digested in lysosomes. Hemoglobin is degraded and iron is liberated through the action of heme oxygenase-1 (HO-1). Iron export from the lysosome is potentially mediated by DMT1 or NRAMP1 (a member of the divalent metal ion transporter family) and may be used by the cell, stored, or exported [45, 46]. Ferroportin and HO-1 are both upregulated after erythrophagocytosis suggesting that liberated iron can be exported via ferroportin [47]. Figure 3 presents a schematic of the proteins involved in iron recycling from senescent red blood cells by macrophages of the reticuloendothelial system.

Systemic coordination of iron metabolism

Iron uptake, redistribution and storage in the body are coordinated by hepcidin, a 25 amino acid peptide concurrently identified by three groups and subsequently shown to be the protein product of the *HAMP* gene [48–50]. Under conditions of iron excess, hepcidin production is stimulated in the liver and limits iron uptake through the gut while also causing iron sequestration

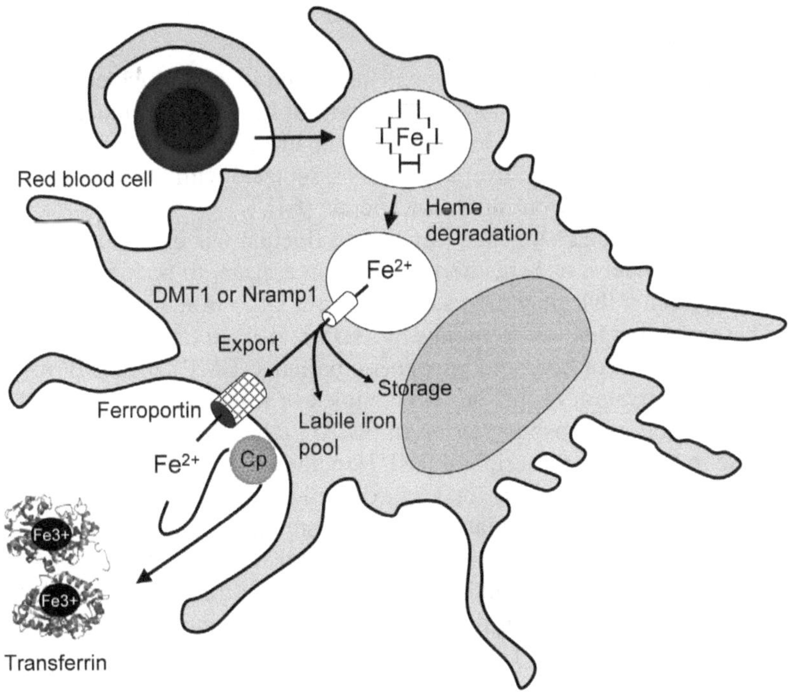

Figure 3. Iron recycling from red blood cells by macrophages of the reticuloendothelial system. See text for further explanation.

in tissues [51, 52]. Conversely, under conditions of iron limitation, hepcidin concentrations are lowered and gut uptake and release from the tissues is increased [53]. Hepcidin mediates these effects by binding to ferroportin and stimulating its internalization and degradation [54]. In macrophages of the reticuloendothelial system, degradation of ferroportin results in iron sequestration. In the gut, hepcidin activity inhibits release of dietary iron into the circulation.

Hepcidin is expressed mainly in the liver, with some expression in the heart, kidney, adipose tissue, spinal cord, pancreas, hematopoietic cells, retina and brain [49, 55–59]. Hepcidin was originally cloned as the product of a gene up-regulated by iron in mice [50]. It was identified in parallel by other groups as a protein present in blood ultrafiltrate from patients with renal failure [48] and in the urine of healthy donors [49]. Insight into its controlling role in iron regulation came from the fortuitous observation that mice with a disrupted *USF2* gene also lacked expression of hepcidin in the liver [60]. As a result, the mice were unable to regulate iron uptake, and they developed an iron loading condition indistinguishable from hereditary hemochromatosis in humans. Subsequent studies using a hepcidin-specific knockout mouse confirmed that hepcidin alone can account for the original phenotype [61].

Transgenic overexpression of hepcidin in mouse liver resulted in severe iron-deficiency anemia [62].

Mutations in the human hepcidin gene have been identified which result in juvenile onset hemochromatosis, a severe form of hereditary hemochromatosis resulting in extensive iron loading early in life [63]. Hepcidin has been shown to be overexpressed in liver adenomas associated with glycogen storage disease and linked to an iron-refractory anemia [64].

Hepcidin expression changes in response to fluctuations in amounts of systemic iron. Modulation of hepcidin transcription appears to be mediated by a number of proteins that may associate to form a large iron sensing complex [65]. These proteins include hemojuvelin (HJV), HFE, transferrin receptor 2 (TfR2), and a subset of bone morphogenetic proteins (BMP) and BMP receptors [66–69]. Activation of this protein complex or a subset of the proteins in this complex induces phosphorylation of SMAD family members, leading to an increase in hepcidin transcription [68]. Hepcidin expression is also regulated by infection, inflammation, or erythropoietic demand [69, 70]. Several inflammatory cytokines upregulate hepcidin expression, including interleukin (IL)-6 and IL-1 [71, 72]. Studies using IL-6 indicate that activation of the JAK/STAT3 pathway is involved in triggering hepcidin transcription [73]. The proteins involved in sensing erythropoietic demand that may in turn influence hepcidin expression have yet to be determined but are the subject of much attention. Hepcidin regulation by erythropoietic demand is distinct from regulation by iron or inflammation as erythropoietic demand downregulates hepcidin transcription whereas increases in iron or inflammation upregulate hepcidin transcription. Much of the understanding of hepcidin regulation has been pieced together over the last few years. The next few years may bring more understanding of how these different types of regulation fit together to control normal and abnormal iron metabolism.

Disorders of iron metabolism

Iron overload disorders

Iron overload can occur through two main mechanisms. Hereditary iron overload (hereditary hemochromatosis) is caused by genetic mutations affecting iron homeostasis. Secondary iron overload is generally a consequence of iron deposits from red blood cell transfusion. Iron loading occurs first in the liver, causing cirrhosis and subsequent end-stage liver disease [74]. Other organs affected are the pancreas, heart, and anterior pituitary, leading to diabetes, cardiomyopathy, and hypogonadism, respectively. Both hereditary and secondary types of iron overload lead to an increased mortality risk due to iron-induced cardiomyopathy, which can be illustrated most clearly in thalassemia major patients with secondary iron overload, who in the absence of iron chelation therapy have a life expectancy of less than 20 years [75].

Hereditary hemochromatosis

Hereditary hemochromatosis results from a number of genetic mutations leading to dysregulated iron uptake and distribution within the body. Increased uptake results in gradual iron loading over time, and is generally diagnosed when serum iron concentrations are inappropriately high and symptoms have arisen due to organ damage caused by excess iron deposition.

There are five types of hereditary hemochromatosis: 1, 2a, 2b, 3 and 4, which correspond to genetic defects in HFE, HJV, hepcidin, TfR2, and ferroportin, respectively. As described above, HFE, HJV, and TfR2 are all suggested to be components of the iron-sensing complex that regulates hepcidin transcription. As would be expected from defects in the proteins involved in hepcidin transcriptional induction, hereditary hemochromatosis types 1 to 3 are all characterized by inappropriately low levels of hepcidin relative to the degree of systemic iron overload [76].

Hereditary hemochromatosis type 1 (*HFE* mutation) is the most widespread form of hereditary hemochromatosis. Approximately 15% of individuals of northern European descent are heterozygous for defects in HFE and 0.5% are homozygous [77]. Despite the number of individuals carrying these genetic lesions, only a very small percent of homozygous individuals (approximately 1%) develop symptoms severe enough to be detected. If disorders do present, the age of onset is 40–60 years for men and after menopause for women [78].

Hereditary hemochromatosis types 2a (*HJV* mutation) and 2b (hepcidin mutation) are the most severe forms of hereditary hemochromatosis [79]. Individuals carrying these mutations start to experience the effects of iron loading within the second or third decade of life. Due to the early age of onset, these conditions are also known as juvenile hemochromatosis. Unlike hereditary hemochromatosis type 1, juvenile hemochromatosis affects both males and females equally [80].

There are few examples of individuals with hereditary hemochromatosis type 3 (*TfR2* mutation); however, the clinical features appear to be similar or perhaps slightly more severe than those observed for type 1 hereditary hemochromatosis and if present with mutations in *HFE* can lead to juvenile hemochromatosis [81].

Hereditary hemochromatosis type 4 (ferroportin mutation) is also termed ferroportin disease. Ferroportin mutations yield two distinct functional outcomes: loss-of-function mutations result in loss of iron export capabilities, and gain-of-function mutations result in uncontrolled iron export due to loss of hepcidin responsiveness. Iron uptake and deposition characteristics vary according to the type of mutation. Loss-of-function mutations lead to iron loading in macrophages, whereas gain-of-function mutations lead to iron loading in parenchymal cells, similar to hereditary hemochromatosis types 1–3.

Diagnosis and treatment of hereditary hemochromatosis

The first indicator of hereditary hemochromatosis types 1–3 is an increase in the percentage of transferrin occupied by iron (transferrin saturation or Tsat). If Tsat is >60% in men or >50% in women and there is no other cause for increased serum iron (e.g., medication or ingested iron), there is a high probability of hemochromatosis. Diagnosis can be confirmed by measurement of serum ferritin, which can be used as a surrogate of body iron stores [82]. Normal ferritin values are <300 ng/mL for men and <150 ng/mL for women. Iron excess can be qualified in terms of serum ferritin values as mild (between normal values and 500 ng/mL), moderate (500–1,000 ng/mL) or severe (>1,000 ng/mL). Treatment of hereditary hemochromatosis types 1–3 and disease caused by gain-of-function ferroportin mutations involves removal of 500 mL of blood once or twice weekly, depending on the degree of iron overload [82]. Each treatment removes approximately 250 mg of iron. Iron overloaded individuals may have an excess iron burden of 30–40 g, requiring several years of regular treatment to bring iron levels into the normal range. Iron chelation is not typically used for the treatment of hereditary hemochromatosis.

Secondary iron overload

Secondary iron overload after transfusion occurs in numerous conditions including thalassemia, sickle cell disease, myelodysplastic syndrome, and hemolytic and sideroblastic anemias. It can also be a significant issue in anemic patients with cancer who require multiple transfusions. In patients with secondary iron overload, iron is derived from transfused red blood cells, which are degraded in macrophages at the end of their lifespan. Because the liberated iron cannot be used to make new red blood cells due the underlying disease, the transfusion-derived iron is stored. The requirement for multiple transfusions causes accumulation of iron to toxic levels that leads to organ damage. On average, each unit of blood introduces 200–250 mg of iron [83] and no mechanism exists to upregulate iron excretion to compensate.

Treatment of secondary iron overload

Secondary iron overload cannot, in general, be addressed by phlebotomy since the patients are anemic. For this reason, the only current treatment option is chelation therapy. Although iron chelation can dramatically increase life expectancy, it has some drawbacks in terms of dosage, compliance, and toxicity. Given the fact that one iron ion requires 1, 2, or 3 molecules of chelator to remove it, daily doses are very high. Side effects differ with the chelator, but examples include neutropenia and agranulocytosis, gastrointestinal side effects (nausea, vomiting, diarrhea), hearing loss, retinal damage, and arthritis [84].

As proteins involved in iron transport and distribution are identified and characterized, novel therapeutics to modulate the activity of these proteins and alleviate secondary iron overload without the use of chelators are being explored.

Iron limitation, anemia, and ESAs

Iron limitation is a frequent cause of anemia and is quoted as the most common cause of hyporesponsiveness to erythropoietic stimulating agents (ESAs) [85]. Approximately 150 mg of iron is necessary to fuel an increase of 1 g/dL in hemoglobin and hence an adequate and continuous supply of iron is required. Red blood cell production may be limited by either absolute iron deficiency, which occurs when body iron stores are exhausted and dietary uptake is insufficient, or functional iron deficiency, which occurs when iron cannot be mobilized from iron stores. Patients with iron-limited anemia may have one or both of these conditions. Although related in outcome, these types of iron deficiency have very different presentations and etiologies. They may also benefit from different treatments.

Absolute iron deficiency

Absolute iron deficiency is a function of increased iron loss (most commonly blood loss), increased iron demand (e.g., growth, pregnancy), or limited iron supply (e.g., poor nutrition, malabsorption). It is an endemic problem in all areas of the world, affecting approximately 20% of the population [86]. Iron deficiency is associated with reduced motor and mental development in children and frequently manifests as anemia in adults. Hemodialysis patients are at particularly high risk for iron deficiency due to blood loss from repeated blood draws, blood loss during dialysis, gastrointestinal bleeding, and poor uptake from the diet. These losses may total 1–3 g/year. Other disorders with occult blood loss such as cancer also have a high incidence of iron deficiency anemia.

Diagnosis of iron deficiency anemia

Before automated blood counters and clinical laboratory analyzers were available, diagnosis of iron deficiency was based primarily on identification of abnormal red blood cell morphology on a peripheral blood smear. The first visual sign of iron deficiency anemia is the appearance of hypochromic (reduced hemoglobin content) and microcytic (smaller in size) erythrocytes mixed with normochromic and normocytic erythrocytes. The iron deficiency anemia diagnosis could be confirmed by examination of stainable iron stores in bone marrow biopsy, if necessary.

With the advent of automated blood cell counting, parameters such as mean corpuscular volume (MCV) and mean corpuscular hemoglobin (MCH) can be accurately quantified, producing standardization in diagnosis and reducing reliance on manual techniques. Because these indices measure the mean values of the circulating red blood cell population and red blood cells have an average life expectancy of 120 days, the patient may be iron deficient for several weeks before a change in MCV or MCH can be detected. To obtain a more real-time assessment of the health of the bone marrow and the state of erythropoiesis, reticulocyte parameters can be evaluated.

Immature red blood cells that are newly-released into the circulation (reticulocytes) contain residual RNA that can be detected by staining techniques. Automated cell counters can be used to measure the number and characteristics of these red blood cell precursors. Reticulocyte numbers have been used to account for reticulocyte growth, a first step in erythropoiesis [87] and reticulocyte mean corpuscular hemoglobin (CHr) is gaining acceptance as a means of detecting iron limited erythropoiesis [88]. Other reticulocyte parameters, such as reticulocyte mean corpuscular volume (MCVr), may also be useful but they have not been explored. Reticulocyte testing provides an additional cost burden to diagnosis of iron deficiency. For this reason, identification of a parameter which could supply similar information but does not require reticulocyte detection would preferable. One potential candidate for this is the percentage of hypochromic red blood cells (%HYPO). Because this parameter evaluates the number of cells at an extreme of the distribution curve as opposed to evaluating the mean of the population, it is easier to detect small changes. Hence, although %HYPO does not give information on the age of the iron deficient cells, it may provide a good assessment of an ongoing mild iron deficiency in a cost-effective way. This test is not used in routine clinical practice but is being tested in exploratory settings [89, 90].

Serum iron parameters can be used to predict iron deficiency and are routinely used in combination with red blood cell parameters. The serum iron indices most commonly used are Tsat and serum ferritin, with Tsat serving as a transient measure of available iron and ferritin, in this case, reflecting accumulated iron stores. Based on data from the Centers for Disease Control and Prevention, normal ranges of serum ferritin vary from 12–300 ng/mL for males and 12–150 ng/mL for females and normal ranges of Tsat vary from 20–50%. Absolute levels of these parameters used to diagnose iron deficiency anemia may vary. According to the Kidney Disease Outcomes Quality Initiative and the National Comprehensive Cancer Network guidelines, an individual is iron deficient when serum ferritin is <100 ng/mL and Tsat is <20%.

Another serum iron parameter that may aid diagnosis of iron deficiency anemia is the concentration of soluble transferrin receptor (sTfR), which is believed to be shed from cells with increased TfR production due to intracellular iron deficit. This parameter is rarely used in isolation, but can be used in combination with serum ferritin (sTfR/ferritin ratio, or R/F ratio) to evaluate body iron stores and determine whether iron supplementation is required [91].

Treatment of iron deficiency anemia

The first line of treatment for iron deficiency is oral iron administration (150–250 mg of elemental iron daily). If oral iron is not efficacious, intravenous iron can be considered. Both forms of iron have associated complications. Oral iron is poorly absorbed and may cause gastrointestinal disturbances. Intravenous iron may induce immune complications, myalgias, arthralgias, headache, and a rare incidence of anaphylaxis. It has also been speculated to increase risk of infection, is associated with increased oxidative products, and may in itself be a risk factor for some disorders such as atherosclerosis [92, 93].

In classical iron deficiency anemia with no inflammation, the amounts of erythropoietin are normal or increased, making ESA treatment unnecessary; however, in patients with iron deficiency anemia who are receiving hemodialysis and who do not have increased amounts of erythropoietin due to impaired or absent kidney function, combination treatment of an ESA and intravenous iron has been shown to improve patient response [94]. In this setting, oral iron has been shown not to be efficacious, presumably due to iron malabsorption as a function of kidney disease.

Functional iron deficiency

Functional iron deficiency occurs when circulating iron is insufficient to satisfy demand. In contrast to absolute iron deficiency where body stores are entirely absent, functional iron deficiency can occur in the presence of adequate iron stores. Functional iron deficiency can result from ESA treatment where demand is greatly increased over a short time frame. ESA treatment can boost the rate of erythropoiesis two- to three-fold [95], and under these conditions, the iron available in serum cannot be replenished quickly enough to fully meet demand and erythropoiesis may be restricted by lack of iron. This type of functional iron deficiency is transient. To compensate, iron uptake and mobilization are induced by suppressing hepcidin levels [69].

Another cause of functional iron deficiency is inflammation. Patients with inflammation show decreased serum iron and increased amounts of hepcidin [96, 97]. Mouse models lacking hepcidin do not show a serum iron decrease when stimulated with inflammation, suggesting that hepcidin is required for induction of functional iron deficiency by inflammation [69]. It has been speculated that functional iron deficiency during inflammation is a means of depriving invading organisms of iron [98–100]. Functional iron deficiency is believed to be a contributing or causative factor in the anemia of inflammation. Anemia of inflammation, sometimes referred to as anemia of chronic disease, is a broad term used to describe anemia which is inflammatory in origin.

Inflammation-induced functional iron deficiency may limit response to ESAs. In patients with chronic kidney disease, the study of inflammatory markers has demonstrated that a link exists between inflammation, anemia, and relative resistance to ESA therapy [101–107]. Other inflammatory patient populations, such as patients diagnosed with autoimmune conditions, show a correlation between inflammatory status and response to ESA therapy [108]. A subset of patients with cancer, chronic heart failure [109] and those receiving treatment in intensive care units [110] are also predicted to have an anemia of inflammatory origin. Definitive studies correlating inflammation with ESA responsiveness have not been conducted in these patient populations.

Diagnosis of anemia of inflammation

Distinguishing anemia of inflammation from iron deficiency anemia and mixed anemia (containing components of both) is complicated since most of the commonly used diagnostic parameters are influenced by inflammation (present as part of the acute phase response). The ratio of sTfR concentration (a marker of erythroid activity) to the log of the ferritin (Ft) concentration has been described as a means of distinguishing between inflammatory and iron deficiency anemia [111]. Either sTfR or ferritin alone can give information about iron stores or iron mobilization, but both are impacted by inflammation. sTfR concentrations are high in patients with active erythropoiesis or iron deficiency but low or normal in patients with inflammation (blunted erythropoiesis). Conversely, ferritin can be increased due to inflammation or to increased iron stores. Although both ferritin and sTfR are confounded by multiple interpretations when considered independently, they can effectively differentiate between iron deficiency and inflammation when used in combination. Anemia of inflammation is reported to be characterized by a low sTfR/log Ft ratio (values <1), while a high ratio is indicative of iron deficiency anemia [111, 112].

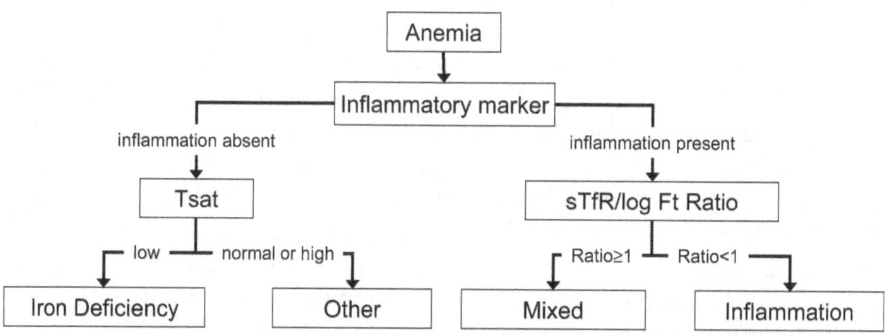

Figure 4. A potential diagnostic scheme to identify patients with iron deficiency anemia (IDA), anemia of inflammation (AI), anemia of mixed origin (some components of IDA and AI) and anemia of other origins (other).

A diagnostic scheme using inflammatory status, Tsat, and sTfR/log Ft (modified from that proposed by [112]) is presented in Figure 4. This scheme illustrates how existing tests could be used to distinguish anemia of inflammation from iron deficiency anemia and mixed anemia. This diagnostic decision tree relies on first detecting the presence or absence of inflammation and then the presence of iron deficiency. A good candidate for the inflammatory marker may be C-reactive protein, a liver-expressed protein that is upregulated in response to inflammation. Another potential candidate is serum hepcidin although a broadly-available and technically accessible assay would first be required [113, 114]. Further distinction between anemia of inflammation and mixed anemia is achieved using sTfR/log ferritin. Iron deficiency is identified in noninflamed patients using Tsat. This type of approach has not been validated and substantial patient data with treatment outcomes will be needed to define the cutoffs for each of the parameters used in the scheme. Tests for other markers clarifying the relationship between iron metabolism and anemia of inflammation are actively sought, including the measurement of hepcidin, which may simplify diagnosis of anemia of inflammation [113].

Treatment of anemia of inflammation

Anemia of inflammation is a multifactorial disease: in addition to functional iron deficiency, it is characterized by inflammatory impairment of endogenous erythropoietin production, reduced responsiveness of marrow cells to ESAs, and shortened red blood cell lifespan [115, 116]. Administration of ESAs can compensate for decreased EPO production and decreased progenitor responsiveness; however, the reticuloendothelial iron block remains as an obstacle to a complete therapeutic response. Treatment with oral iron provides little benefit as uptake is blocked by the increased hepcidin in these patients. The use of intravenous iron may generate short-term response to ESAs, but the bulk of the parenteral infusion may be rapidly sequestered in the reticuloendothelial system potentially creating reactive iron stores in macrophages. Some studies have suggested that intravenous iron treatment and sequestration may be a risk factor for cardiovascular disease and other disorders [117].

New therapeutic strategies for the manipulation of iron

As understanding of the molecules and mechanisms involved in iron metabolism increases, a number of therapeutic opportunities may emerge. Treatment options for iron excess and deficiency have changed little over the years and equate to either removing (phlebotomy, chelation) or administering iron. These treatments have a number of drawbacks including poor patient compliance, undesirable side effects, and limited efficacy. The hope of future therapeutics is the advent of molecular therapies that will enable the controlled redistribution of iron, which would prevent inappropriate accumulation in conditions of iron excess and allow adequate mobilization to satisfy demand in conditions of iron deficiency. It would also avoid iron sequestration in potentially dangerous locations. Potential therapeutics for hereditary hemochromatosis may include antagonism of enterocyte transport proteins or chelation of iron in the gastrointestinal tract to inhibit iron uptake from the diet. Administration of hepcidin or a hepcidin agonist could potentially reverse or limit progression of hereditary hemochromatosis as has been suggested by mouse models [118]. For iron overload disorders in general, modulation of proteins involved in iron sensing or transport may provide therapeutic potential.

Treatment opportunities to address functional iron deficiency include direct neutralization of hepcidin or modulation of proteins that regulate hepcidin expression such as HFE, HJV, TfR2, BMPs, or inflammatory cytokines. The ability to downregulate IL-6-induced hepcidin expression in multicentric Castleman's disease (an IL-6 driven disorder) has been demonstrated using a humanized anti-IL-6 antibody [119]. In addition, hypoxia-inducible factor prolyl hydroxylase (HIF-PH) inhibitors are reported to suppress hepcidin expression in addition to stimulating EPO expression [120]. Development of even a few of these possibilities would change the treatment paradigm for many iron metabolism disorders and potentially provide patients with safer and more efficacious treatment.

References

1 Gunshin H, Mackenzie B, Berger UV et al. Cloning and characterization of a mammalian proton-coupled metal-ion transporter. *Nature* 1997;388:482–488.
2 Liuzzi JP, Aydemir F, Nam H et al. Zip14 (slc39a14) mediates non-transferrin-bound iron uptake into cells. *Proc Natl Acad Sci USA* 2006;103:13612–13617.
3 McKie AT, Barrow D, Latunde-Dada GO et al. An iron-regulated ferric reductase associated with the absorption of dietary iron. *Science* 2001;291:1755–1759.
4 Gunshin H, Starr CN, DiRenzo C et al. Cybrd1 (duodenal cytochrome b) is not necessary for dietary iron absorption in mice. *Blood* 2005;106:2879–2883.
5 Shayeghi M, Latunde-Dada GO et al. Identification of an intestinal heme transporter. *Cell* 2005;122:789–801.
6 Qiu A, Jansen M, Sakaris A et al. Identification of an intestinal folate transporter and the molecular basis for hereditary folate malabsorption. *Cell* 2006;127:917–928.
7 Parmley RT, Barton JC, Conrad ME et al. Ultrastructural cytochemistry and radioautography of hemoglobin – iron absorption. *Exp Mol Pathol* 1981;34:131–144.
8 Donovan A, Lima CA, Pinkus JL et al. The iron exporter ferroportin/slc40a1 is essential for iron homeostasis. *Cell Metab* 2005;1:191–200.
9 Vulpe CD, Kuo Y-M, Murphy TL et al. Hephaestin, a ceruloplasmin homologue implicated in intestinal iron transport, is defective in the sla mouse. *Nat Genet* 1999;21:195–199.
10 Harris ZL, Durley AP, Man TK et al. Targeted gene disruption reveals an essential role for ceruloplasmin in cellular iron efflux. *Proc Natl Acad Sci USA* 1999;96:10812–10817.
11 Sharp P. The molecular basis of copper and iron interactions. *Proc Nutrition Soc* 2004;63:563–569.
12 Aisen P, Leibman A, Zweier J. Stoichiometric and site characteristics of the binding of iron to human transferrin. *J Biol Chem* 1978;253:1930–1937.
13 Wada HG, Hass PE, Sussman HH. Transferrin receptor in human placental brush border membranes. Studies on the binding of transferrin to placental membrane vesicles and the identification

of a placental brush border glycoprotein with high affinity for transferrin. *J Biol Chem* 1979;254:12629–12635.
14 Paterson S, Armstrong NJ, Iacopetta BJM et al. Intravesicular ph and iron uptake by immature erythroid cells. *J Cell Physiol* 1984;120:225–232.
15 Graham RM, Chua AC, Herbison CE et al. Liver iron transport. World. *J Gastroenterol* 2007;13:4725–4736.
16 Jandl JH, Katz JH. The plasma-to-cell cycle of transferrin. *J Clin Invest* 1963;42:314–326.
17 Trinder D, Morgan E. Uptake of transferrin-bound iron by mammalian cells; in Templeton DM (ed.) *Molecular and cellular iron transport. Toronto*, Marcel Dekker, Inc., 2002;427–442.
18 Brissot P, Pigeon C, Loreal O. Regulation of systemic iron transport and storage; in Templeton DM (ed.) *Molecular and cellular iron transport. Toronto*, Marcel Dekker, Inc., 2002;597–609.
19 Ma Y, Yeh M, Yeh K-Y et al. Iron imports. V. Transport of iron through the intestinal epithelium. *Am J Physiol Gastrointest Liver Physiol* 2006;290:G417–422.
20 Arosio P, Levi S. Ferritin, iron homeostasis, and oxidative damage. *Free Radical BiolMed* 2002;33:457–463.
21 Cragg SJ, Wagstaff M, Worwood M. Detection of a glycosylated subunit in human serum ferritin. *Biochem J* 1981;199:565–571.
22 Halliday JW, Powell LW. Ferritin metabolism and the liver. *Sem Liver Dis* 1984;4:207–216.
23 Blight GD, Morgan EH. Ferritin and iron uptake by reticulocytes. *Br J Haematol* 1983;55:59–71.
24 Loken MR, Shah VO, Dattilio KL et al. Flow cytometric analysis of human bone marrow: I. Normal erythroid development. *Blood* 1987;69:255–263.
25 Smith DW. The molecular biology of mammalian hemoglobin synthesis. *Ann Clin Lab Sci* 1980;10:116–122.
26 Shaw GC, Cope JJ, Li L et al. Mitoferrin is essential for erythroid iron assimilation. *Nature* 2006;440:96–100.
27 Napier I, Ponka P, Richardson DR. Iron trafficking in the mitochondrion: Novel pathways revealed by disease. *Blood* 2005;105:1867–1874.
28 Quigley JG, Yang Z, Worthington MT et al. Identification of a human heme exporter that is essential for erythropoiesis. *Cell* 2004;118:757–766.
29 Jonker JW, Buitelaar M, Wagenaar E et al. The breast cancer resistance protein protects against a major chlorophyll-derived dietary phototoxin and protoporphyria. *Proc Natl Acad Sci USA* 2002;99:15649–15654.
30 Shirihai OS, Gregory T, Yu C et al. Abc-me: A novel mitochondrial transporter induced by gata-1 during erythroid differentiation. *EMBO J* 2000;19:2492–2502.
31 Allen DW, Jandl JH. Kinetics of intracellular iron in rabbit reticulocytes. *Blood* 1960;15:71–81.
32 Noyes WD, Hosain F, Finch CA. Incorporation of radioiron into heme. *J Lab Clin Med* 1964;64:574–580.
33 Fontenay M, Cathelin S, Amiot M et al. Mitochondria in hematopoiesis and hematological diseases. *Oncogene* 2006;25:4757–4767.
34 Beutler E. Hemolytic anemia due to chemical and physical agents. In Beutler E, Coller BS, Lichtman MA, Kipps TJ, Seligsohn U (eds): Williams hematology, McGraw-Hill, 2001;629–632.
35 Bowman WDJ. Abnormal ("Ringed") sideroblasts in various hematologic and non-hematologic disorders. *Blood* 1961;18:662–671.
36 Bekri S, May A, Cotter PD et al. A promoter mutation in the erythroid-specific 5-aminolevulinate synthase (alas2) gene causes x-linked sideroblastic anemia. *Blood* 2003;102:698–704.
37 Taketani S, Adachi Y, Nakahashi Y. Regulation of the expression of human ferrochelatase by intracellular iron levels. *Eur J Biochem* 2000;267:4685–4692.
38 Tahara T, Sun J, Igarashi K et al. Heme-dependent up-regulation of the [alpha]-globin gene expression by transcriptional repressor bach1 in erythroid cells. *Biochem Biophys Res Commun* 2004;324:77–85.
39 Alter BP, Goff SC. Globin synthesis in mouse erythroleukemia cells *in vitro*: A switch in beta chains due to inducing agent. *Blood* 1977;50:867–876.
40 Cotter PD, Baumann M, Bishop DF. Enzymatic defect in "X-linked" Sideroblastic anemia: Molecular evidence for erythroid {delta}-aminolevulinate synthase deficiency. *Proc Natl Acad Sci USA* 1992;89:4028–4032.
41 Cotter PD, May A, Li L et al. Four new mutations in the erythroid-specific 5-aminolevulinate synthase (alas2) gene causing x-linked sideroblastic anemia: Increased pyridoxine responsiveness after removal of iron overload by phlebotomy and coinheritance of hereditary hemochromatosis.

Blood 1999;93:1757–1769.
42 Bottomley S, Tanaka M, Everett M. Diminished erythroid ferrochelatase activity in protoporphyria. *J Lab Clin Med* 1975;86:126–131.
43 Borgna-Pignatti C, Galanello R. Thalassemias and related disorders: Quantitative disorders of hemoglobin synthesis.; *Wintrobe's clinical hematology*. Philadephia, Lippincott, Williams and Wilkins, 2004;1319–1365.
44 Bratosin D, Mazurier J, Tissier JP et al. Cellular and molecular mechanisms of senescent erythrocyte phagocytosis by macrophages. A review. *Biochimie* 1998;80:173–195.
45 Biggs TE, Baker ST, Botham MS et al. Nramp1 modulates iron homoeostasis *in vivo* and *in vitro*: Evidence for a role in cellular iron release involving de-acidification of intracellular vesicles. *Eur J Immunol* 2001;31:2060–2070.
46 Jabado N, Canonne-Hergaux F, Gruenheid S et al. Iron transporter nramp2/dmt-1 is associated with the membrane of phagosomes in macrophages and sertoli cells. *Blood* 2002;100:2617–2622.
47 Delaby C, Pilard N, Puy H et al. Sequential regulation of ferroportin expression after erythrophagocytosis in murine macrophages: Early mrna induction by haem, followed by iron-dependent protein expression. *Biochem J* 2008;411:123–131.
48 Krause A, Neitz S, Magert HJ et al. Leap-1, a novel highly disulfide-bonded human peptide, exhibits antimicrobial activity. *FEBS Lett* 2000;480:147–150.
49 Park CH, Valore EV, Waring AJ et al. Hepcidin, a urinary antimicrobial peptide synthesized in the liver. *J Biol Chem* 2001;276:7806–7810.
50 Pigeon C, Ilyin G, Courselaud B et al. A new mouse liver-specific gene, encoding a protein homologous to human antimicrobial peptide hepcidin, is overexpressed during iron overload. *J Biol Chem* 2001;276:7811–7819.
51 Laftah AH, Ramesh B, Simpson RJ et al. Effect of hepcidin on intestinal iron absorption in mice. *Blood* 2004;103:3940–3944.
52 Yeh KY, Yeh M, Glass J. Hepcidin regulation of ferroportin 1 expression in the liver and intestine of the rat. *Am J Physiol GastroLiver Physiol* 2004;286:G385–394.
53 Frazer DM, Wilkins SJ, Becker EM et al. Hepcidin expression inversely correlates with the expression of duodenal iron transporters and iron absorption in rats. *Gastroenterology* 2002;123:835–844.
54 Nemeth E, Tuttle MS, Powelson J et al. Hepcidin regulates cellular iron efflux by binding to ferroportin and inducing its internalization. *Science* 2004;306:2090–2093.
55 Kulaksiz H, Theilig F, Bachmann S et al. The iron-regulatory peptide hormone hepcidin: Expression and cellular localization in the mammalian kidney. *J Endocrinol* 2005;184:361–370.
56 Bekri S, Gual P, Anty R et al. Increased adipose tissue expression of hepcidin in severe obesity is independent from diabetes and nash. *Gastroenterology* 2006;131:788–796.
57 Ilyin G, Courselaud B, Troadec MB et al. Comparative analysis of mouse hepcidin 1 and 2 genes: Evidence for different patterns of expression and co-inducibility during iron overload. *FEBS Lett* 2003;542:22–26.
58 Peyssonnaux C, Zinkernagel AS, Datta V et al. Tlr4-dependent hepcidin expression by myeloid cells in response to bacterial pathogens. *Blood* 2006;107:3727–3732.
59 Gnana-prakasam JP, Martin PM, Mysona BA et al. Hepcidin expression in mouse retina and its regulation via lipopolysaccharide/toll-like receptor-4 pathway independent of hfe. *Biochem J* 2008;411:79–88.
60 Nicolas G, Bennoun M, Devaux I et al. Lack of hepcidin gene expression and severe tissue iron overload in upstream stimulatory factor 2 (usf2) knockout mice. *Proc Natl Acad Sci USA* 2001;98:8780–8785.
61 Lesbordes-Brion JC, Viatte L, Bennoun M et al. Targeted disruption of the hepcidin 1 gene results in severe hemochromatosis. *Blood* 2006;108:1402–1405.
62 Nicolas G, Bennoun M, Porteu A et al. Severe iron deficiency anemia in transgenic mice expressing liver hepcidin. *Proc Natl Acad Sci USA* 2002;99:4596–4601.
63 Roetto A, Papanikolaou G, Politou M et al. Mutant antimicrobial peptide hepcidin is associated with severe juvenile hemochromatosis. *Nature Gen* 2003;33:21–22.
64 Weinstein DA, Roy CN, Fleming MD et al. Inappropriate expression of hepcidin is associated with iron refractory anemia: Implications for the anemia of chronic disease. *Blood* 2002;100:3776–3781.
65 Schmidt PJ, Huang FW, Wrighting DM et al. Hepcidin expression is regulated by a complex of hemochromatosis-associated proteins. *Blood* 2006;108:abstract 267

66 Schmidt PJ, Toran PT, Giannetti AM et al. The transferrin receptor modulates hfe-dependent regulation of hepcidin expression. *Cell Metab* 2008;7:205–214.
67 Goswami T, Andrews NC. Hereditary hemochromatosis protein, hfe, interaction with transferrin receptor 2 suggests a molecular mechanism for mammalian iron sensing. *J Biol Chem* 2006;281:28494–28498.
68 Babitt JL, Huang FW, Wrighting DM et al. Bone morphogenetic protein signaling by hemojuvelin regulates hepcidin expression. *Nature Gen* 2006;38:531–539.
69 Nicolas G, Chauvet C, Viatte L et al. The gene encoding the iron regulatory peptide hepcidin is regulated by anemia, hypoxia, and inflammation. *J Clin Invest* 2002,110:1037–1044.
70 Pak M, Lopez MA, Gabayan V et al. Suppression of hepcidin during anemia requires erythropoietic activity. *Blood* 2006;108:3730–3735.
71 Nemeth E, Rivera S, Gabayan V et al. Il-6 mediates hypoferremia of inflammation by inducing the synthesis of the iron regulatory hormone hepcidin. *J Clin Invest* 2004;113:1251–1253.
72 Lee P, Peng H, Gelbart T et al. Regulation of hepcidin transcription by interleukin-1 and interleukin-6. *Proc Natl Acad Sci USA* 2005;102:1906–1910.
73 Wrighting DM, Andrews NC. Interleukin-6 induces hepcidin expression through stat3. *Blood* 2006;108:3204–3209.
74 Jean G, Terzoli S, Mauri R et al. Cirrhosis associated with multiple transfusions in thalassaemia. *Arch Dis Child* 1984;59:67–70.
75 Kushner JP, Porter JP, Olivieri NF: Secondary iron overload. *Hematology Am Soc Hematol Educ Program* 2001:47–61.
76 Nemeth E, Ganz T. Regulation of iron metabolism by hepcidin. *Annu Rev Nutr* 2006;26:323–342.
77 Waalen J, Beutler E. Hereditary hemochromatosis: Screening and management. *Curr Hematol Rep* 2006;5:34–40.
78 Yen AW, Fancher TL, Bowlus CL et al. Revisiting hereditary hemochromatosis: Current concepts and progress. *Am J Med* 2006;119:391–399.
79 De Gobbi M, Roetto A, Piperno A et al. Natural history of juvenile haemochromatosis. *Br J Haematol* 2002;117:973–979.
80 Lamon JM, Marynick SP, Roseblatt R et al. Idiopathic hemochromatosis in a young female. A case study and review of the syndrome in young people. *Gastroenterology* 1979;76:178–183.
81 Pietrangelo A, Caleffi A, Henrion J et al. Juvenile hemochromatosis associated with pathogenic mutations of adult hemochromatosis genes. *Gastroenterology* 2005;128:470–479.
82 Brissot P, de Bels F. Current approaches to the management of hemochromatosis. Hematol *Am Soc Hematol Educ Program* 2006:36–41.
83 Jabbour E, Kantarjian HM, Koller C et al. Red blood cell transfusions and iron overload in the treatment of patients with myelodysplastic syndromes. *Cancer* 2008;112:1089–1095.
84 Beutler E, Hoffbrand AV, Cook JD. Iron deficiency and overload. Hematol *Am Soc Hematol Educ Program* 2003;2003:40–61.
85 Besarab A. Resolving the paradigm crisis in intravenous iron and erythropoietin management. *Kidney Int Suppl* 2006;101:S13–18.
86 Cook JD, Skikne BS, Baynes RD. Iron deficiency: The global perspective. *Adv Exp Med Biol* 1994;356:219–228.
87 d'Onofrio G, Kuse R, Foures C et al. Reticulocytes in haematological disorders. *Clin Lab Haematol* 1996;18:29–34.
88 Brugnara C. Reticulocyte cellular indices: A new approach in the diagnosis of anemias and monitoring of erythropoietic function. *Crit Rev Clin Lab Sci* 2000;37:93–130.
89 Thomas C, Thomas L. Biochemical markers and hematologic indices in the diagnosis of functional iron deficiency. *Clin Chem* 2002;48:1066–1076.
90 Katodritou E, Terpos E, Zervas K et al. Hypochromic erythrocytes (%): A reliable marker for recognizing iron-restricted erythropoiesis and predicting response to erythropoietin in anemic patients with myeloma and lymphoma. *Ann Hematol* 2007;86:369–376.
91 Brugnara C. Iron deficiency and erythropoiesis: New diagnostic approaches. *Clin Chem* 2003;49:1573–1578.
92 Sengoelge G, Sunder-Plassmann G, Hörl WH. Potential risk for infection and atherosclerosis due to iron therapy. *J Ren Nutr* 2005;15:105–110.
93 Burns DL, Pomposelli JJ. Toxicity of parenteral iron dextran therapy. *Kidney Int Suppl* 1999;69:S119–124.
94 Besarab A, Amin N, Ahsan M et al. Optimization of epoetin therapy with intravenous iron

therapy in hemodialysis patients. *J Am Soc Nephrol* 2000;11:530–538.
95 Eschbach JW, Egrie JC, Downing MR et al. Correction of the anemia of end-stage renal disease with recombinant human erythropoietin. Results of a combined phase I and II clinical trial. *N Engl J Med* 1987;316:73–78.
96 Kemna E, Pickkers P, Nemeth E et al. Time-course analysis of hepcidin, serum iron, and plasma cytokine levels in humans injected with lps. *Blood* 2005;106:1864–1866.
97 Nemeth E, Valore EV, Territo M et al. Hepcidin, a putative mediator of anemia of inflammation, is a type ii acute-phase protein. *Blood* 2003;101:2461–2463.
98 Roy CN, Weinstein DA, Andrews NC. The molecular biology of the anemia of chronic disease: A hypothesis. *Ped Res* 2003;53:507–512.
99 Ganz T. Hepcidin, a key regulator of iron metabolism and mediator of anemia of inflammation. *Blood* 2003;102:783–788.
100 Andrews NC. Anemia of inflammation: The cytokine-hepcidin link. *J Clin Invest* 2004; 113:1251–1253.
101 Horl WH, Jacobs C, Macdougall IC et al. European best practice guidelines 17–18: Adverse effects. *Nephrol Dial Transplant* 2000;15 S4:51–56.
102 Horl WH, Jacobs C, Macdougall IC et al. European best practice guidelines 14–16: Inadequate response to epoetin. *Nephrol Dial Transplant* 2000;15 S4:43–50.
103 Barany P, Divino-Filho JC, Bergstrom J. High c-reactive protein is a strong predictor of resistance to erythropoietin in hemodialysis patients. *Am J Kidney Dis* 1997;29:565–568.
104 Qureshi AR, Alvestrand A, Danielsson A et al. Factors predicting malnutrition in hemodialysis patients: A cross-sectional study. *Kidney Int* 1998;53:773–782.
105 Yeun JY, Levine RA, Mantadilok V et al. C-reactive protein predicts all-cause and cardiovascular mortality in hemodialysis patients. *Am J Kidney Dis* 2000;35:469–476.
106 Zimmermann J, Herrlinger S, Pruy A et al. Inflammation enhances cardiovascular risk and mortality in hemodialysis patients. *Kidney Int* 1999;55:648–658.
107 Fine A. Relevance of c-reactive protein levels in peritoneal dialysis patients. *Kidney Int* 2002;61:615–620.
108 Bertero MT, Caligaris-Cappio F. Anemia of chronic disorders in systemic autoimmune diseases. *Haematology* 1997;82:375–381.
109 Nanas JN, Matsouka C, Karageorgopoulos D et al. Etiology of anemia in patients with advanced heart failure. *J Am Coll Cardiol* 2006;48:2485–2489.
110 Pieracci FM, Barie PS. Diagnosis and management of iron-related anemias in critical illness. *Crit Care Med* 2006;34:1898–1905.
111 Punnonen K, Irjala K, Rajamaki A. Serum transferrin receptor and its ratio to serum ferritin in the diagnosis of iron deficiency. *Blood* 1997;89:1052–1057.
112 Weiss G, Goodnough LT. Anemia of chronic disease. *N Engl J Med* 2005;352:1011–1023.
113 Kemna EHJM, Tjalsma H, Willems HL et al. Hepcidin: From discovery to differential diagnosis. *Haematol* 2008;93:90–97.
114 Ganz T. Molecular control of iron transport. *J Am Soc Nephrol* 2007;18:394–400.
115 Cartwright GE. The anemia of chronic disorders. *Semin Hematol* 1966;3:351–375.
116 Means RT Jr, Advances in the anemia of chronic disease. *Int J Hematol* 1999;70:7–12.
117 Yuan XM, Li W, Yuan X-M et al. The iron hypothesis of atherosclerosis and its clinical impact. *Ann Med* 2003;35:578–591.
118 Viatte L, Nicolas G, Lou D-Q et al. Chronic hepcidin induction causes hyposideremia and alters the pattern of cellular iron accumulation in hemochromatotic mice. *Blood* 2006;107:2952–2958.
119 Kawabata H, Tomosugi N, Kanda J et al. Anti-interleukin 6 receptor antibody tocilizumab reduces the level of serum hepcidin in patients with multicentric castleman's disease. *Haematologica* 2007;92:857–858.
120 Lagsetmo I, Young B, Zhang W et al. Effect of fg-2216 on anemia and iron transport in a rat model of anemia of chronic disease. *J Am Soc Nephrol* 2004;15:548A.

Erythropoietins, Erythropoietic Factors and Erythropoiesis
Edited by S.G. Elliott, M.A. Foote, and G. Molineux

Nonhematopoietic effects of erythropoiesis-stimulating agents

Wolfgang Jelkmann[1], Reinhard Depping[1] and Eric Metzen[2]

[1] *Institute of Physiology, University of Lübeck, Lübeck, Germany*
[2] *Institute of Physiology, University Duisburg-Essen, Essen, Germany*

Introduction

Erythropoietin (EPO) was originally considered a hormone with action restricted to erythrocytic progenitors. The colony-forming units–erythroid (CFU–E) possess abundant EPO receptor (EPOR) molecules, and EPO is essential for their survival, proliferation, and differentiation. The mature functional human erythropoietic EPOR is a homodimer of two 484 amino acids transmembrane glycoproteins of approximately 60 kDa. On binding of 1 EPO molecule, the EPOR dimer undergoes a conformational change that is transmitted to the cytosolic domain resulting in the autophosphorylation of EPOR-associated Janus protein tyrosine kinases 2 (JAK-2). The tyrosine phosphorylated EPOR dimer exhibits intracellular docking sites for signaling proteins containing SRC homology 2 (SH2) domains [1]. The specific roles of the various then-activated enzymes, such as phosphatidylinositol-3 kinase (PI-3K)/Akt, mitogen-activated protein kinase (MAPK), protein kinase C (PKC), and transcription factors, such as signal transducer and activator of transcription 5 (STAT-5), with respect to survival, proliferation, and differentiation are only partly understood. Clearly, however, the action of EPO requires the presence of functional EPOR and of appropriate EPOR signaling molecules. Activation of the EPOR is terminated by JAK-2 dephosphorylation and internalization of the EPO/EPOR complex [2].

Studies suggest that EPOR is expressed in a variety of nonhematopoietic tissues, including the brain, the cardiovascular bed, and the kidneys [3, 4]; however, the physiologic role of EPO/EPOR system in nonhematopoietic tissues deserves further clarification for several reasons. First, EPOR mRNA expression does not prove the translation of functional EPOR molecules. Unfortunately, the detection of EPOR protein has been rendered difficult due to the lack of appropriate antibodies for use in immunohistochemistry and Western blotting [5]. Second, while EPO/EPOR system is essential for erythropoiesis, nonhematopoietic EPOR may be dispensible under normal conditions. Transgenic mice with a constitutive genetic deletion of *EPOR* are not viable, but the animals develop normally and are fertile after tissue-specific, re-expression of EPOR in the erythrocytic lineage [6]. Distinct tissue-protective effects of EPO outside the bone marrow, however, are probably relevant under pathophysiologic conditions of organ injury. Signaling through EPOR can prevent apoptotic cell death caused by reactive O_2 species, Ca^{2+} mobilization, and mitochondrial dysfunction resulting from stressors such as hypoxia/reoxygenation, cytotoxic drugs, or mediators of inflammation. The pharmacological use of recombinant human EPO (rHuEPO), of analogous erythropoiesis-stimulating agents (ESA) and of nonerythropoietic derivatives thereof for nonhematopoietic indications, such as neuro-, cardio-, and nephroprotection, is in the focus of nonclinical and clinical research (Fig. 1). In considering EPO a pleiotropic cell survival factor, however, the question of whether administration of the drugs may promote tumor growth [5, 7] must be addressed.

We summarize *in vitro* and *in vivo* studies of the role of EPOR in nonhematopoietic tissues. The available data are considered with respect to the following questions:

- Is EPOR mRNA detectable?
- Is EPOR protein detectable?
- Can EPOR signaling be demonstrated?
- Do effective ESA concentrations match the physiological and/or pharmacologic levels attained in patients treated with ESA?

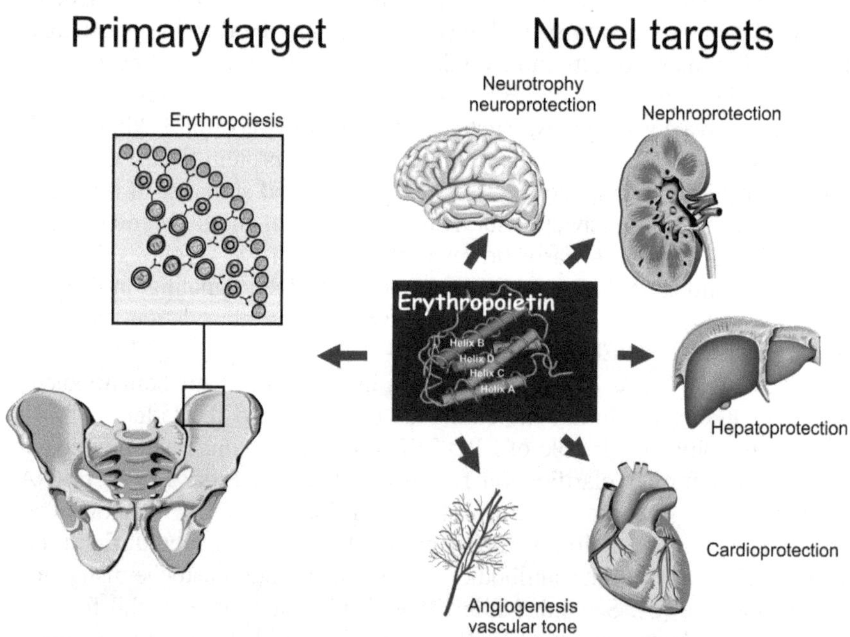

Figure 1. Targets of EPO, respectively its derivatives and analogs. Apart from stimulating erythropoiesis, EPO has tissue protective potential in various other organs.

Effects of ESA in the central nervous system

First evidence for the expression of functional EPOR by neuronal cells was provided 15 years ago with the findings that rat PC12 pheochromocytoma and mouse basal forebrain SN6 cells have EPO-binding sites, and that EPO stimulates the release of monoamines from PC12 cells, augments choline acetyltransferase activity in primary cultures of mouse neurons, and supports the survival of septal cholinergic neurons in rats with fimbria-fornix transactions [8, 9]. Subsequently, investigators detected EPOR mRNA and EPOR protein in defined areas of the brain of rodents, monkeys, and humans, with EPO-binding sites being detectable mainly in the hippocampus, capsula interna, cortex, and midbrain [10]. EPOR mRNA is expressed in the central nervous system during human fetal development at about 7 weeks, and increases from 8–24 weeks [11]. Simultaneously, the *EPO* gene is expressed in brain, where EPO can exert cytoprotective functions separate from the remainder of the body [12]. EPO has been reported to stimulate the generation of neurons from mammalian forebrain neuronal stem cells [13]. In addition to neurons and astrocytes, brain capillary endothelial cells express EPO-binding sites [14]. It was earlier claimed that EPOR can be detected within and around human brain capillaries [15], but the anti-EPOR antiserum used for the study was later shown to be unspecific [16–19]. Thus, the localization of EPOR protein in the brain requires reinvestigation [18].

Most *in vitro* studies aimed at demonstrating effects of EPO on neuronal cells were performed with very high EPO concentrations (≥ 0.5 IU/mL) with few reports showing positive effects at physiologically low EPO concentrations [20–22]. The concentration of endogenous EPO in the cerebrospinal fluid is normally <0.001 U/mL [23], but the actual concentration in the brain microenvironment is not known, as EPO derived from neuronal and glial cells may act in an autocrine and paracrine way. This fact must be remembered when proposing that EPO can protect neuronal cells against toxic agents and hypoxia/reoxygenation injury. EPO rescues hippocampal and cerebral rat cortical neurons in primary culture from nitric oxide (NO)-induced death [24]. Preconditioning with EPO protects neurons from NO and N-methyl-D-aspartate (NMDA) injury [25]. Preconditioning is not necessarily required for neuroprotection, because EPO prevents apoptosis of neurons when applied simultaneously with hypoxic stress [26]. Inhibition of glutamate-induced damage of cultured neurons may be the primary mechanism by which EPO protects hippocampal and cerebral cortical neurons from death due to hypoxia [27]. Of potential clinical interest is the finding that EPO can protect rat primary hippocampal neuronal cells in culture from damage induced by β-amyloid [28], as this protein contributes to Alzheimer's disease. The neuroprotective action of EPO involves the JAK-2, PI-3K/Akt, MAPK, and NFκB pathways [25, 28] (Fig. 2).

In vivo evidence for the neuroprotective potential of the EPO/EPOR system was provided with the findings that the infusion of EPO into the lateral ventricles of Mongolian gerbils rescues hippocampal CA1 neurons from ischemic damage, while infusion of EPOR causes neuronal degeneration and impaired

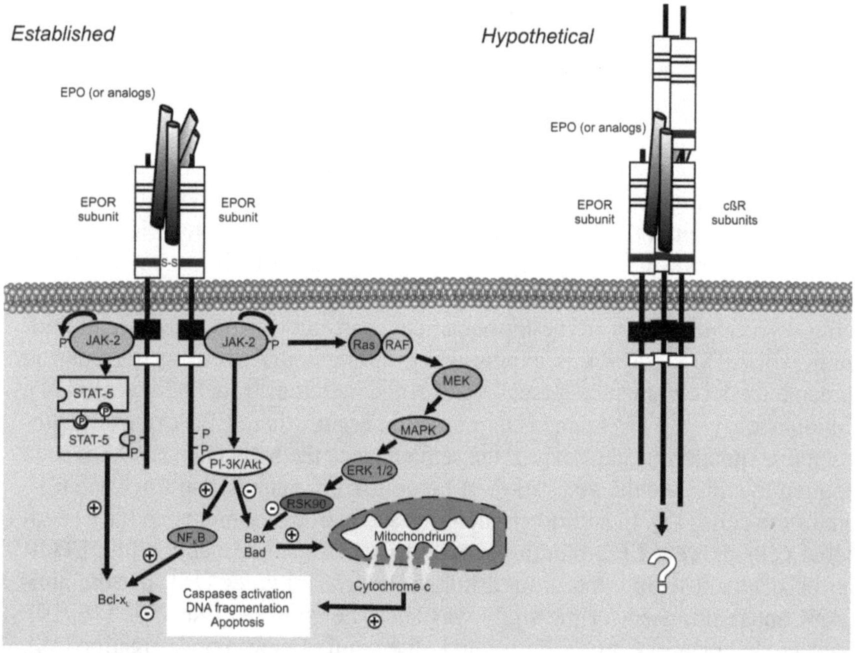

Figure 2. Scheme of the established homodimeric EPO receptor (EPOR) and the heteromeric receptor proposed to function in nonhematopoietic tissues. The antiapoptotic action of EPO on binding to EPOR involves autophosphorylation of JAK-2 (Janus kinase 2), tyrosine phosphorylation of EPOR, homodimerization of STAT-5 (signal transducer and activator of transcription 5), induction of Bcl-x_L, activation of PI-3K (phosphatidylinositol-3 kinase), the G-protein Ras and the sequential activation of the serine kinase RAF, MEK (syn. MAPKK) and MAPK (mitogen activated protein kinase). As a result Bax and Bad are prevented from increasing mitochondrial membrane permeability and cytochrome c release and the activities of caspase 3 and other caspases otherwise executing apoptosis are silenced. Whether EPO and its derivatives and analogs can exert anti-apoptotic effects on binding to a putative heteromeric receptor composed of EPOR and cβR (common β-subunit of the IL-3, IL-5, and GM-CSF receptor) in nonhematopoietic tissues is under debate.

learning ability [24]. EPOR mRNA concentrations increase in the periphery of the cerebrocortical infarct, the so-called ischemic penumbra, in rats with permanent middle cerebral artery occlusion [29]. In this model, the infusion of EPO into the cerebral ventricles reduces the ischemia-induced place navigation disability, cortical infarction and thalamic degeneration. The protective effect of EPO involves the induction of Bcl-x_L, an anti-apoptotic member of the Bcl-2 family [30] (Fig. 2). Using mice with a conditional *EPOR* knockdown, it has been shown that the brain-specific deletion of *EPOR* is associated with reduced cell proliferation in the subventricular zone and a specific deficit in post-stroke neurogenesis through impaired migration of neuroblasts to the peri-infarct cortex [31]. Neuronal cells from mice with a partial *EPOR* knockout restricted to organs outside the hematopoietic tissues are more sensitive to hypoxia/glutamate toxicity than normal neurons *in vitro* and *in vivo* [32]. Thus, EPOR appears to be required for the development and for repair processes of the central nervous system. Less conclusive evidence is available in proof of the idea that these effects are mediated by a putative heteroreceptor consisting of EPOR and the common cytokine beta receptor (βcR) [33] (Fig. 2). EPOR/βcR hypothesis is mainly based on the observation that erythropoietically inactive carbamoylated EPO (CEPO) provides neuroprotection in animal models [34].

The brain EPO/EPOR system initially was thought to be completely separate from blood-borne EPO [4], because it seemed unlikely that a large glycoprotein like EPO could readily cross the blood-brain barrier. Brines et al. [15] showed that the systemic administration of rHuEPO reduces brain lesions due to focal ischemia, trauma, or experimental autoimmune encephalitis in rodents. Furthermore, the authors reported that 5 h after intraperitoneal administration of biotinylated rHuEPO, biotin was detectable by immunohistochemistry, which was initially localized around capillaries and later (after 17 h) to scattered neurons [15]. Note, however, that the possibility cannot be excluded that the biotin-labeled EPO was degraded in endothelial cells leaving the liberated biotin to move into the brain. In another study, the systemic application of rHuEPO (5,000 IU/kg, administered intraperitoneally) reduced the volume of infarction in rats 24 h after middle cerebral artery occlusion [35]. The neuroprotective effects of systemically administered rHuEPO in experimental animals with acute cerebral ischemia, crush injuries or encephalomyelitis has been confirmed in several other studies [12, 36].

With respect to the potential use of rHuEPO or its derivatives and analogs as neuroprotective agents in humans, the question as to what extent the drugs can cross the blood-brain barrier has remained a critical issue [37]. Possibly, the drugs get easier access to the brain after a hypoxic lesion of the blood-brain barrier. Another point of interest relates to likely differences in the transfer of drugs with a different molecular size. When either epoetin alfa (5,000 IU/kg) or an equivalent dose of its hyperglycosylated analog darbepoetin alfa (25 μg/kg) were administered intravenously to rats, both drugs were measurable in the cerebrospinal fluid for up to 8 h with the rate of penetration amounting to 0.09% for epoetin alfa and 0.08% for darbepoetin alfa [38]. Interestingly, intravenously administered asialo-rHuEPO, which has a very short plasma half-life of 1.4 min and does not stimulate erythropoiesis, exerts neuroprotective effects in cerebral ischemia, spinal compression, and sciatic nerve crush in rodents [39]. Whether the small cyclic EPO mimetic peptides (EMP; composed of about 20 amino acids) are of therapeutic value in brain injury has not been disclosed. Such studies may be promising, because EMP may cross the blood-brain barrier more readily, and one compound (EMP1) has been shown to inhibit Ca^{2+}-induced glutamate release from cultured rat cerebral cells of hippocampal neurons [27].

The fact that systemically administered EPOR ligands are neuroprotective in experimental animals has led to consideration of rHuEPO or derivatives thereof as therapeutic option in human brain diseases. A pioneering double-blind, placebo-controlled, randomized, proof-of-concept study on 40 patients with acute stroke used magnetic resonance tomography to show a strong trend for reduction in infarct size on rHuEPO treatment [23]. After intravenous administration of 33,000 IU rHuEPO, plasma EPO concentrations amounted to about 5 IU/mL after 3 h, while in the cerebrospinal fluid, the concentration was about 0.02 IU/mL (normal < 0.001 IU/mL). The rHuEPO-treated patients exhibited a markedly improved neurological recovery 1 month after the stroke [23]. 39 men diagnosed with chronic schizophrenia were recruited for a double-blind, placebo-controlled, randomized, multicenter EMP proof-of-principle (Phase II) study to test the hypothesis that EPO is useful as a neuroprotective add-on strategy in schizophrenia. The patients were treated with 40,000 IU rHuEPO intravenously or placebo weekly for 12 weeks. Compared with baseline, the schizophrenia-relevant cognitive function score improved in both groups, presumably reflecting the nonspecific stimulatory effect of the study environment. However, the improvement was significantly greater in the rHuEPO than in the placebo group. The hemoglobin values were relatively stable at a low rate of venesections, indicating a small hematopoietic response [40]. rHuEPO, respectively its derivatives and analogs, are presently considered for neuroprotection in other neurological diseases such as multiple sclerosis, Alzheimer's disease, or Parkinson's disease [37, 41]. A placebo-controlled study of the effects of rHuEPO (40,000 IU) on self-reported mood and cognitive function in healthy volunteers has resulted in behavioral effects similar to those seen on acute administration of serotoninergic antidepressants [42].

In conclusion, EPO appears to provide neuroprotection by two means: systemic EPO stimulates erythropoiesis and, thereby, increases the blood O_2-carrying capacity and brain oxygenation [12]; and brain EPO promotes neurogenesis and the survival of neurons on hypoxic and toxic stress [36]. The local production of EPO in the brain will allow for sufficient bioavailability of the survival factor independent of the transfer across the blood-brain barrier. The neuroprotective action of EPO has prompted clinical trials to use ESA in humans with stroke, schizophrenia, and neurodegenerative diseases [37].

Effects of ESA in the cardiovascular system

Effects of ESA on vascular cells

Although the rHuEPO therapy-induced increase in arterial blood pressure seen in some patients with chronic renal failure results primarily from the increased blood viscosity and the abolishment of tissue hypoxia-associated vasodilation, it is clear that EPO can stimulate the production of endothelial cell-derived modulators of the vascular tone, including both vasoconstrictive mediators, such as endothelin, and vasorelaxant products, such as NO. EPOR mRNA is expressed in human vascular endothelial cells in culture [43]. EPO stimulates the *in vitro* proliferation and migration of endothelial cells, and it induces a proangiogenic phenotype of endothelial cells and neovascularization [44, 45].

The angiogenic potential of EPO may be disadvantageous under certain conditions. For example, the question has been raised whether EPO could promote the retinopathy in diabetes mellitus patients [46]. On the other hand, EPO itself is an important survival factor for photosensor cells [47].

Like erythropoietic cells, cultured human umbilical vein endothelial cells (HUVEC) respond to rHuEPO with the rapid tyrosine phosphorylation of cytosolic proteins and the translocation of STAT-5 [48]. A differential display analysis has shown that rHuEPO upregulates four groups of genes in HUVEC, including genes that encode proteins controlling vascular function (thrombospondin-1), gene transcription (c-myc purine-binding transcription factor PuF), mitochondrial function (cytochrome c oxidase subunit 1) and regulators of signal transduction [49]. High doses of EPO (≥100 IU/mL) have been shown to reduce the incidence of apoptosis in bovine pulmonary artery endothelial cultures treated with bacterial lipopolysaccharide (LPS) [50]. EPO has been reported to protect rat cerebral microvascular endothelial cells in primary culture from apoptosis due to O_2 and glucose deprivation [51]. The protective effect requires PI-3K activation, and it uses the phosphorylation of STAT-3, STAT-5, MAPK, and the fork head transcription factor FOXO3a. Note that in all of the cited endothelial cell culture studies, the concentration of EPO greatly exceeded that reached in the blood of patients treated with rHuEPO.

In addition to acting directly on the resident vasculature, EPO promotes the mobilization of myeloid $CD34^+$ hematopoietic and endothelial progenitor cells into the blood stream. The primary role of endothelial progenitor cells has been demonstrated in studies on transgenic mice lacking *EPOR* in nonerythroid lineages. On exposure of the mice to normobaric hypoxia (10% O_2 for 3 weeks), mobilization of endothelial progenitor cells and incorporation into the pulmonary endothelium was greatly impaired, which resulted in pulmonary hypertension [52]. The increased number of circulating cells detected after the administration of rHuEPO or darbepoetin alfa in humans may prove useful for augmenting the neovascularisation of ischemic tissues [53, 54].

Vascular smooth muscle cells also express EPOR mRNA *in vitro* [55]. When applied in concentrations that are by several orders of magnitude higher than those reached in the clinical setting, rHuEPO induces *in vitro* vascular smooth muscle contraction [56] and vasoconstriction [57]. EPO signaling in vascular smooth muscle cells is Ca^{2+}-dependent and involves the activation of the phospholipase C cascade and of oncogenes (Myc, jun, fos) [58] that promote DNA replication and cellular growth. However, it seems unlikely that EPO plays a major role in vascular smooth muscle biology *in vivo*, because circulating EPO will rarely penetrate through intact endothelium.

Effects of ESA on the heart

EPOR mRNA and EPOR protein have been detected in cultures of human [59] and rat [60] cardiomyocyte cell lines and in samples of human heart [61]. With respect to reports of EPOR protein in the human heart it must be noted, however, that for Western blotting and immunohistochemistry, antiEPOR antibody was used that was subsequently shown to be unspecific and to interact with several other proteins including heat shock proteins [16–19]. On the other hand, animal studies have provided strong evidence that EPO activates cardiac protein kinases [62], opens potassium channels [63], and inhibits apoptosis of cardiac myocytes and fibroblasts [64]. When human atrial trabeculae isolated from patients undergoing coronary artery bypass grafting were exposed to 90 min hypoxia followed by 120-min reoxygenation the addition of rHuEPO (≥5 IU/mL) to the reperfusion medium increased the force of contraction, which required PI-3K and MAPK activities and was associated with reduced caspase 3 activity [65]. Transgenic mice with a tissue-specific EPOR-knockout outside the bone marrow exhibit greater myocardial damage on ischemia/reperfusion than wild-type mice, suggesting that the endogenous EPO/EPOR system plays a protective role in the heart [66]. Treatment of experimental animals with ESA reduces the myocardial infarct volume and improves contractile properties after ischemia reperfusion injury [60, 67–70]. Moreover, the administration of rHuEPO (1,000 IU/kg 3 times a week) for 4 weeks after coronary artery constriction has been shown to prevent cardiac remodeling and dysfunction, and to improve hemodynamic parameters [71]. CEPO, which is devoid of erythropoietic activities but appears to confer cytoprotection outside the bone marrow, has been assigned cardioprotective properties on ischemia/reperfusion *in vivo* [72]. Further nonclinical studies of the cardioprotective potential of EPO and its derivatives and analogs have been summarized elsewhere [73].

Well-designed large clinical trials are required to demonstrate benefits of the administration of rHuEPO, derivatives or analogues thereof, in humans with acute or chronic coronary syndrome and cardiac failure. In a single-center, investigator-initiated, prospective study on 10 patients with a first acute myocardial infarction, the administration of a single dose (300 µg) darbepoetin alfa was found to be safe and well tolerated [74]. Treatment with drug produced a strong increase in the number of circulating $CD34^+/CD45^-$ cells within 72 h. The left ventricular ejection fraction did not differ when compared with a control group of patients [74]. Previous studies were performed on anemic patients with congestive heart failure, and the improved condition after ESA therapy has been primarily related to the increased red blood cell concentration and, hence, tissue oxygenation [75].

Effects of ESA on renal cells

EPOR mRNA has been detected in tubular and mesangial cells from human and rodent kidneys [76]. Binding studies with radiolabeled EPO provided some evidence for the presence of EPOR, although unspecific antiEPOR antibody was used for Western blotting. In addition, EPO was shown to stimulate

DNA synthesis and cell proliferation in cultures of murine proximal tubular cells [76]. High concentrations of rHuEPO (200 IU/mL) protect human proximal tubular epithelial cells (PTEC) in primary culture from hypoxia-induced apoptosis [77]. Furthermore, rHuEPO (≥10 IU/mL) has been shown to prevent the apoptotic cell death associated with oxidative stress in the human PTEC cell line HK-2 [78]. EPO inhibits tubular cell caspase-3 activation, with upregulation of Bcl-x_L and XIAP (X-linked inhibitor of apoptosis protein) in tubular cells. These effects are dependent on JAK-2 signaling and the phosphorylation of Akt by PI-3K [78].

Single injections of high doses of rHuEPO (3,000 IU/kg) have been shown to protect against subsequent ischemia/reperfusion injury of the kidneys in rats [79]. rHuEPO treatment accelerates renal tubular regeneration with cisplatin-induced acute renal failure [80]. When rHuEPO was given subcutaneously as either a 3-day pretreatment (1,000 IU/kg and day) or as a single bolus (1,000 IU/kg) upon reperfusion to mice subjected to bilateral renal artery occlusion for 30 min, the protection was greater by the pretreatment regime than by the single bolus at the time of reperfusion [81]. The protective potential of darbepoetin alfa was studied in a classic remnant kidney rat model [82]. Animals were subjected to 5/6 nephrectomy followed by either saline or 0.1 μg/kg darbepoetin alfa once weekly. The authors observed a persistent activation of the pro-survival Akt signaling pathway in endothelial and glomerular epithelial cells in the darbepoetin-treated animals, which was accompanied by a significant reduction of apoptotic cell death in renal tissue [82]. Darbepoetin alfa treatment ameliorated endothelial damage, attenuated the composite tissue injury score, which includes vascular sclerosis, glomerulosclerosis, tubulo-interstitial damage, and preserved renal function. Based on the observation that low-dose darbepoetin alfa treatment confers vascular and tissue protection it has been suggested that the use of rHuEPO or analogs may have utility in preventing ischemia-related progressive renal vascular injury and organ failure [82].

Effects of ESA on hepatic cells

Similar to their effects on kidneys ESA appear to protect the liver from ischemia/reperfusion injury. When rats were pretreated with 1,000 IU rHuEPO intravenously before liver ischemia (30 or 45 min) followed by reperfusion, there was less hepatic apoptosis on EPO treatment compared with control animals [83, 84]. Thus, it may be worthy to evaluate the therapeutic properties of ESA in preventing an ischemia/reperfusion injury of the liver, including the preservation of livers for transplantation [85]. Furthermore, ESA are apparently hepatoprotective in acute liver failure due to inflammation, as has been demonstrated in darbepoetin alfa-treated mice challenged with D-galactosamine and *E coli* LPS [86]. Although the site of action of darbepoetin alfa was not clear from this study (hepatocytes?, endothelial cells?, leukocytes?), the observed reduction in liver damage and improved recovery call for a clinical trial with ESA in patients with acute liver injury.

Effects of ESA on tumor growth

Relevance of tumor oxygenation

EPO may influence the growth of solid tumors directly by stimulating erythropoiesis, thereby increasing tumor oxygenation, and indirectly by modulating local immune responses, angiogenesis and cell proliferation (Fig. 3).

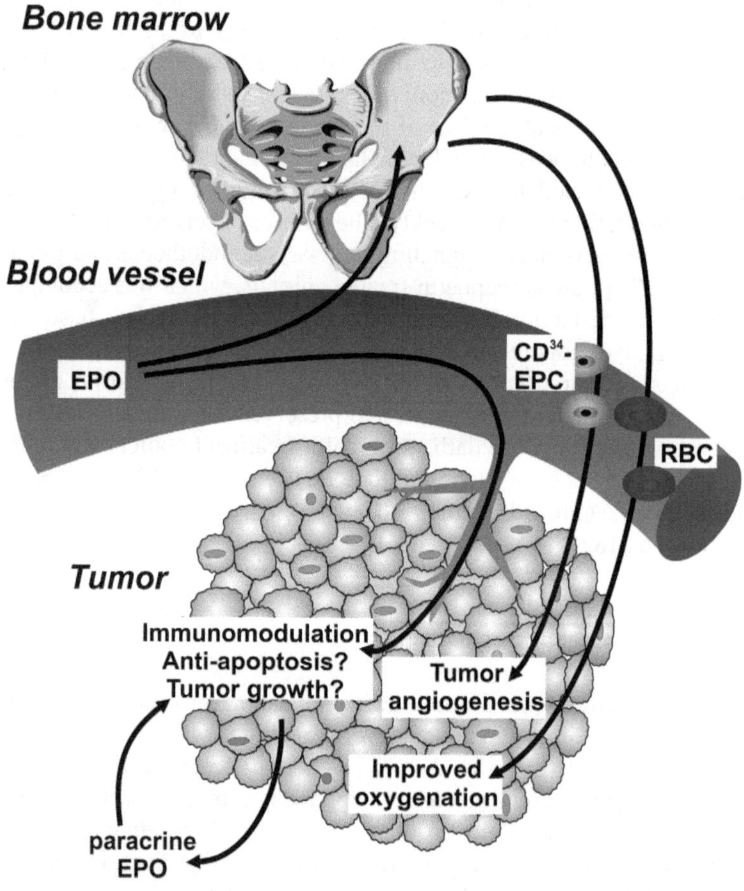

Figure 3. Possible role of EPO in tumor biology. EPO increases red blood cell concentration, blood O_2 capacity and tumor oxygenation. EPO and its pharmacological analogs may stimulate angiogenesis through mobilization of CD34-positive endothelial progenitor cells (CD^{34}-EPC) and through local effects on the endothelium. Whether endogenous EPO and exogenous ESA (erythropoiesis stimulating agents) can promote tumor growth directly via functional EPOR is a controversial issue.

Oxygen availability is central in controlling energy homeostasis, tumor cell survival and metastasis [87]. Several studies have considered the importance of the correction of anemia in tumor patients which leads to an increase in the O_2 supply and, thus, the sensitivity towards chemo- and radiotherapy [88]. For example, the cytotoxic efficacy of cyclophosphamide is greater in rats treated with carboplatin, if anemia is prevented by the administration of rHuEPO [89]. rHuEPO has been shown to suppress the growth of Lewis lung carcinoma in synergy with cisplatin, though not with mitomycin C or cyclophosphamide [90]. Darbepoetin alfa has been proven to increase blood hemoglobin concentration and tumor oxygenation and to reduce tumor mass in murine models of Lewis lung carcinoma in synergy with cisplatin, although darbepoetin alfa does not modulate tumor growth directly [91].

Direct effects of ESA on tumor cells

Because EPOR mRNA is present in essentially every organ, it is also detectable in cancer tissue. A study has shown that *EPOR* is not overexpressed in human cancers. A gene amplification analysis of 1,083 human solid tumors has revealed that amplification of the *EPOR* locus is rare with frequencies similar to other nononcogenes [92]. In addition, *EPOR* transcript amounts in human tumors are low compared with bone marrow and equivalent to, or lower than, the one in matched nonmalignant tissues [92].

Another important question is whether tumor cells translate EPOR protein and whether the EPOR is functional. Since specific anti-EPOR antibodies suitable for immunohistochemistry or Western blotting are not available at present [16–19, 93], an encyclopedic survey of the numerous investigations utilizing such antibodies [5, 7] is not provided here. In brief, most reports claiming the presence of EPOR protein were based on studies with the commercial polyclonal antibody C-20 (Santa Cruz) that was raised against the 20 *C*-terminal amino acids of the human EPOR. The C-20 antibody binds to proteins of 35, 66 and 100 kDa that are clearly unrelated to the EPOR. In all likelihood, the 66 kDa band seen on Western Blotting is heat shock protein HSP70 [16, 17, 94] and the 100 kDa band is HSP90 [19]. Another commonly used polyclonal antibody, M-20 (Santa Cruz), which is raised against the 20 *C*-terminal amino acids of murine EPOR, reacts with murine and human EPOR but also with several other proteins [17]. Both the C-20 and the M-20 antibody falsely produce stain signals in EPOR knockout mouse tissues [17]. Other antibodies used for detection of EPOR on human tumor tissue have included the polyclonal anti-human EPOR antibody H194 (Santa Cruz), the polyclonal anti-murine EPOR antibody 07-311 (Upstate) and the monoclonal anti-human EPOR antibody MAB307 (R&D Systems). All of these antibodies crossreact with multiple proteins different from the EPOR as predicted by the apparent molecular weight [17–19]. When the *EPOR* gene was knocked out by small interfering RNA technique (RNAi) in *EPOR* transfected human osteosarcoma cells, only the 60 kDa protein was lost, while several other bands still stained with the C-20 antibody [19]. Thus, the significance of published immunohistochemical and blotting studies of EPOR protein in biopsies from human tumors or human cancer cell lines has to be called in question.

A number of *in vitro* studies using a variety of tumor cell lines have shown that EPO does not induce EPOR signaling and does not stimulate cell proliferation [5, 7]). In contrast, some investigators have reported growth stimulating effects [5, 7, 95]. Growth stimulation was generally observed only with very high doses exceeding the concentrations measured in the plasma of healthy non-anemic humans (approximately 0.02 IU/mL) or of patients after subcutaneous administration of rHuEPO in clinical practice (approximately 0.15 IU/mL). For example, 250 IU/mL were required to show significant effects on protein phosphorylation, DNA synthesis, proliferation and migration in breast cancer cell cultures [96, 97]. Somewhat lower concentrations of EPO (0.5–1 IU/mL) sufficed to stimulate the growth of distinct human renal [98] or prostate cancer cell lines [99]. It should be noted that the negative and positive responses of tumor cells on the addition of EPO are not related to the use of serum-containing or of serum-free media. In addition, even if significant effects of EPO were shown, these were generally very small, in absolute terms. It was reported that rHuEPO (10 IU/mL) induces signaling (activation of STAT-5, Akt, MAPK) in the nonsmall-cell lung carcinoma cell line H838, but no proliferative response [100]. In contrast to erythropoietic cells, H838 cells do not ubiquitinate and degrade EPOR on binding of EPO. Even in tumor cell lines expressing EPOR mRNA, no EPO binding sites were detectable on the cell surface as assessed by application of radiolabeled rHuEPO [92, 101], which may be due to the lack of EPO mRNA translation or lack of cell-surface-trafficking factors such as JAK-2.

Liu et al. [102] have raised the question as to whether ESA may not influence the basal viability of tumor cells but may protect the cells from the cytotoxic effect of drugs like cisplatin. rHuEPO has been reported to cause resistance to cisplatin in HELA [103], U87 glioma, and HT100 cervical cancer cells [104], and to dacarbazine in melanoma cells [105]. When ovarian cancer cells (line A2780) were treated with rHuEPO in the long term a phenotype developed that exhibited enhanced EPOR signaling and increased resistance to paclitaxel [106]. This effect was drug-specific, since no change in cisplatin or carboplatin sensitivity was observed. The increased resistance towards paclitaxel was explained by a reduced expression of the pro-apoptotic proteins Bcl-2 and Bcl-10 [106]. In contrast to this observation, Carvalho et al. [107] have reported that renal carcinoma cells display a high rate of apoptosis on treatment with a combination of daunorubicine or vinblastine and rHuEPO than with either of these agents alone. Still other investigators have reported that rHuEPO treatment fails to interfere with the cytotoxicity of chemotherapeutics such as adriamycin, taxol or tamoxifen in breast cancer cell lines [108].

Since most *in vitro* investigations on tumor cell growth were performed with permanent cell lines, earlier studies by Bauer et al. [109] are of major importance. These scientists investigated the effects of rHuEPO (up to 400 IU/mL) on the *in vitro* clonal growth of 53 human primary tumor specimens (mainly of renal and colorectal origin) in a soft agar cloning system. rHuEPO treatment was without effect in 47 specimens while stimulation was observed in two, and inhibition in five specimens [109]. Others investigated the effects of epoetin alfa, epoetin beta, and darbepoetin alfa on the *in vitro* proliferation of hematologic tumor cells from eight patients with B-cell chronic lymphocytic leukemia (B-CLL), three patients with mantle-cell lymphoma (MCL), and four patients with multiple myeloma (MM) [110]. None of the ESA induced proliferation in any of the cultures. Thus, neither cancer nor lymphoid malignant or myeloma cells in primary culture are responsive to ESA, in general.

Effects of EPO on the local immune response

Studies of the effects of EPO on whole blood cell cultures from rHuEPO treated patients with chronic kidney disease [111] and on cultures of peripheral polymorphonuclear leukocytes have suggested direct immunomodulating effects of EPO [112]. rHuEPO augments B-cell responses, manifested by stimulation of immunoglobulin production and LPS-induced proliferation of splenocytes [113]. A T cell-mediated tumor-specific immune reaction to myeloma cells has been assumed from studies showing that rHuEPO treatment induces complete tumor regression in 30–60% of mice with syngeneic myeloma [114]. This interpretation has been confirmed in subsequent studies using two lymphoproliferative murine models (MOPC-315 MM and BCL1 B-cell leukemia lymphoma) in which rHuEPO treatment reduced tumor cell growth [115]. The action of EPO as an immunomodulator is incompletely understood, however. Investigators have explained the positive effect of rHuEPO therapy on the survival of mice carrying a subclone of colon 26 adenocarcinoma by a decrease in the production of the cachexia-inducing cytokine interleukin 6 (IL-6) [116].

Effects of EPO on tumor angiogenesis

Several lines of evidence suggest that endogenous EPO may be an important survival factor for the endothelium of tumor vessels. Yasuda et al. [117] have shown that the application of anti-EPO antibody or of soluble EPOR in the transplants of uterine or ovarian tumors in nude mice results in a reduction of tumor size. In subsequent studies it was found that the intraperitoneal administration of a synthetic EPOR blocking peptide (EMP9) in mice with subcutaneous xenografts of human stomach choriocarcinoma (SHC) or melanoma (P39) inhibits tumor angiogenesis and opposes tumor cell viability, whereas the application of an EPO mimetic peptide (EMP1) promotes tumor angiogenesis and tumor cell survival [118]. Furthermore, the application of either anti-EPO antibody, soluble EPOR or an inhibitor of JAK-2 delays tumor growth in a tumor-Z chamber model with rat syngeneic mammary adenocarcinoma cells [119]. Studies utilizing the dorsal skin-fold window chamber technique, which allows one to study tumor morphology by intravital microscopy, have shown that the coinjection of mammary carcinoma cells with either soluble EPOR or anti-EPO antibody results in reduced tumor angiogenesis and tumor growth in mice [120]. Transgenic expression of an EPO-antagonist protein (R103A-EPO) likewise inhibits tumor angiogenesis and tumor growth whereas expression of a constitutively active *EPOR (EPOR-R129C)* in the tumor cells is associated with enhanced tumor angiogenesis and growth [120]. Since tumor cells can endogenously express the *EPO* gene, tumor cell-derived EPO may promote angiogenesis *in vivo* [121]. With respect to the administration of ESA to tumor patients a deeper insight into the role of EPO for tumor angiogenesis is obviously required.

Conclusions

rHuEPO and its hyperglycosylated analog darbepoetin alfa are routinely used for the treatment of the anemias associated with chronic kidney disease or chemotherapy. Primarily in preclinical studies, rHuEPO and its derivatives and analogs prevented apoptosis and provided protection against damage caused by hypoxia, ischemia/reperfusion or cytotoxic agents in several nonhematopoietic organs and tissues, including the brain, the heart, the kidneys and the liver. Reportedly, certain EPO derivatives that are devoid of erythropoietic activity, such as carbamoylated EPO, can still confer cytoprotection in nonhematopoietic tissues.

However, many of the enthusiastic reports of the nonhematopoietic cytoprotective potential of EPO and its derivatives have remained phenomenological, in that statistically significant effects were deduced on the addition – often in very high doses – of the drugs to more or less artificial experimental setups. Furthermore, the presence of EPOR mRNA does not equal the presence of functional cell surface EPOR protein. EPOR protein was often not demonstrated, primarily because specific anti-EPOR antibodies for immunodetection are not available. This problem also holds true when reports proposing growth promoting effects of EPO on tumor tissue are critically assessed. EPOR transcripts are not increased in tumors, and ligand/EPOR binding and EPOR signaling has rarely been proven.

It remains to be investigated whether the promising results of the use of EPO in neuronal, myocardial, renal and hepatic cell culture or whole organ studies and in animal models can be reproduced in clinical practice. Properly designed, carefully blinded and placebo-controlled, clinical trials are needed to assess the efficacy and safety of these interventions in patients.

References

1. Barber DL, Beattie BK, Mason JM et al. A common epitope is shared by activated signal transducer and activator of transcription-5 (STAT5) and the phosphorylated erythropoietin receptor: implications for the docking model of STAT activation. *Blood* 2001;97:2230–2237.
2. Walrafen P, Verdier F, Kadri Z et al. Both proteasomes and lysosomes degrade the activated erythropoietin receptor. *Blood* 2005;105:600–608.
3. Maiese K, Li F, Chong ZZ. New avenues of exploration for erythropoietin. *JAMA* 2005;293:90–95.
4. Sasaki R, Masuda S, Nagao M. Erythropoietin: multiple physiological functions and regulation of biosynthesis. *Biosc Biotechnol Biochem* 2000;64:1775–1793.
5. Sinclair AM, Todd MD, Forsythe K et al. Expression and function of erythropoietin receptors in tumors: implications for the use of erythropoiesis-stimulating agents in cancer patients. *Cancer* 2007;110:477–488.
6. Suzuki N, Ohneda O, Takahashi S et al. Erythroid-specific expression of the erythropoietin receptor rescued its null mutant mice from lethality. *Blood* 2002;100:2279–2288.
7. Jelkmann W, Bohlius J, Hallek M et al. The erythropoietin receptor in normal and cancer tissues. *Crit Rev Oncol Hematol* 2008;67:39–61
8. Konishi Y, Chui DH, Hirose H et al. Trophic effect of erythropoietin and other hematopoietic factors on central cholinergic neurons *in vitro* and *in vivo*. *Brain Res* 1993;609:29–35.
9. Masuda S, Nagao M, Takahata K et al. Functional erythropoietin receptor of the cells with neural characteristics. Comparison with receptor properties of erythroid cells. *J Biol Chem* 1993;268:11208–11216.
10. Digicaylioglu M, Bichet S, Marti HH et al. Localization of specific erythropoietin binding sites in defined areas of the mouse brain. *Proc Natl Acad Sci USA* 1995;92:3717–3720.
11. Juul SE, Anderson DK, Li Y et al. Erythropoietin and erythropoietin receptor in the developing human central nervous system. *Pediat Res* 1998;43:40–49.
12. Jelkmann W. Effects of erythropoietin on brain function. *Curr Pharmaceut Biotechnol* 2004;6:65–79.
13. Shingo T, Sorokan ST, Shimazaki T et al. Erythropoietin regulates the *in vitro* and *in vivo* production of neuronal progenitors by mammalian forebrain neural stem cells. *J Neurosci* 2001;21:9733–9743.
14. Yamaji R, Okada T, Moriya M et al. Brain capillary endothelial cells express two forms of erythropoietin receptor mRNA. *Eur J Biochem* 1996;239:494–500.
15. Brines ML, Ghezzi P, Keenan S et al. Erythropoietin crosses the blood-brain barrier to protect against experimental brain injury. *Proc Natl Acad Sci USA* 2000;97:10526–10531.
16. Brown WM, Maxwell P, Graham ANJ et al. Erythropoietin receptor expression in non-small cell lung carcinoma: a question of antibody specificity. *Stem Cells* 2007;25:718–722.
17. Elliott S, Sinclair AM, Begley CG. Anti-Epo receptor antibodies do not predict Epo receptor expression. *Blood* 2006;107:1892–1895.
18. Kirkeby A, Nielsen J, Leist M et al. Functional and immunochemical characterisation of different antibodies against the erythropoietin receptor. *J Neurosci Methods* 2007;164:50–58.
19. Laugsch M, Metzen E, Svensson T et al. Lack of functional erythropoietin receptors of cancer cell lines. *Int J Cancer* 2008;122:1005–1011.
20. Koshimura K, Murakami A, Sohmiya M et al. Effects of erythropoietin on neuronal activity. *J Neurochem* 1999;72:2565–2572.
21. Morishita E, Masuda S, Nagao M et al. Erythropoietin receptor is expressed in rat hippocampal and cerebral cortical neurons, and erythropoietin prevents *in vitro* glutamate-induced neuronal death. *Neuroscience* 1997;76:105–116.
22. Sinor AD, Greenberg DA. Erythropoietin protects cultured cortical neurons, but not astroglia, from hypoxia and AMPA toxicity. *Neurosci Lett* 2000;290:213–215.
23. Ehrenreich H, Hasselblatt M, Dembrowski C et al. Erythropoietin therapy for acute stroke is both safe and beneficial. *Mol Med* 2002;8:495–505.
24. Sakanaka M, Wen TC, Matsuda S et al. *In vivo* evidence that erythropoietin protects neurons from ischemic damage. *Proc Natl Acad Sci USA* 1998;95:4635–4640.
25. Digicaylioglu M, Lipton SA. Erythropoietin-mediated neuroprotection involves cross-talk between Jak2 and NF-kappaB signalling cascades. *Nature* 2001;412:641–647.
26. Lewczuk P, Hasselblatt M, Kamrowski-Kruck H et al. Survival of hippocampal neurons in culture

upon hypoxia: effect of erythropoietin. *Neuroreport* 2000;11:3485–3488.
27. Kawakami M, Sekiguchi M, Sato K et al. Erythropoietin receptor-mediated inhibition of exocytotic glutamate release confers neuroprotection during chemical ischemia. *J Biol Chem* 2001;276:39469–39475.
28. Chong ZZ, Li F, Maiese K. Erythropoietin requires NF-kappaB and its nuclear translocation to prevent early and late apoptotic neuronal injury during beta-amyloid toxicity. *Curr Neurovasc Res* 2005;2:387–399.
29. Sadamoto Y, Igase K, Sakanaka M et al. Erythropoietin prevents place navigation disability and cortical infarction in rats with permanent occlusion of the middle cerebral artery. *Biochem Biophys Res Commun* 1998;253:26–32.
30. Wen TC, Sadamoto Y, Tanaka I et al. Erythropoietin protects neurons against chemical hypoxia and cerebral ischemic injury by up-regulating Bcl-xL expression. *J Neurosci Res* 2002;67:795–803.
31. Tsai PT, Ohab JJ, Kertesz N et al. A critical role of erythropoietin receptor in neurogenesis and post-stroke recovery. *J Neurosci* 2006;26:1269–1274.
32. Chen ZY, Asavaritikrai P, Prchal JT et al. Endogenous erythropoietin signaling is required for normal neural progenitor cell proliferation. *J Biol Chem* 2007;282:25875–25883.
33. Brines M, Cerami A. Emerging biological roles for erythropoietin in the nervous system. *Nat Rev Neurosci* 2005;6:484–494.
34. Leist M, Ghezzi P, Grasso G et al. Derivatives of erythropoietin that are tissue protective but not erythropoietic. *Science* 2004;305:239–242.
35. Siren AL, Fratelli M, Brines M et al. Erythropoietin prevents neuronal apoptosis after cerebral ischemia and metabolic stress. *Proc Natl Acad Sci USA* 2001;98:4044–4049.
36. Noguchi CT, Asavaritikrai P, Teng R et al. Role of erythropoietin in the brain. *Crit Rev Oncol Hematol* 2007;64:159–171.
37. Ehrenreich H, Degner D, Meller J et al. Erythropoietin: a candidate compound for neuroprotection in schizophrenia. *Mol Psychiatr* 2004;9:42–54.
38. Jumbe NL. Erythropoietic agents as neurotherapeutic agents: what barriers exist? *Oncology* 2002;16:91–107.
39. Erbayraktar S, Grasso G, Sfacteria A et al. Asialoerythropoietin is a nonerythropoietic cytokine with broad neuroprotective activity *in vivo*. *Proc Natl Acad Sci USA* 2003;100:6741–6746.
40. Ehrenreich H, Hinze-Selch D et al. Improvement of cognitive functions in chronic schizophrenic patients by recombinant human erythropoietin. *Mol Psychiatr* 2006;12:206–220.
41. Assaraf MI, Diaz Z, Liberman A et al. Brain erythropoietin receptor expression in Alzheimer disease and mild cognitive impairment. *J Neuropathol Exp Neurol* 2007;66:389–398.
42. Miskowiak K, Inkster B, Selvaraj S et al. Erythropoietin improves mood and modulates the cognitive and neural processing of emotion 3 days post administration. *Neuropsychopharmacology* 2008;33:611–618.
43. Anagnostou A, Liu Z, Steiner M et al. Erythropoietin receptor mRNA expression in human endothelial cells. *Proc Natl Acad Sci USA* 1994;91:3974–3978.
44. Carlini RG, Reyes AA, Rothstein M. Recombinant human erythropoietin stimulates angiogenesis *in vitro*. *Kidney Int* 1995;47:740–745.
45. Jaquet K, Krause K, Tawakol-Khodai M et al. Erythropoietin and VEGF exhibit equal angiogenic potential. *Microvasc Res* 2002;64:326–333.
46. Watanabe D, Suzuma K, Matsui S et al. Erythropoietin as a retinal angiogenic factor in proliferative diabetic retinopathy. *N Engl J Med* 2005;353:782–792.
47. Grimm C, Wenzel A, Groszer M et al. HIF-1-induced erythropoietin in the hypoxic retina protects against light-induced retinal degeneration. *Nat Med* 2002;8:718–724.
48. Haller H, Christel C, Dannenberg L et al. Signal transduction of erythropoietin in endothelial cells. *Kidney Int* 1996;50:481–488.
49. Fodinger M, Fritsche-Polanz R, Buchmayer H et al. Erythropoietin-inducible immediate-early genes in human vascular endothelial cells. *J Investig Med* 2000;48:137–149.
50. Carlini RG, Alonzo EJ, Dominguez J et al. Effect of recombinant human erythropoietin on endothelial cell apoptosis. *Kidney Int* 1999;55:546–553.
51. Chong ZZ, Maiese K. Erythropoietin involves the phosphatidylinositol 3-kinase pathway, 14-3-3 protein and FOXO3a nuclear trafficking to preserve endothelial cell integrity. *Brit J Pharmacol* 2007;150:839–850.
52. Satoh K, Kagaya Y, Nakano M et al. Important role of endogenous erythropoietin system in

recruitment of endothelial progenitor cells in hypoxia-induced pulmonary hypertension in mice. *Circulation* 2006;113:1442–1450.
53. Bahlmann FH, De Groot K, Spandau JM et al. Erythropoietin regulates endothelial progenitor cells. *Blood* 2004;103:921–926.
54. Heeschen C, Aicher A, Lehmann R et al. Erythropoietin is a potent physiologic stimulus for endothelial progenitor cell mobilization. *Blood* 2003;102:1340–1346.
55. Ammarguellat F, Llovera M, Kelly PA et al. Low doses of EPO activate MAP kinases but not JAK2-STAT5 in rat vascular smooth muscle cells. *Biochem Biophys Res Commun* 2001;284:1031–1038.
56. Morakkabati N, Gollnick F, Meyer R, Fandrey J et al. Erythropoietin induces Ca^{2+} mobilization and contraction in rat mesangial and aortic smooth muscle cultures. *Exp Hematol* 1996;24:392–397.
57. Heidenreich S, Rahn KH, Zidek W. Direct vasopressor effect of recombinant human erythropoietin on renal resistance vessels. *Kidney Int* 1991;39:259–265.
58. Gogusev J, Zhu DL, Herembert T, Ammarguellat F et al. Effect of erythropoietin on DNA synthesis, proto-oncogene expression and phospholipase C activity in rat vascular smooth muscle cells. *Biochem Biophys Res Commun* 1994;199:977–983.
59. Tramontano AF, Muniyappa R, Black AD et al. Erythropoietin protects cardiac myocytes from hypoxia-induced apoptosis through an Akt-dependent pathway. *Biochem Biophys Res Commun* 2003;308:990–994.
60. Wright GL, Hanlon P, Amin K et al. Erythropoietin receptor expression in adult rat cardiomyocytes is associated with an acute cardioprotective effect for recombinant erythropoietin during ischemia-reperfusion injury. *FASEB J* 2004;18:1031–1033.
61. Depping R, Kawakami K, Ocker H et al. Expression of the erythropoietin receptor in human heart. *J Thorac Cardiovasc Surg* 2005;130:877–878.
62. Rafiee P, Shi Y, Su J et al. Erythropoietin protects the infant heart against ischemia-reperfusion injury by triggering multiple signaling pathways. *Basic Res Cardiol* 2005;100:187–197.
63. Shi Y, Rafiee P, Su J et al. Acute cardioprotective effects of erythropoietin in infant rabbits are mediated by activation of protein kinases and potassium channels. *Basic Res Cardiol* 2001;99:173–182.
64. Parsa CJ, Kim J, Riel RU et al. Cardioprotective effects of erythropoietin in the reperfused ischemic heart: a potential role for cardiac fibroblasts. *J Biol Chem* 2004;279: 20655–20662.
65. Mudalagiri NR, Mocanu MM, Di SC et al. Erythropoietin protects the human myocardium against hypoxia/reoxygenation injury via phosphatidylinositol-3 kinase and ERK1/2 activation. *Brit J Pharmacol* 2008;153:50–56.
66. Asaumi Y, Kagaya Y, Takeda M et al. Protective role of endogenous erythropoietin system in nonhematopoietic cells against pressure overload-induced left ventricular dysfunction in mice. *Circulation* 2007;115:2022–2032.
67. Cai Z, Manalo DJ, Wei G et al. Hearts from rodents exposed to intermittent hypoxia or erythropoietin are protected against ischemia-reperfusion injury. *Circulation* 2003;108:79–85.
68. Calvillo L, Latini R, Kajstura J et al. Recombinant human erythropoietin protects the myocardium from ischemia-reperfusion injury and promotes beneficial remodeling. *Proc Natl Acad Sci USA* 2003;100:4802–4806.
69. Moon C, Krawczyk M, Ahn D et al. Erythropoietin reduces myocardial infarction and left ventricular functional decline after coronary artery ligation in rats. *Proc Natl Acad Sci USA* 2003;100:11612–11617.
70. Parsa CJ, Matsumoto A, Kim J et al. A novel protective effect of erythropoietin in the infarcted heart. *J Clin Invest* 2003;112:999–1007.
71. Nishiya D, Omura T, Shimada K et al. Effects of erythropoietin on cardiac remodeling after myocardial infarction. *J Pharmacol Sci* 2006;101:31–39.
72. Fiordaliso F, Chimenti S, Staszewsky L et al. A nonerythropoietic derivative of erythropoietin protects the myocardium from ischemia-reperfusion injury. *Proc Natl Acad Sci USA* 2005;102:2046–2051.
73. Lipsic E, Schoemaker RG, van der MP et al. Protective effects of erythropoietin in cardiac ischemia: from bench to bedside. *J Am Coll Cardiol* 2006;48:2161–2167.
74. Lipsic E, van der MP, Voors AA et al. A single bolus of a long-acting erythropoietin analogue darbepoetin alfa in patients with acute myocardial infarction: a randomized feasibility and safety study. *Cardiovasc Drug Therapy* 2006;20:135–141.

75. Schwartzenberg S, Ben-Shoshan J, Keren G et al. The role of erythropoietin in myocardial protection: potential mechanisms and applications. *Expert Rev Cardiovasc Ther* 2006;4:41–50.
76. Westenfelder C, Biddle DL, Baranowski RL. Human, rat, and mouse kidney cells express functional erythropoietin receptors. *Kidney Int* 1999;55:808–820.
77. Vesey DA, Cheung C, Pat B et al. Erythropoietin protects against ischaemic acute renal injury. *Nephrol Dial Transplant* 2004;19:348–355.
78. Sharples EJ, Patel N, Brown P et al. Erythropoietin protects the kidney against the injury and dysfunction caused by ischemia-reperfusion. *J Am Soc Nephrol* 2004;15: 2115–2124.
79. Yang CW, Li C, Jung JY et al. Preconditioning with erythropoietin protects against subsequent ischemia-reperfusion injury in rat kidney. *FASEB J* 2003;17:1754–1755.
80. Vaziri ND, Zhou XJ, Liao SY. Erythropoietin enhances recovery from cisplatin-induced acute renal failure. *Am J Physiol* 1994;266:F360–F366.
81. Patel NS, Sharples EJ, Cuzzocrea S et al. Pretreatment with EPO reduces the injury and dysfunction caused by ischemia/reperfusion in the mouse kidney *in vivo*. *Kidney Int* 2004;66:983–989.
82. Bahlmann FH, Song R, Boehm SM et al. Low-dose therapy with the long-acting erythropoietin analogue darbepoetin alpha persistently activates endothelial Akt and attenuates progressive organ failure. *Circulation* 2004;110:1006–1012.
83. Schmeding M, Neumann UP, Boas-Knoop S et al. Erythropoietin reduces ischemia-reperfusion injury in the rat liver. *Eur Surg Res* 2007;39:189–197.
84. Sepodes B, Maio R, Pinto R et al. Recombinant human erythropoietin protects the liver from hepatic ischemia-reperfusion injury in the rat. *Transplant Int* 2006;19:919–926.
85. Seehofer D, Neumann UP, Schirmeier A et al. Synergistic effect of erythropoietin but not G-CSF in combination with curcumin on impaired liver regeneration in rats. *Langenbecks Arch Surg* 2008;393:325–332.
86. Le MK, Klemm K, Abshagen K et al. Attenuation of inflammation and apoptosis by pre- and posttreatment of darbepoetin-alpha in acute liver failure of mice. *Am J Pathol* 2007;170:1954–1963.
87. Pouyssegur J, Dayan F, Mazure NM. Hypoxia signalling in cancer and approaches to enforce tumour regression. *Nature* 2006;441:437–443.
88. Vaupel P, Mayer A, Höckel M. Tumor hypoxia and malignant progression. *Methods Enzymol* 2004;381:335–354.
89. Thews O, Kelleher DK, Vaupel P. Erythropoietin restores the anemia-induced reduction in cyclophosphamide cytotoxicity in rat tumors. *Cancer Res* 2001;61:1358–1361.
90. Sigounas G, Sallah S, Sigounas VY. Erythropoietin modulates the anticancer activity of chemotherapeutic drugs in a murine lung cancer model. *Cancer Lett* 2004;214:171–179.
91. Shannon AM, Bouchier-Hayes DJ, Condron CM et al. Correction of anaemia through the use of darbepoetin alfa improves chemotherapeutic outcome in a murine model of Lewis lung carcinoma. *Br J Cancer* 2005;93:224–232.
92. Sinclair AM, Rogers N, Busse L et al. Erythropoietin receptor transcription is neither elevated nor predictive of surface expression in human tumour cells. *Br J Cancer* 2008;98:1059–1067.
93. Hoogsteen IJ, Peeters WJ, Marres HA et al. Erythropoietin receptor is not a surrogate marker for tumor hypoxia and does not correlate with survival in head and neck squamous cell carcinomas. *Radiother Oncol* 2005;76:213–218.
94. Della RF, Cucciolla V, Borriello A et al. Erythropoietin receptors on cancer cells: a still open question. *J Clin Oncol* 2007;25:1812–1813.
95. Hardee ME, Arcasoy MO, Blackwell KL et al. Erythropoietin biology in cancer. *Clin Cancer Res* 2006;12:332–339.
96. Acs G, Acs P, Beckwith SM et al. Erythropoietin and erythropoietin receptor expression in human cancer. *Cancer Res* 2001;61:3561–3565.
97. Lester RD, Jo M, Campana WM et al. Erythropoietin promotes MCF-7 breast cancer cell migration by an ERK/mitogen-activated protein kinase-dependent pathway and is primarily responsible for the increase in migration observed in hypoxia. *J Biol Chem* 2005;280:39273–39277.
98. Westenfelder C, Baranowski RL. Erythropoietin stimulates proliferation of human renal carcinoma cells. *Kidney Int* 2000;58:647–657.
99. Feldman L, Wang Y, Rhim JS et al. Erythropoietin stimulates growth and STAT5 phosphorylation in human prostate epithelial and prostate cancer cells. *Prostate* 2006;66:135–145.
100. Dunlop EA, Percy MJ, Boland MP et al. Induction of signalling in non-erythroid cells by phar-

macological levels of erythropoietin. *Neurodegener Dis* 2006;3:94–100.
101. LaMontagne KR, Butler J, Marshall DJ et al. Recombinant epoetins do not stimulate tumor growth in erythropoietin receptor-positive breast carcinoma models. *Mol Cancer Ther* 2006;5:347–355.
102. Liu WM, Powles T, Shamash J et al. Effect of haemopoietic growth factors on cancer cell lines and their role in chemosensitivity. *Oncogene* 2004;23:981–990.
103. Acs G, Zhang PJ, McGrath CM et al. Hypoxia-inducible erythropoietin signaling in squamous dysplasia and squamous cell carcinoma of the uterine cervix and its potential role in cervical carcinogenesis and tumor progression. Am. *J Pathol* 2003;162.1789–1806.
104. Belenkov AI, Shenouda G, Rizhevskaya E et al. Erythropoietin induces cancer cell resistance to ionizing radiation and to cisplatin. *Mol Cancer Ther* 2004;3:1525–1532.
105. Kumar SM, Acs G, Fang D et al. Functional erythropoietin autocrine loop in melanoma. Am. *J Pathol* 2005;166:823–830.
106. Solar P, Feldman L, Jeong J-Y et al. Erythropoietin treatment of human ovarian cancer cells results in enhanced signaling and a paclitaxel-resistant phenotype. *Int J Cancer* 2007;122:281–288.
107. Carvalho G, Lefaucheur C, Cherbonnier C et al. Chemosensitization by erythropoietin through inhibition of the NF-kappaB rescue pathway. *Oncogene* 2005;24:737–745.
108. Gewirtz DA, Di X, Walker TD et al. Erythropoietin fails to interfere with the antiproliferative and cytotoxic effects of antitumor drugs. *Clin Cancer Res* 2006;12:2232–2238.
109. Bauer E, Danhauser-Riedl S, De Riese W et al. Effects of recombinant human erythropoietin on clonogenic growth of primary human tumor specimens *in vivo*. *Onkologie* 1992;15:254–258.
110. Kokhaei P, Abdalla AO, Hansson L et al. Expression of erythropoietin receptor and *in vitro* functional effects of epoetins in B-cell malignancies. *Clin Cancer Res* 2007;13:3536–3544.
111. Bryl E, Mysliwska J, Abska-Slizien A et al. The influence of recombinant human erythropoietin on tumor necrosis factor alpha and interleukin-10 production by whole blood cell cultures in hemodialysis patients. *Artif Organs* 1998;22:177–181.
112. Kimata H, Yoshida A, Ishioka C et al. Effect of recombinant human erythropoietin on human IgE production *in vitro*. *Clin Exp Immunol* 1991;83:483–487.
113. Katz O, Gil L, Lifshitz L et al. Erythropoietin enhances immune responses in mice. *Eur J Immunol* 2007;37:1584–1593.
114. Mittelman M, Neumann D, Peled A et al. Erythropoietin induces tumor regression and antitumor immune responses in murine myeloma models. *Proc Natl Acad Sci USA* 2001;98:5181–5186.
115. Katz O, Barzilay E, Skaat A et al. Erythropoietin induced tumour mass reduction in murine lymphoproliferative models. *Acta Haematol* 2005;114:177–179.
116. Kanzaki M, Soda K, Gin PT et al. Erythropoietin attenuates cachectic events and decreases production of interleukin-6, a cachexia-inducing cytokine. *Cytokine* 2005;32:234–239.
117. Yasuda Y, Musha T, Tanaka H et al. Inhibition of erythropoietin signalling destroys xenografts of ovarian and uterine cancers in nude mice. *Br J Cancer* 2001;84:836–843.
118. Yasuda Y, Fujita Y, Matsuo T et al. Erythropoietin regulates tumour growth of human malignancies. *Carcinogenesis* 2003;24:1021–1029.
119. Arcasoy MO, Amin K, Karayal AF et al. Functional significance of erythropoietin receptor expression in breast cancer. *Lab Invest* 2002;82:911–918.
120. Hardee ME, Cao Y, Fu P et al. Erythropoietin blockade inhibits the induction of tumor angiogenesis and progression. *PLoS ONE* 1007;2:e549.
121. Ribtti D, Marzullo A, Gentile A et al. Erythropoietin/erythropoietin-receptor system is involved in angiogenesis in human hepatocellular carcinoma. *Histopathology* 2007;50:591–596.

Index

The MDT-Series
Milestones in Drug Therapy

The discovery of drugs is still an unpredictable process. Breakthroughs are often the result of a combination of factors, including serendipidity, rational strategies and a few individuals with novel ideas. *Milestones in Drug Therapy* highlights new therapeutic developments that have provided significant steps forward in the fight against disease. Each book deals with an individual drug or drug class that has altered the approach to therapy. Emphasis is placed on the scientific background to the discoveries and the development of the therapy, with an overview of the current state of knowledge provided by experts in the field, revealing also the personal stories behind these milestone developments. The series is aimed at a broad readership, covering biotechnology, biochemistry, pharmacology and clinical therapy.

Forthcoming titles

PEGylated Protein Drugs: Basic Science and Clinical Applications, F.M. Veronese, M.J. Bossard (Editors), 2009
Bortezomib in the Treatment of Multiple Myeloma, K.C. Anderson, P.G. Richardson, I. Ghobrial (Editors), 2009
Influenza Virus Sialidase – A Drug Discovery Target, M. von Itzstein (Editor), 2010

Published volumes

Bipolar Depression: Molecular Neurobiology, Clinical Diagnosis and Pharmacotherapy, C.A. Zarate, H.K. Manji (Editors), 2009
Treatment of Psoriasis, J.M. Weinberg (Editor), 2008
Aromatase Inhibitors, 2nd revised edition, B.J.A. Furr (Editor), 2008
Pharmacotherapy of Obesity, J.P.H. Wilding (Editor), 2008
Entry Inhibitors in HIV Therapy, J.D. Reeves, C.A. Derdeyn (Editors), 2007
Drugs affecting Growth of Tumours, H.M. Pinedo, C. Smorenburg (Editors), 2006
TNF-alpha Inhibitors, J.M. Weinberg, R. Buchholz (Editors), 2006
Aromatase Inhibitors, B.J.A. Furr (Editor), 2006
Cannabinoids as Therapeutics, R. Mechoulam (Editor), 2005
St. John`s Wort and its Active Principles in Anxiety and Depression, W.E. Müller (Editor), 2005
Drugs for Relapse Prevention of Alcoholism, R. Spanagel, K. Mann (Editors), 2005
COX-2 Inhibitors, M. Pairet, J. Van Ryn (Editors), 2004

Calcium Channel Blockers, T. Godfraind (Author), 2004
Sildenafil, U. Dunzendorfer (Editor), 2004
Hepatitis Prevention and Treatment, J. Colacino, B.A. Heinz (Editors), 2004
Combination Therapy of AIDS, E. De Clercq, A.M. Vandamme (Editors), 2004
Cognitive Enhancing Drugs, J. Buccafusco (Editor), 2004
Fluoroquinolone Antibiotics, A.R. Ronald, D. Low (Editors), 2003
Erythropoietins and Erythropoiesis, G. Molineux, M. Foote, S. Elliott (Editors), 2003
Macrolide Antibiotics, W. Schönfeld, H. Kirst (Editors), 2002
HMG CoA Reduktase Inhibitors, G. Schmitz, M. Torzewski (Editors), 2002
Antidepressants, B.E. Leonard (Editor), 2001
Recombinant Protein Drugs, P. Buckel (Editor), 2001
Glucocorticoids, N. Goulding, R.J. Flower (Editors), 2001
Modern Immunosuppressives, H.-J. Schuurman (Editor), 2001
ACE Inhibitors, P. D'Orleans-Juste, G. Plante (Editors), 2001
Atypical Antipsychotics, A.R. Cools, B.A. Ellenbroek (Editors), 2000
Methotrexate, B.N. Cronstein, J.R. Bertino (Editors), 2000
Anxiolytics, M. Briley, D. Nutt (Editors), 2000
Proton Pump Inhibitors, L. Olbe (Editor), 1999
Valproate, W. Löscher (Editor), 1999

Kolibri 360 GmbH
Ferdinand-Jühlke-Straße 7
99095 Erfurt, Deutschland
produktsicherheit@kolibri360.de